Numanities - Arts and Humanities in Progress

Volume 16

Series Editor

Dario Martinelli, Kaunas University of Technology, Kaunas, Lithuania

The series originates from the need to create a more proactive platform in the form of monographs and edited volumes in thematic collections, to contribute to the new emerging fields within art and humanistic research, and also to discuss the ongoing crisis of the humanities and its possible solutions, in a spirit that should be both critical and self-critical.

"Numanities" (New Humanities) aim at unifying the various approaches and potentials of arts and humanities in the context, dynamics and problems of current societies.

The series, indexed in Scopus, is intended to target an academic audience interested in the following areas:

- Traditional fields of humanities whose research paths are focused on issues of current concern;
- New fields of humanities emerged to meet the demands of societal changes;
- Multi/Inter/Cross/Transdisciplinary dialogues between humanities and social and/or natural sciences;
- Humanities "in disguise", that is, those fields (currently belonging to other spheres), that remain rooted in a humanistic vision of the world;
- Forms of investigations and reflections, in which the humanities monitor and critically assess their scientific status and social condition;
- Forms of research animated by creative and innovative humanities-based approaches;
- Applied humanities.

More information about this series at http://www.springer.com/series/14105

Roberto Marchesini · Marco Celentano

Critical Ethology and Post-Anthropocentric Ethics

Beyond the Separation between Humanities and Life Sciences

Springer

Roberto Marchesini
School of Human-Animal Interaction
and Centre Study for Post-human
philosophy
Bologna, Italy

Marco Celentano
Università di Cassino e del Lazio
Meridionale
Terni, Italy

ISSN 2510-442X ISSN 2510-4438 (electronic)
Numanities - Arts and Humanities in Progress
ISBN 978-3-030-74202-7 ISBN 978-3-030-74203-4 (eBook)
https://doi.org/10.1007/978-3-030-74203-4

This Springer imprint is published by the registered company Springer Nature Switzerland AG
The registered company address is: Gewerbestrasse 11, 6330 Cham, Switzerland

Introduction

This volume offers an updated analysis and discussion on some of the most debated theoretical, epistemological, social and ethical issues within that very young field of study called "philosophy of ethology" or "philosophical ethology" (Lestel 2001; Lestel, Brunois, Gaunet 2006; Chrulew 2014; Buchanan, Bussolini, Chrulew 2014; Bussolini 2016; Marchesini 2016; Celentano 2018).

As its name suggests, this field of investigation compares philosophical reflection and ethological research into the questions raised by one of the most important scientific breakthroughs of our times: the discovery that, besides Man, many other animals can think, invent new techniques, develop cultural traditions and manifest individual differences in their aesthetic sensibility and expressive ability.

This book has three main aims:

- to contribute to the development of a philosophy of ethology and an ethical reflection based on a both post-genocentric and post-anthropocentric approach;
- to promote an overcoming of the traditional division between life sciences and humanities;
- to attempt a critical (and self-critical) integration of these two scientific traditions.

We believe it is necessary to develop ethology as a comparative study of the behaviour of all the living species (not just animals), aimed at fully overcoming the mechanistic and predominantly nativist setting of classical ethology and develop a critical approach to the study of psychic and cultural phenomena that could archive the prejudices of anthropocentrism and anthropodenial (de Waal 2016) and open up to a horizontal (non-hierarchical) comparison between human minds and cultures and all other, past or present animal minds and cultures.

The first part of the book which presents three essays by Celentano, illustrates this approach by likening it to the historical progression from Darwinian proto-ethology to cognitive and cultural ethology, and from neo-Darwinian formulations of a theory of evolution to its most recent renditions in light of the discovery of epigenetic inheritance and the birth of Evolutionary Developmental (Evo-Devo) Biology.

The first chapter reconstructs the process which led from Darwin's revolution to contemporary ethology, emphasizing the topicality and fruitfulness of the pluralistic approach to the explanation of the animal and human mental evolution that Darwin assumed from the 1870s.

The second chapter focuses on the analysis of the affinities and bonds between the two different attempts to reformulate Darwinian theory in a "cognitive" way which have profoundly renewed evolutionary studies: Evolutionary Epistemology, first promoted by ethologist Konrad Lorenz in the 1970s, and Extended Synthesis, or the post-genocentric synthesis of Darwinism, towards which a large part of contemporary evolutionary biology is converging. An ongoing synthesis to which, in the last three decades, authoritative members of the Konrad Lorenz Institute for Evolution and Cognition Research (Klosterneuburg—Austria), such as Werner Callebaut, Gerd Müller and Massimo Pigliucci have made important contributions.

The third chapter illustrates the projects of *Interspecific Cultural Studies*: a meta-disciplinary area which aims to contribute to a self-critical refoundation of both humanities and behavioural sciences, as well as to a new organization of university and post-university education and basic and applied research enabling it to offer the development of skills that are transversal to the traditional bipartition between human and life sciences. The first goal of Interspecific Cultural Studies is in fact to train generations of scholars capable of profoundly renewing the Humanities, critically inserting the brief history of human cultures into the much longer history of animal ones, which has endured for hundreds of millions of years (differentiated cultural traditions are in fact already observable in bony fish that have existed for more than 400 million years). In various respects, this is a project very close to the radical self-reform of humanities proposed in Martinelli's *Manifesto of Numanities* (Martinelli 2016) that, precisely because of this affinity, in our opinion finds its ideal location, and a precious source of comparison, in this Springer editorial series specifically dedicated to the Numanities, to which we are very proud to contribute.

The second part of the book entitled *Knowledge, Subjectivity, and Intelligence in Non-Human Animals*, presents three essays by Marchesini and tries to clarify the methods and contents of a holistic and post-mechanistic approach to the study and understanding of animal minds. The chapter delves into three deep fundamental aspects of animal life: (a) learning process, (b) subjectivity and (c) intelligence. The phenomenon of learning represents one of the most important processes of behavioural identity development in animals. Starting from the mid-nineteenth century, there have been many explicative proposals and models aimed at describing this process. Nowadays we are faced with a number of models, often forcedly juxtaposed even when incompatible: this was the case with associationist, psychoenergetic and cognitive models. The question is whether this abundance could be replaced with a unique model able to subsume the different occurrences and resolve the inconsistencies still present in all these explanations. As a second aspect, this chapter wants to show how the issue of animal subjectivity has been addressed in many ways over time: the different explanations had come from the anthropomorphic assimilative interpretation, supporting a projective view by which animality is just a declination of being human, to the break operated by René Descartes, who assimilated the animal condition to mere *res extensa*, thereby annihilating any hint of its subjectivity and depicting the non-human as a mass of deterministic mechanisms. I would like to deal with the issue of animal subjectivity avoiding any recourse to a mere projection through which conscience is considered the decisive element that causes subjectivity to emerge out of nothing, like a rabbit out of the hat. Finally, we come to intelligence: despite its multiple possible interpretations, Western tradition has restricted the act of knowing to a principle of disjunction from corporeality, in a sort of detachment and contemplation of the world ascribable to the res cogitans and to the exclusion of mere fruition. Knowledge, understood as a neutral and objectivizing act, as something neither participatory nor emerging from the relational predicate, has engendered a diaspora between the knowing and the known: a dichotomy that reflects in a fractal producing other dichotomies in turn. The author tries to develop the last path described here, referring to Lorenz's view on the adherence of cognitive tools to the configuration of reality. The issue is also to try to understand how to reconcile an interpretation based on an animal being's subjectivity, full of perspective protagonism in any interaction with the world, with an epistemological framework that accounts for the phylogenetic declination as a given dimension.

The third part of the book discusses the contributions of ethological research and behavioural epigenetics to furthering the exploration of two topics that the authors consider equally important for contemporary ethical debate. One is the *critique of the anthropocentric moral tradition* by contemporary anti-speciesist ethics; the other is the refutation of the theoretical presuppositions of *behavioural, social and moral determinism* re-proposed in recent years by leading exponents of evolutionary psychology, now sanctioned by the developments of the research on epigenetic inheritance.

The aims to which the studies here presented are oriented imply, in fact, together with the effort to understand both human and non-human minds and cultures in a non-anthropocentric and non-ideological way, the commitment to defend their autonomous existence. In other words, the comparative study of human and non-human animal traditions and forms of thought requires, as one of its indispensable correlates, *an active participation in the struggle to protect them and the natural environments in which they have evolved from the destruction and extinction* that many of them are undergoing due to anthropic impact and dominant destructive forms of economic exploitation of both human communities and natural resources.

The essay opening this part (chapter 7), written by Celentano and entitled *Contributions of Ethology to the Birth of a Post-Anthropocentric Ethics*, documents the role that ethological research has played in promoting the birth of post-anthropocentric and anti-speciesist ethical movement taking on this fight. Therefore, this chapter highlights the historical and cultural link between two ongoing revolutions one scientific, introduced by ethology and one ethical, proper of current anti-speciesism.

The eighth chapter, written by Marchesini and entitled *A Re-evaluation of animal interests starting from a critique of Maslow's Pyramid*, shall demonstrate that understanding the subject of interest as an intentional entity is not enough to infer sentiency neither from welfare parameters nor Maslow's Pyramid, and compassion or sympathy are not even enough, interpreted in the etymological way of "being of the same dispositional feeling". In order to preserve the interests of the non-human animal as a subject, it is indispensable to: (1) accept the existentiality of him/her and avoid the mechanicism of heritage; (2) carry out an empathic approach, that is the ability to reproduce a condition or an inclination different from ours, that requires a suspension of the anthropomorphism. It is possible to know the intentional dimension of non-human animals if we apply Darwin's criterion of adaptative resemblance and distinction. But in order to do that it is necessary to strengthen scientific knowledge and ethical reflection because neither approach explains how the interest subjectivity emerges.

The ninth and final chapter, written by Celentano and titled *Behavioral and Cultural Epigenetics. Social Biologisms Refuted by the Developments in Biology*, aims to clarify that the theoretical framework of the contemporary evolutionary biology, and the experimental evidence on which it is based, allows us to definitively refute and dismiss all approaches to the study of animal and human behaviour based on genetic determinism. In fact, much data accumulated over the past thirty years show that, in the course of phylogenesis, three kinds of selection, heredity and variation, respectively, *epigenetic, behavioural and cultural* ones operated alongside the slow processes of genetic variation, and much faster than it, in tight conjunction with environmental inputs.

As far as our species is concerned, this data shows that, as already understood by Darwin (Darwin 1871a), for a long time it has been human *social* history that has guided and shaped human biological history, not vice versa. The analysis proposed in this chapter shows that it is now anachronistic to hypothesize a "human nature" rigidly codified at the genetic level and substantially unchangeable in its fundamental mental and behavioural propensities, as some exponents of evolutionary psychology still do (Pinker 2005; Tibayrenc and Ayala 2016).

A large amount of experimental evidence in fact demonstrates that social context can, through experience and its epigenetic, behavioural and cultural inheritance, either inhibit or implement the cognitive potentials and behavioural attitudes of its members with effects that are transmitted from generation to generation.

In other words, we can finally *refute old and new social biologisms*, or ancient and re-emerging genetic determinisms, *with the tools of biology itself*.

The authors wish to conclude this introduction by expressing their deepest gratitude to Springer editions, to Dario Martinelli, director of the Numanities Arts and Humanities in Progress series, and to the scientific and editorial board of the series for having considered this fruit of their work worthy of publication.

Marco Celentano
Roberto Marchesini

References

Bjorklund, David and Pellegrini, Anthony. 2002. The origins of human nature: Evolutionary developmental psychology. Washington (DC): American Psychological Association.

Buchanan, Brett, Bussolini, Jeffrey, Chrulew, Matthew. 2014. General introduction: Philosophical ethology. *Angelaki. Journal of the Theoretical Humanities*. 19 (3): 1–3.

Bussolini, Jeffrey. 2016. The philosophical ethology of Roberto Marchesini. *Angelaki. Journal of the Theoretical Humanities*. 21 (1): 17–38.

Celentano, Marco. 2018. *Interspecific cultural convergences (ICC) and Interspecific cultural studies (ICS)*. In Ceccarelli, David, Frezza, Giulia (eds.). *Predicatbility and the unpredictable*. Roma: CNR Edizioni.

Chrulew, Matthew. 2014. The philosophical ethology of Dominique Lestel. *Angelaki. Journal of the Theoretical Humanities*. 19 (3): 17–44.

Darwin, Charles. 1871. *The descent of man*. London: John Murray.

de Waal, Frans, 2016. *Are we smart enough to know how smart animals are?* London: Granta.

Lestel, Dominique. 2001. *Les origines animales de la culture*. Paris: Flammarion.

Lestel, Dominique, Brunois, Florence, Gaunet, Florence. 2006. Towards an etho-ethnology and an ethno-ethology. *Social Sciences Information*. 45: 155–177.

Marchesini, Roberto. 2016. Philosophical ethology and animal subjectivity. *Angelaki. Journal of the Theoretical Humanities*. 21 (1): 237–252.

Tibayrenc, Michel and Ayala, Francisco (eds). 2016. *On Human Nature: Biology, Psychology, Ethics, Politics, and Religion. Amsterdam, Boston, Heidelberg, London, New York: Elsevier.*

Contents

Part I
Towards a Critical Re-Foundation of Cultural Studies

Chapter 1
From the Darwinian to the Ethological Revolutions: An Ongoing Process

Marco Celentano

Abstract Like Darwin's works and theories, the studies and discoveries produced by ethology (a research area of which the great naturalist was the first promoter) inspired not one, but *two* scientific revolutions. The second is still in progress. As with the *first Darwinian revolution*, some of the theoretical, social and ethical implications of the *first ethological revolution* have long been distorted, partly even by its own promoters. It has been arbitrarily used to support forms of behavioural determinism according to which all aspects of animal and human minds and activities are substantially regulated by hereditary mechanisms that are scarcely modifiable through experience, education, culture and socio-environmental stimuli. One of the goals of this book is to demonstrate that this form of ethological mechanicism and social biologism can now be refuted with the theoretical and methodological, empirical and experimental tools of biology and ethology themselves. To this aim the present chapter contributes through a critical review of the two Darwinian revolutions, of the first ethological revolution, and of some of their interpretations that had a wide echo. It also introduces an analysis of some aspects of the second, ongoing ethological revolution, and of contemporary evolutionary studies, which are further examined in the following sections of the book, showing that developments in both these areas are converging towards a post-mechanistic model of animal behaviour and a post-genocentric explanation of evolutionary processes. In this chapter I try to show that, with respect to these developments, the *Darwinism of Darwin* is demonstrating a fruitfulness, a resilience, and an attitude to frame phenomena that at the time of its formulation were unknown, far superior to that of all the "*neo-Darwinian*" models which predominated in evolutionary biology after Darwin. That is to say that, at least since August Weismann's *Germinal Selection* (Weismann 1896), to Jacques Monod's *Le hasard et la nécessité* (Monod 1970), Darwin's Darwinism, although focused on the concept of natural selection, implied an explanatory pluralism and a series of (albeit critical and cautious) openings to the possibility of "Lamarckian" forms of inheritance rejected by subsequent neo-Darwinist models in defence of a supposed Darwinian "orthodoxy" only to be once again re-evaluated by contemporary epigenetics. In the following pages I attempt to summarize the outcomes both of the two Darwinian and the first ethological revolutions, highlighting their nature as flows of ensuing scientific-cultural events, the implications of which are in many respects still at stake, open-ended and ongoing. The scientific revolutions discussed

R. Marchesini and M. Celentano, *Critical Ethology and Post-Anthropocentric Ethics*, Numanities - Arts and Humanities in Progress 16,
https://doi.org/10.1007/978-3-030-74203-4_1

in these pages are in fact only mere stages of a long, single, internally conflicting and composite revolutionary process that leads from Darwin's proto-ethology to contemporary ethology.

1.1 Introduction

Over the last sixty years many scholars and disseminators have referred to the concept of a "Darwinian revolution", but Patrick Tort, one of the most authoritative historians of Darwin and Darwinism has since the 1980s pointed out that it is more appropriate to talk about *not one, but two Darwinian revolutions.*[1]

In this essay I welcome Tort's suggestion and draw on his extensive reconstruction of Darwin's research path and the social processes that influenced his reception. Nevertheless, I will advance some criticisms of Tort's interpretation of the "second Darwinian revolution" and the risks of an *idealization* or a "monumental" reconstruction of the "*civilization*" process that in my opinion, it presents.

Like any other scientist and human being, Charles Darwin was of course not immune to the ideological conditioning and social prejudices of his time and social environment. Thus his theories are not lacking in limits, inadequacies, fluctuations or ambiguities.

However, among his contemporaries disseminating and renewing "transformism", he was at once the most sober, radical, coherent and far-sighted. A man inclined to subscribe to an optimistic faith in progress typical of his time, but also one of the most lucid scholars in glimpsing crucial issues arising from the social and theoretical implications of the genealogical perspective.

In a nutshell, as Karl Marx wrote to Ferdinand Lassalle in 1861, the first Darwinian revolution, consisting in the detailed exposition of the theory of natural selection contained in *The Origin of Species*, gave "a mortal blow to teleology" (Marx in Marx and Engels 1975–2004, 41: 246–247), making explainable the origin of all the living species without resorting to finalistic principles.

With his second revolution, of which the works *The Descent of Man* (1871) and *The Expression of Emotions in Man and Animals* (1872) were the heralds, Darwin obtained at least three important results:

[1]Curiously, the wide diffusion of the concept of a "Darwinian revolution", recorded in the last sixty years, took its cue in 1959 from the title of a biography of Darwin, written by the historian Gertrude Himmelfarb (1959), which radically opposed to the theory of natural selection and tried to refute it. Since then, the formula "Darwinian revolution" has been taken up by numerous scholars and advisers, appearing in the titles of many volumes and articles, predominantly but not exclusively aimed at emphasizing the scientific relevance and the still current aspects of Darwin's theories and studies. I will limit myself to recall: Michael Ruse (1979), Patrick Tort (1983, 1992); *Journal of the History of Biology*, 38 (1) 2005, entirely devoted to this theme with contributions pro and against Darwin written by many well-known scholars, and David Sloan Wilson (2019).

- the abolishment of the traditional metaphysical man-animal dichotomy, having shown that almost all mental capacities for millennia considered exclusive to man are also widespread in other animal species;
- the foundation of a new field of research: the comparative study of animal expression, behaviours and abilities, which also includes the human species, and their evolutionary history;
- a critique and an overcoming of the *ethological determinism*, improperly referred to as a "social Darwinism" (Tort 1999), which had become very widespread in the previous decade.[2] Darwin rejected the Spencerian belief that human moral and social traditions were the product of a natural selection that only preserves what is really "useful", proposing the alternative hypothesis that, in the history of human customs, *education and social control* have supplanted and replaced "natural selection", becoming largely autonomous from it (Darwin 1871: 404).

Of course, both Darwinian revolutionary turns left some problems open. Undoubtedly, however, as a whole, Darwin's work opened a new phase in Western thought, helping to demolish prejudices rooted for centuries and, in some cases, millennia.

In fact, the Darwinian revolutions produced the effect of reconnecting humans to other animals and to their natural history, introducing a change no less radical than that caused by the Galilean revolution, reconnecting Earth and sky.

Daughters of the Darwinian ones have been the two ethological revolutions that crossed the twentieth century:

- The first goes from the birth of classical ethology founded in the 1930s to human and cognitive ethology, which arose in the Sixties and Seventies.
- The second, still in progress, is the transition to a post-genocentric and post-anthropocentric turning point that in the last three decades has led to a new "philosophy of ethology", to important developments in cultural and cognitive ethology, and to the emergence of new areas of research such as behavioural and cultural epigenetics.

As is documented in the concluding essay of this volume, these changes, inherent to behavioural sciences, converge with the concurrent developments in evolutionary studies. Both indeed move towards a vision of evolution that is not only characterized by external selection and genetic mutations, but also by an organisms' active search for more suitable living conditions in driven by epigenetic, behavioural and cultural inheritance forces of evolutionary processes.

Both in evolutionary and ethological studies, evolution is today conceived as a selective process in which organisms are protagonists, and animals are considered,

[2]For a critical approach to "social Darwinism" and an analysis of its multifaceted character, see: La Vergata (1999, 2001, 2005). La Vergata shows how social Darwinism became a pseudo-scientific justification of different ideological positions. Among these, a prevalence was granted to those supporting the "elimination of the unfit", providing biologistic and pseudo-naturalistic justifications to political cynicism and to economic exploitation of men and nature, but other brands of social Darwinism existed, such as "a liberalist Social Darwinism, a statist-conservative one, a militaristic one, and then one pacifist, one socialist, one anarchist" (La Vergata 2005: 21).

not as Cartesian "machines" or Dawkinsian genetic "robots", but as sentient and intelligent beings who learn from experiences, transmit them, and actively transform their environments, orienting their ontogenetic and phylogenetic history.

The scientific revolutions discussed in these pages are actually stages of one, ongoing process: a single, long, revolution, internally conflicting and composite like any revolutionary process.

Developments are ongoing because of the immense scope of the yet to be studied phenomena concerned and the ethical and social implications they bring about, too. In fact, we now have a possibility that was previously not available: conducting an empirical, experimental, theoretical and historical *refutation of both anthropocentrism and gene-centrism*, using instruments offered by developments in the very same biological and behavioural sciences.

While we discover phenomena unsuspected until a few decades ago, such as the existence of millenary and/or secular cultural traditions in other species,[3] or the complexity that animal thought can reach, we also live in an age characterized by the daily devastation of ecosystems in which all wild animal species live perpetrated by an anthropic development guided by a single logic: that of immediate profit. Enormous industrial apparatuses linked to intensive breeding of animals for meat production significantly contribute to pollution, foolish consumption and environmental catastrophes. These phenomena thus pose new important ethical, social and ecological challenges.

As shown in greater detail in the following essays of this volume, these are historical passages that call upon both human and natural sciences to undertake paths of critical re-foundation of their own educational and research methods, calling for epochal changes overcoming the traditional bipartition between humanities and life sciences, creating scientific and professional training courses offering skills that are transversal to these two blocks.

1.2 Darwinian Revolutions and Their Emancipatory Effects

At the end of *The Origin of Species*, Darwin wrote: "When the views entertained in this volume on the origin of species, or when analogous views are generally admitted,

[3] As already clarified in the third chapter, various sites and finds, discovered in the last two decades, attest to the existence of very ancient animal traditions. In 2007, in Côte d'Ivoire, a coconut crushing site that had been used by local chimpanzee populations for no less than 4300 years was discovered (Mercader et al. 2007). The use of stone tools has also been observed in some anthropoid monkeys and, in 2016, a site for crushing cashew nuts, used by local communities of striped cebi (Sapajus libidinosus) for over 700 years was found in Brazil, in the National Park of Serra da Capivara, (Haslam et al. 2016). Moreover, in 2014, an article by Catherine Hobaiter and her collaborators, published in PLoS Biology, for the first time documented a phenomenon of transmission of a cultural innovation consisting in the invention of a sponge made with leaves and mosses among a group of wild chimpanzees (Hobaiter et al. 2014).

we can dimly foresee that there will be a considerable revolution in natural history" (Darwin 1859: 484).

He was well aware that his revolution in natural history would trigger a domino effect in the whole domain of sciences, challenging beliefs rooted in a millenarian tradition and forcing a drastic redefinition of the distinction on which the whole Western system of knowledge was based: that between *natural sciences and humanities*.

Darwin was of course not the first to refute the belief in the fixity of species, to conceive living beings as products of an historical process and to affirm that man descends from other animals. The same road had already been taken by Diderot, Buffon, Saint Hilaire, Erasmus Darwin and other, more or less renowned scholars since the seventeenth century. Thanks to Lamark's theory and Spencer's from the 1850s, an evolutionary perspective was rather common for the first decades of the nineteenth century. Furthermore, in parallel with Darwin, Wallace had also independently conceived a theory of natural selection.

Darwin's approach was, however, unlike other any previous evolutionary models. His was immediately perceived as subversive by the cultivated classes of that time. Other evolutionists spoke of "vital forces" or of an "essential irreducibility" of human mind to its material components. Herbert Spencer, the most notorious among them, hypothesized an allegedly necessary "law of progress", operating at each and every level of reality (Spencer 1857). "A panoply of concepts that traditional Christianity could accept in compromise, for they permitted a Christian God to work by Evolution instead of Creation" (Gould 1977: 24–25). In fact, such models, although rejecting some traditional religious dogmas such as the fixity of the species and the theory of "separate creations", still re-launched and strengthened other important aspects of the Western traditions, among which the teleological (and at times explicitly theological) approach to the studying of natural phenomena and the anthropocentrism re-launched by the image of Man as the maximum height of evolution. The difficulty to attempt, in this historical phase, a passage from models that were clearly suspended half-way between innovation and tradition to a rigorous genealogical approach is testified by the fact that Alfred Russel Wallace himself, the joint discoverer of natural selection with Darwin, made ample concessions to Christian dogmatism, describing the human mind as "the only divine contribution to the history of life", and human evolution as a process led by a "superior intelligence" (Wallace 1870: 360).

Darwin took a firm position against him on this ground, although he himself had not been completely immune to some lexical concessions to the religious orthodoxy. In the final chapter of *The Origin*, for example, there is a reference to the moment when "the first creature [...] was created" (Darwin 1859: 488), though the Creator's possible role is confined to the appearance of the earlier living forms, in direct opposition to the traditional hypothesis of the "separate creations".[4]

---

[4]One of the explicit goals set by Darwin in *The Origin of Species* was to demonstrate the unsustainability of the dogma of separate creations, i.e. of the conviction, based on biblical sources but debated at length in the sixteenth and seventeenth century, that each living species had been called into being by God through a separate act of creation. Such a doctrine is of course irreconcilable

However, aside from this terminological concession to creationism in relation to *the origin of life on Earth*, a topic that was not covered in the work, Darwin's theory explained the existence of all present species, including humans, as deriving from one or a few common progenitors through a process of selection and conservation of the variants which were most adaptable to the environment. The strength and weakness, simplicity and intricateness of the Darwinian concept of "natural selection" derives precisely from this fact: it is presented by the author as a principle that is both *negative*, i.e., privative, and positive, i.e., cumulative., thus being a principle capable of giving rise to new useful solutions. Natural selection is the gradual elimination of the less suitable, but also the "*accumulation and strengthening of advantageous variations*" (La Vergata 2001: 208).

In short Darwin's theory represented an explanatory model that, for the first time, did not resort to making any allowances for the intervention of divine forces or mysterious progressive tendencies in inherent biological and human evolution.

Exactly for this reason, as Marx and Engels pointed just a few months after the publication of *The Origin*, Darwin's approach inflicted "a fatal blow to teleology" (Marx in Marx and Engels 1975–2004, 41: 246–247).[5]

As Friedrich Nietzsche (1868) and Ernst Haeckel (1868) reiterated a few years later, Darwin, with his theory of "natural selection", had invalidated the assertion made by Kant in section 75 of the *Critique of Judgment*, according to which: "This is so certain that we can boldly say that it would be absurd for humans even to make such an attempt or to hope that there may yet arise a Newton who could make comprehensible even the generation of a blade of grass according to natural laws that no intention has ordered" (Kant 1790 [2005]: 185). In other words, Darwin paved the way to a radical *secularization* of the problem of the descent of living species and man. Human history was reunited with animal history, producing a paradigmatic change no less dramatic than the Copernican revolution which had reunited sky and Earth.

After Darwin, not only the traditional (implicitly or explicitly) theological and teleological presuppositions of natural sciences, but also the anthropocentric prejudices on which human sciences had been founded for millennia, and the whole traditional philosophical field, from the theory of knowledge to ethics, was profoundly and radically problematized. The investigation of man's "spiritual" activities, emotions, feeling and knowledge, as well as of human expressiveness and language, took a different direction from that moment onwards. Without Darwin, many milestones of Western culture would have simply not existed: from Nietzsche and Freud's revolutionary approaches to the exploration of the subconscious and the problem of "discontents of civilization", to the birth of new research fields like classical, human,

---

with Darwin's genealogical perspective, according to which all existing species derive from few common progenitors.

[5]The passage is taken from a letter written by Marx to Lassalle (January 1, 1861). Writing on the same topic to Engels, a few days earlier (December 19, 1860) Marx had noted: "Although it is developed in a crude English style, here is the book that contains the natural-history foundation of our point of view" (Marx in Marx and Engels 1975–2004, 41: 231–232).

cognitive and cultural ethology with the inclusion of non-human animals, or a part of them, among subjects considered to have an intrinsic ethical value.

For his part Darwin himself, in the course of his entire scientific activity, assigned a central place to the problem of the origin of animal and human "mental faculties", trying to construct a genealogical theory to explain the processes leading from the appearance of the earliest organisms to the development of the human species with their species-specific features.

Still as a young naturalist, since 1838, he conceived in his Notebooks (Darwin *Posthumous* 2009) the ambitious project of a theory capable of explaining both the origins of the *anatomical, morphological* and *physiological* features of living beings and the appearance and transformations of the *behavioural* and *mental* animal and human traits, freeing these research domains from theology and teleology.[6] This polemic motivation was at the origin of an extensive research project, which he later abandoned, but continued to provide a general framework for his later studies.

While studies of morphological, anatomical and functional differentiation between species were integrated twenty years later in *The Origin of Species*, the behavioural parts were incorporated in Chapters 3, 4, 5 and 21 of *The Descent of Man* (Darwin 1871) and in *The Expression of Emotions in Man and Animals* (Darwin 1872), these are now rightly considered forerunner texts of modern ethology. At least from a general theoretical perspective, it is with these studies that Darwin completed his revolutionary enterprise, obtaining an extremely shocking triple effect for the culture of the time, proving that:

(a) it is indeed possible to explain the evolution of living organisms, from its first steps to the appearance of man without resorting to any extra-natural, teleological, or aprioristic factor;
(b) the so-called "superior" abilities that were traditionally exclusively attributed to man are at least in part found in other animals and depend on organs and apparatuses that we share with many other species.
(c) the theory of natural selection does not imply that human social behavior is determined by hereditary factors in a non-modifiable way, because it is fully compatible with the finding that, in human history, social environment and education have gradually become more powerful selective factors than external environmental selection.

1.3 Social Darwinism as a "Conservative Revolution"

He who proclaims a new idea never gets away with it. Moreover, if this idea is the Darwinian doctrine of evolution, which, since the second half of the nineteenth century has become a

[6] In a letter to Wallace in 1867 Darwin wrote: "I want anyhow to upset [the] idea [...] that certain muscles have been given to man solely that he can reveal to other men his feelings. I want to try & show how expressions have arisen" (Ch. Darwin to A.R. Wallace, [12–17] March [1867], in Burkhardt and Secord 2005: 141).

> field for recurring polemics (the periodical character of which calls for deeper analysis) for the simple fact that it challenged a body of dominant conceptions, one faces a twofold risk: either having it repressed it in its entirety, or having it reabsorbed within that very system of representations it had intended to overcome and demolish. (Tort 2000: 19)[7]

How was the critical and revolutionary potential of Darwinian theory channeled and controlled? In which of the ideological trends associated to that period was the reception of Darwinism, at least partially, reabsorbed? On which *internal* elements of the theory did such attempts leverage and which explicit Darwinian positions instead had to be arbitrarily twisted or misrepresented in order to achieve these results?

There is no doubt that some theoretical elements that Darwin assumed in *The Origin of Species* derived from the classics of liberal and liberalist thought. The concept of evolution as a gradual ascending progress, very appealing during the Victorian age and already asserted as a scientific certainty by Lamarck and Spencer, was certainly present, though restrained in Darwin's work. Nevertheless, it is honoured in the concluding pages of the *The Origin of Species* where Darwin wrote: "Hence we may look with some confidence to a secure future of equally inappreciable length. And as natural selection works solely by and for the good of each being, all corporeal and mental endowments will tend to progress towards perfection" (Darwin 1859: 489). But most of all, from a classic of the late seventeenth century liberalist literature, Malthus's *Essay on the Principle of Population* (Malthus 1798), Darwin drew a model which assumed an important role in his theory of natural selection: the model asserted that the human population, in the absence of obstacles, tends to increase more rapidly than the livelihoods it is able to produce, due to its geometrical increase (1–2–4–8–etc.), whereas the latter increased arithmetically (1–2–3–4–etc.). On the basis of some aspects already found in Malthus, Darwin extended this principle to all living species and concluded that an insufficiency of resources would ineluctably induce a "struggle for survival" among individuals of the same species and between antagonist species.

The observations made as a naturalist during his voyage on the Beagle, the studies on artificial selection carried out by livestock breeders and farmers and the works of Adam Smith, another classical exponent of liberal thought, all suggested to Darwin the idea that this struggle could gradually lead to a "selection of the fittest".[8] An idea also supported by Spencer: a differential reproduction, favourable, within a species or population, or between competing species, to the individuals or species best at exploiting their environmental conditions. A process of adaptation that led to a slow modification of the species and to the advent of all past and present species out of a limited number of simple, primordial ancestors.

The concepts of natural selection and struggle for existence presented by Darwin in 1859 undoubtedly indicated the liberal optimism about the regulatory effects of a "free" competition for the hoarding of resources that was so widespread during

[7] This and all the other quotations from essays that have not been translated into English, contained in this chapter, are my translations.

[8] Darwin adopted the expression "selection of the fittest" from Spencer, starting from the third edition of *The Origin*.

his times. It also favoured a utilitarian view of organisms, in which the behaviour that every living being has to follow, in order to remain alive, is "conceived as a variant of competitive, acquisitive, «egoistic» and calculating behavior that is attributed to «rationality» *tout court* by the liberal theorists of classical and neo-classical economics" (Cavazzini 2009: 5). However, Darwin's approach changed, at least in part in *The Descent of Man*, published eleven years and three months after *The Origin*.

In fact, in this work Darwin argued that the moral rules oriented to mutual solidarity and to support the weakest had evolved in human societies from forms of parental care and "social instincts" present in all the gregarious animals, to be later rewarded by natural selection having proved useful in strengthening the group (Darwin 1871: 166). In other words, according to the Darwin of 1871, in the most recent stages of human history, social selection has increasingly taken precedence over natural selection. It acts through the rules, traditions and educational processes and has become the main driving force of conservation or changes in customs and behaviour (Darwin 1871, I).

Assuming this hypothesis, Darwin distanced himself not only from Spencer, who had criticized public aid to the less well-off by justifying the system of competition between classes, nations, economic groups and individuals as an inescapable law of nature, but he also distanced himself from the positions of all the other main evolutionism exponents of the time, including his friend, T.H. Huxley who postulated a radical break between the moral sphere and the natural sphere. Neither did Darwin endorse the biologistic justifications of eugenics advanced by his cousin Galton (to which, however, especially in the final pages of *The Descent of Man*, Darwin made some concessions[9]), nor the racist and colonialist ideology of the German "mastiff" of Darwinism, E. Haeckel.

But despite this, the interpretation given by most of Darwin's contemporaries misunderstood the meaning and field of application of concepts such as "struggle for existence" and "selection of the fittest", arbitrarily extending their use to the analysis of human social history and economic reality. As Tort observes, "the mainframe of European, and later American, interpretations of Darwin after 1860 is always

[9]Although fiercely taking position against slavery and other forms of social discrimination and exploitation, Darwin was not entirely immune to eugenic concerns and not completely averse to the promotion of some positive and negative eugenic measures. For example, in chapter V of *The Descent of Man*, he writes: "We must bear without complaining the undoubtedly bad effects of the weak surviving and propagating their kind; but there appears to be at least one check in steady action, namely the weaker and inferior members of society not marrying so freely as the sound" (Darwin 1871, I: 169). He adds then on the same page, a few lines later, this consideration: "In all civilized countries man accumulates property and bequeaths it to his children. I know that they are children in the same country. But this is far from an unmixed evil; for the capital the arts could not progress; and it is chiefly through their power that the civilized races have extended, and now they are everywhere, their range, so to take the place of the lower races" (Darwin 1871, I: 169).

constant: themes such as competition, struggle for life, survival of the fittest, cumulative transmission of benefits, elimination of the less fit and negative selection are always underscored and applied to human societies" (Tort 2000: 19).

The model that inspired this so called "social Darwinism" was actually Spencer's evolutionary metaphysics and not Darwin's theory, anthropology or political convictions.[10]

At that time, the influence of this biologistic approach, which should be more correctly called "social Spencerism", became so pervasive, and widespread in so many different currents of thought and disciplinary fields that we could compare its success to a sort of "conservative revolution" *ante litteram*.[11] Through this process of interpretative distortion and theoretical flattening, Darwin's theories of natural selection were arbitrarily equated with the metaphysical principles of the gradual emergence of the fittest and of the gradual progress towards the best, which the Spencerism applied at a cosmological level.[12]

---

[10]Of course, emphasizing the differences between Darwin and the social Darwinists (even those, such as Haeckel and Darwin's cousin Galton, who were closest to him) I do not intend to present Darwin as a man who was above all the prejudices of his time and cultural environment. As I tried to show, the mentality then dominant in Great Britain and Europe was variously reflected in the works of Darwin. However I find it necessary not to lose sight of the differences that allowed the Darwinian theory of descent with modifications, to impose itself against earlier contemporary and later genealogical models, for its superior scientific rigor, for the unprecedented attempt to do away with theology, metaphysics and teleology and, finally, for its internal consistence, its adherence to the observed phenomena and its explanatory power.

[11]As it is well known, between 1918 and 1932 the German culture was greatly influenced by some theorists explicitly referring to Hugo von Hoffmanstahl's idea of a "conservative revolution". They propounded the rediscovery of Germanic national traditions, an anti-modernism and an elitism of a romantic brand, keen on theories of racial discrimination, and the exaltation of the heroic and tragic element of life. Among the best-known exponents of this wave were the philosopher O. Spengler and the philosopher/writer E. Jünger. Culturally close to the ideologies of part of the nascent Nazi regime ideology, with which they initially collaborated, the main representatives of the "conservative revolution" remained marginal after Hitler's advent. They were close to some aspects of the theories of the fathers of "conservative revolution" and to other intellectuals, such as C. Schmitt, A. Moeller van den Bruck, M. Heidegger, Th. Mann, W. Sombart, M. Scheler and the philosopher-psychologist L. Klages.

[12]In the books *Progress, Its Law and Cause* (1857) and *A System of Synthetic Philosophy: First Principles* (1862), Herbert Spencer had theorized the existence of an evolutionary "law" of the "selection of the fittest", or "law of progress", operating as a universal principle at all levels of reality: cosmic, biological, social and moral. In fact, according to the author, evolution can gradually originate a growing amount of "happiness", and this "law" acts identically both in nature and in human societies. Economic and social differences, as the differences in development among different cultures, are to be intended as the outcome of differential adaptability at an individual and group level. On these grounds, and following Malthus, Spencer fiercely criticized the "Law for the Poor", or the earliest forms of social assistance in Britain, as well as the religious practice of charity, specifically addressing the taxing of the rich in order to alleviate the sorrow of the poor; he interpreted them as "obstacles" to the survival of the fittest (Spencer 1887). According to Spencer, in fact, it is from the death of the "unfit" that Evolution receives its ascending character and those who survive must in any case "be the chosen of their generation" (Spencer 1887).

In this sense, natural selection suffered a fate analogous to that which, a few decades later, was to fall upon the Nietzschean theory of the "Wille zur Macht", distorted by the promoters of the so-called "conservative revolution", and later by Nazism, in an exaltation of nationalism, an apology of expansionist policies and a political anti-Semitism that were foreign to Nietzsche.

Apologists of this tendency were, together with Herbert Spencer, a large number of European and American epigones of his approach (notable for his radicalization of the Spencerian doctrine was William Graham Sumner), according to whom the criterion of *laissez faire* should have ruled all aspects of social life and also provided a model for governmental policies. They believed, in fact, that only a free competition of forces would have led to a gradual elimination of the "unfittest" and to a world in constant progress towards the better. The Darwinian "struggle for existence" in which Kropotkin rightly saw both co-operation and competition among individuals and species, was transformed by this radically reductionist interpretation into an equivalent of Hobbes' "*bellum omnium contra omnes*".

Patrick Tort rightly emphasized the fact that Darwin, with his second revolution, distanced himself from Spencerism and various other forms of "social Darwinism" which had begun to spread after 1959. In fact, Darwin makes it clear in this work that, in his opinion, since ancient times and then in an increasingly incisive way in modern ones, behavioural, cognitive, social and "moral" human propensities have been conditioned and oriented *by a social selection and not by the natural one*, by "education" and social control, not by hereditary factors.

As Tort reiterates, from this point of view, Darwin's anthropology achieves almost a double revolutionary result: firstly, to abolish every metaphysical break between human and other animals, rejecting the hypothesis that to explain the origin of our mental and "moral" characteristics it is necessary to postulate the action of extra-natural factors, as Wallace stated. Secondly, to defend the hypothesis of a direct continuity between animal and human evolution, rejecting at the same time the hypothesis of a "simple continuity" between them, in which natural selection drives both natural and human history. This way Darwin arrived at an epistemological approach that recognized the (at least partial) autonomy of social development from natural selection, thus allowing credit for "the theoretical autonomy of the sciences of man and society without breaking the historical-material continuum between «nature» and «culture»" (Tort 2000: 53). It was a turning point, not less important than that marked by the theory of natural selection. However, in obedience to the historical "law" suggested by Tort (no great theoretical innovation escapes ideological distortion), even in this case the anti-deterministic revolution introduced by Darwin with *The Descent of Man* in some way paid its price to the ideological universe of the time.

1.4 Tort's Interpretation of the Second Darwinian Revolution

As I have already mentioned, in my opinion, the very important critical goal obtained by the second Darwinian revolution presents, already in its original exposition (Darwin 1871), and even in Tort's interpretation, the risks of a "monumental"[13] and idealized reconstruction of the process that they define as "*civilization*".

In a nutshell, Tort claims that according to the conclusions reached by Darwin in *Descent of Man*, "civilization" allowed human societies to gradually escape the laws of survival of the fittest, and therefore the eliminatory function carried out by natural selection in all the other species. In his opinion, in fact, within our species, social "instincts" and behaviours of mutual support had proven, in the long run, more advantageous than those exclusively based on mere individual competition and had consequently been favorably selected. Since then, those groups and individuals capable of promoting the values of "morality, "altruism" and solidarity in society were favoured. This allowed a transition towards a new social effect: assisting the weak instead of eliminating them. According to the thesis that Darwin exposes in chapter V and takes up in various passages of *The Descent of Man*, the attitude of mutual aid, already rooted in the social instincts of our ancestors, offered human communities that practiced it most *as an established custom* a greater cohesion and incisiveness and new opportunities in the struggle for survival.

According to this interpretation, the process of "civilization" (of which Western culture has been the epicentre and driving force) coincides with a gradual imposition of the tendency to extend solidarity to ever wider circles, and finally even beyond the borders of our species. This process created, according to Tort, the conditions for an overcoming or a "reversal" of natural selection, achieving the conditions to remove its eliminatory mechanism. In other words, "civilized" human societies benefitting from social solidarity have overcome the "struggle for existence" which requires the most disadvantaged to succumb, creating rules for coexistence in which "the weak are no longer eliminated (intending here all the individuals whose psychological, psychic or social condition would have condemned them to death under the hegemony of «natural» law, but are instead protected, cared for and defended" (Tort 2000: 25).

According to Tort it is in this reversal of the effects of natural selection that lies the key to human "civilization" and in its identification does "the key to Darwinian anthropology", which was bearer of a "second revolution", even more important than that introduced with *The Origin of Species*, because capable of escaping the traps of social biologism without failing in the rigor of the genealogical perspective.

[13]I use here the adjective "monumental" in the sense that Nietzsche gave to it in the *II Untimely Meditation*, entitled *On the Advantage and Disadvantage of History for Life* (Nietzsche 1874). Nietzsche describes the monumental way of making history as a tendency to reconstruct the past, or determined epochs, in a celebratory way, removing all the aspects that do not give themselves to their idealization, and reducing the narration to a mythicization of some historical phases or characters in which "only a few embellished facts raise themselves up above, like islands" (Nietzsche 1874: Sect. 2. My translation).

Reaffirming that I consider Tort an undisputed master in his historical reconstruction and critical analysis of Darwinism, evolutionism, their reception and ideological implications, and I find his enhancement of the second Darwinian revolution useful and correct in numerous ways, I would like to summarize here some of the perplexities that his reconstruction of the "civilization" process arouses in me. First of all, I think it is appropriate to point out a gap, perhaps slight but not irrelevant, between what Darwin states in this regard in *The Descent of Man* and the generalizing form in which Tort sums up his position.

As a matter of fact, in the fifth section of *The Descent of Man*, Darwin stated that solidarity within the group has been one of the propulsive factors of human social evolution (Darwin 1871, I: 166) and has become a feature of "civilized" societies: "We civilized men, on the other hand, do our utmost to check the process of elimination; we build asylums for the imbecile, the maimed, and the sick; we institute poor-laws; and our medical men exert their utmost skill to save the life of every one to the last moment" (Darwin 1871, I: 168). However, he still recognizes that in many cases this form of solidarity has developed in co-evolution with the activity of war and the cultural dehumanization of other populations, as recent studies seem to confirm (Choi and Bowles 2007). In fact, Darwin stated that in a "primitive" situation "the tribes inhabiting adjacent districts are almost always at war with each other" and "the social instincts never extend to all the individuals of the same species" (Darwin 1871: 85). They are only addressed to community members and, according to the naturalist, for this very reason the greater internal cohesion and spirit of sacrifice of individuals, controlled by social mechanisms such as "praise and blame", offer more opportunities in competition with other communities, towards whose members no solidarity was expressed. It is thus not correct to assimilate this kind of behavior to a generic, and generally universal, principle of solidarity. But without a doubt, as Tort emphasizes, the interpretation of the civilization process as a gradual extension of the "circle of solidarity" is present in the Darwinian text. The great naturalist states, in the fourth chapter of the work, that feelings of sympathy and solidarity of human beings for their fellows have gradually grown "to extend to the men of all races, to the imbecile, the maimed, and other useless members of society, and finally to the lower animals,—so would the standard of his morality rise higher and higher" (Darwin 1871, I: 103).

According to Darwin, in the moral sphere "the elimination of the worst dispositions is always increasing" both in the "lower races" and in the "civilized nations" (Darwin 1871, I: 173). The main driving forces of this process overcoming the elimination of the less adapted have been and are, for him, "the approbation of our fellowmen—the strengthening of our sympathies by habit-example and imitation-reason-experience and even self-interest-instruction during youth, and religious feelings" (Darwin 1871, I: 173).

As I have already explained, it is precisely in this interpretation of the process of civilization that Tort identifies the turning point of Darwin's thought towards a "materialistic" and continuist, but at the same time non-mechanistic and non-biologistic conception of the relationship between organic and cultural evolution, natural and human social selection.

Without wishing to neglect the importance of the causal factors highlighted by Darwin and Tort, it seems to me that this kind of description of the process of "civilization" results inadequate or incomplete, precisely in identifying *the material causes and socio-economic selective pressures* that have oriented its development.

By this I mean that Darwin and Tort, presenting "civilization" as a process originated by motivations and changes of an almost exclusively ethical and intellectual nature, and characterized by a progressive strengthening of sympathy, empathy and altruism, leave in the shade a fundamental detail. The driving forces of that social process which the Western culture has called civilization, in fact, have been rather, in the first place, the brutal forms of exploitation of working classes imposed by the advent of capitalism and phenomena such as colonialism, slavery, imperialism, wars of conquest and the various forms of racial and sexual segregation that have marked the modern and contemporary era. Processes driven by economic and political interests *opposed* to those of mutual support, even if, from the time of forced evangelization of native Americans up to that of the armed "exporting" of democracy, humanitarian values have always been called into question *to ideologically cover* the pursuit of these interests. In short, in my opinion Darwin and Tort, intending "civilization" exclusively as a gradual progress of the spread of solidarity attitudes, propose an optimistic and unilateral, idealized and monumental reading of this historical and social process, purging it from all its "dark" sides.

This is the same dark side of "civilization" that Nietzsche wanted to bring out instead in his *On the Genealogy of Morality* (Nietzsche 1887). It will be enough here to mention only some passages of the second essay of the book in which the author, rejecting the hypothesis that altruistic behaviour developed spontaneously in conjunction with the progress of intellectual faculties and human moral sensibility, reminds us that "perhaps there is nothing more terrible and strange in man's prehistory than his technique of mnemonics. 'A thing must be burnt in so that it stays in the memory: only something that continues to hurt stays in the memory' – that is a proposition from the oldest (and unfortunately the longest-lived) psychology on earth" (Nietzsche 1887: II, 3).

It is not surprising to find a removal of these aspects and an optimistic-idealizing conception of the history of human customs in Darwin, because it was typical of the classical liberalism with which Victorian culture was imbued. But I confess my bewilderment at the fact that it may have been accepted by a very lucid critic of liberal ideology like Tort, and in an era like the early 1980s, in which the neoliberal winds embodied by Reaganism and Thatcherism were already dismantling the system of social rights and protections acquired by the masses in the previous cycles of social struggles.

In my opinion, it can and should be acknowledged that some trends and currents of modern Western culture, in various historical phases and especially in the epochal arc that goes from The Enlightenment to the social struggles of the 1960–1970s, *affirmed the possibility and the hope* for social forms founded on the Kropotkinian principle of mutual support to exist. Although they fought for this goal, this principle has certainly not become the basis of our associated life.

Unfortunately, this is demonstrated in a both brazen and tragic way by our epoch, in which all the so-called "civilized" nations barricade themselves as fortresses against the migratory waves *that they have provoked with their centuries-old predatory interference* in the African, Eastern, and South American countries. An era, indeed, in which all nations flout the same solidarity principles written in their constitutive national and international documents—such as providing aid to those in mortal danger and the reception for those who flee from conditions of humiliation, abuse and exploitation. Therefore, a historical phase in which all the States considered beacons and champions of "civilization" betray and disregard the aspiration that Tort seems to consider by them acquired and placed at the basis of social life: a social and institutional organization in which the "weak [...] are protected, cared for and defended" (Tort 2000: 25).

The global data concerning the variations in the concentration of wealth in the last centuries and decades also denounce a continuous widening of the gap between rich and poor. That is to say: an opposite trend to a progressive enlargement of the circle of human solidarity. Today, according to the data published by Oxfam (2019), the richest 1% of the planet holds almost half of the total net aggregate wealth (47.2%, to be precise), while 3.8 billion people, equal to the poorest half of the planet's inhabitants, can count just on the 0.4% of it.

As for the extension of ethical sensitivity beyond the boundaries of our species, it certainly represents an important phenomenon of our time. But the anti-speciesist movement, despite the generous forces it manages to mobilize, like environmental protection movements, today still only represents a small group going against the flooding river of a global society that is unfortunately responsible for a process of environmental devastation and mass extinction of animals and plants as well as an unprecedented level of exploitation of humans and animals.

Therefore, for the reasons summarized above, the paradigm drafted by Darwin and developed by Tort appears, from my point of view, a largely idealized reconstruction of "civilization" and its ethical, political, and socio-economic outcomes.

Paradoxically, this vision is intrinsically vulnerable to an ideological mechanism that Tort lucidly criticized on many occasions: the presentation of the *formal* equality of all citizens and of the principles of solidarity declared in the liberal-democratic constitutions and international treaties *as if* they were the real fulcrum of the social organization of the "civilized" countries and of their reciprocal relations. As if the private ownership of the means of production, and their use for private purposes by the ruling classes of every level that the current economic regime legitimizes and protects did not render an effective large-scale implementation of those solidarity principles structurally impossible. As if the capitalist economy had not yet been sufficiently proved to be able to create merely temporary (and never guaranteed in the medium or long term) improvement in the living conditions of the lower classes in one area of the world only at the price of intensifying or extending the processes of exploitation and social oppression in other areas of the globe.

The Kantian illusion that the spread of republican institutions would have led to a "progress towards the best", understood as the overcoming of war, social oppression and minority status human beings and the implementation of a "perpetual peace" was demolished by the actual realization of these institutions.

Therefore, at least from my personal point of view, a society of solidarity described by Tort, capable of protecting the weakest rather than exploiting or abandoning them, appears today a very important goal for which it is necessary to fight and not consider an existing result.

1.5 The Debate Between Genetic and Environmental Determinisms and the Birth of Classical Ethology

From the early decades of the 1900s, the debate on the *origin and modifiability of human propensities* was marked by a contrast between evolutionary innatisms and environmental determinisms, "biologisms" and "culturalisms".

A complex of philosophical hypothesis and ideological doctrines focusing on an *evolutionary nativism*, grafting onto a furrow already traced by nineteenth "social Darwinism", contributes to offer a pseudo-scientific basis to the social exploitation of humans and animals, for colonialist policies, racist and anti-Semitic tendencies, with gender and social discriminations.

On the opposite front, the *environmentalist* one, two major schools were developing:

- American behaviourism, which attempts to offer a scientific foundation to "democratic" propaganda in the USA through an approach based on the practices of classic (and later, operating) conditioning.
- "Dialectical materialism", as interpreted in the Soviet Union and in some currents of contemporary Western Marxism.

Although different and indeed opposite in many ways, these two approaches had in common a rigidly "culturalist" position, based on the conviction that the social environment, if subject to a rigid top-down control, could produce radical and positive changes in human psychology and behaviour in the turn of few generations.

At an ideological level, this principle played in both American behaviourism and the "orthodox" interpretation of dialectical materialism a role similar to that of nativism in the racist and authoritarian ideologies of the Nazi and Fascist regimes: it legitimated extremely invasive practices of institutional intrusion in the life of individuals and society. The two major figures of American behaviourism (J. Watson, since 1912–1913, and B. F. Skinner since the 1950s) scientifically validated the attempt to develop a "technology of behaviour" through large-scale top-down programming of activities and human response, based on the conditioning methods. Their goals were ultimately not different from those of Lysenko, supported by Stalin from the 1930s to 1950s, and reflected, in turn, albeit in forms which were different from those dominant in extreme right-wing regimes, a project of totalitarian control of the masses.

From this point of view, the frontal opposition between the different schools of nativists and culturalists/instructivists[14] hid a profound structural resemblance. On the one hand evolutionary nativism served as a biologistic justification of social traditions and political authoritarianism, on the other, culturalism served to justify the "pseudo-democratic" (as Konrad Lorenz called it) stance of American liberalism and the "pseudo-dialectic" one (as Theodor Adorno tagged it) of Soviet Marxism. Contrasting theoretical models led to similar practical outcomes: the justification of an invasive, manipulative, totalitarian use of scientific knowledge against the most basic rights of the human beings.

Classical ethology was born as an independent scientific discipline, in the first decades of the twentieth century, in a cultural climate characterized by these rigid oppositions. How did ethological disciplines fit into this complex debate? What innovations did they bring to it?

At the beginning its pioneers were definitely oriented towards a nativist approach. Their innatism, however, unlike the philosophical one of Spencer, was based on a deep knowledge of the behaviour of many species of wild animals in their natural environment. In this respect, the ethological approach, based on field observation, revealed an unknown world that had always been before our eyes but rarely studied with the passion, diligence and attention necessary to understand it. Daughter or niece of the Darwinian one, the first ethological revolution marked the third decade of the twentieth century.[15]

"The key actors in the founding of ethology as a discipline were Konrad Lorenz and Niko Tinbergen. It was Lorenz who was primarily responsible for laying the field's early conceptual foundations in the 1930s" (Burkhardt 2005: 4). According to Lorenz, the first steps of ethology were the result of a transfer of "Darwinian" theories and methodologies from specific fields such as morphology and anatomy to the comparative study of animal behaviour (Lorenz 1978 [1981]: 3). In the first decades of their scientific activity, both Lorenz and Tinbergen mainly focused their research on the identification and study of the "innate" components of behaviours. In truth, in that phase, Lorenz had already discovered important phenomena of integration between biological heritage and learning, i.e., *imprinting*.[16] But, as he himself admitted, he

---

[14]I here intend instructivism as a not only pedagogical model, but also a political one (relating to the relationship between ruling classes and masses) focused on the role of the instructor conceived as a figure that must assume full control of what is to be learned and of the ways in which it is to be learned, aiming to design the very same personality of the learners, conceiving them as passive receptors of the conditioning programs they are subjected to.

[15]As I previously, the phase I call "first ethological revolution" embraces the arc of time that goes from the foundation of classical ethology which occurred in the 1930s, to the birth of human and cognitive ethology, between the Sixties and Seventies. The "second ethological revolution", still in progress to this day, goes instead from the post-genocentric and post-anthropocentric turning point started in the ethological field in the Nineties, to the profound critical revision of the lexicon, theoretical assumptions and methods of behavioral science that, in the last two decades, led to important developments in cultural and cognitive ethology and to the emerging of promising new research areas such as behavioral and cultural epigenetics.

[16]Imprinting is a kind of early learning that occurs in some "sensitive phases" of individual maturation; it is based on an "open program", or innate program for learning and involves an integration

was only later to grasp its real relevance and significance, possibly as a consequence of his debate with the behaviorists, the British ethological school and some behavioural scientists then operating in the United States such as Donald O. Hebb, Daniel S. Lehrman and Theodore C. Schneirla (Lehrman 1953). This confrontation, and the considerations it engendered, led Lorenz to write *Evolution and modification of behavior* (in English—1965), in which he developed:

- The seminal concept of "innate instructors", "open programs", or "*phylogenetically adapted teaching mechanisms*" (Lorenz 1965: 44; 80–81; 104; 1973 [1977]: 88–96);
- a "real self-criticism", i.e., a criticism of the original underestimation of the learning processes and of their importance, by himself and by the other pioneers of ethology;
- a methodological and theoretical critique of behaviourism;
- a criticism of the theoretical "compromise" which, in his opinion, some authoritative members of the British ethological school (such as Tinbergen and Hinde) had reached with behaviorism regarding the distinction between innate and learned.

Nikolaas Tinbergen concurred with Lorenz on the key stages of classical ethology and the definition of the concept of "instinctive behavior"[17] which played a key role in the theoretical framework of the new discipline. But moving to England, he modified his position in consideration of criticisms of Lorenz's nativism by Hebb, Schneirla, Lehrman and other scholars.[18] According to the "later" Tinbergen, most animal behaviours depend, even in their "elementary units", on both hereditary and learned factors, albeit in greatly varying degrees according to their evolutionary levels, so that the clear distinction between innate and learned behaviour is only to configure "two extreme, however real, cases" (De Crescenzo 1975: 122). According to R. Hinde's somewhat more radical position, all behaviours derive from an inextricable mixture of genetic and environmental influences, which makes it practically impossible to distinguish the innate from the learned. Hinde goes as far as to reject such a distinction, which he considered not only false, but also misleading and basically detrimental

---

between innate and learned information. Its discovery was extensively discussed by Lorenz already in an essay written in 1935, Der Kumpan in der Umwelt des Vogels. This brought him, in the later Taxis und Instinthandlung, written with Tinbergen (Lorenz and Tinbergen 1938) to hypothesize that new motor patterns derive either from a recombination of different orders of hereditary motor units, or from the interlocking of short hereditary motor sequences with short learned sequences (the concept of *Instinkt-Dressurverschraenkung*).

[17]According to classical ethology, "instinctive behaviour" is defined by the correlation of four phases: appetitive behaviour (*Appetenzverhalten*), the flexible component modifiable through experience; the innate releasing mechanism (*IRM*, or *Angeboren ausloesmechanismus, AAM*), species-specific external stimulus; "hereditary motor co-ordination" or "fixed action pattern" (*Erbkoordination*), and "consummatory act" (*Endhandlung*), the latter coinciding with the execution of hereditary co-ordination, or its final part, and producing a discharge of tension.

[18]The first clear signs of this theoretical change are visible in Tinbergen (1955, 1963). The distancing from the Lorenzian positions appears clearer in Tinbergen (1965, 1968), touching themes inherent to human ethology, and to the polemics triggered by Lorenz's essay *On Aggression* (Lorenz 1963). See also De Crescenzo (1975: 119–131), Nisbett (1976: 158).

to comparative research on behaviour.[19] The largest part of these critiques to the innatism of the Austrian ethologists was, in most cases, dictated by common sense. They were intended to mark a distance from the "irrelational"[20] concept of instinct of the earliest Lorenz, i.e. from the idea that "instinctive behaviour" is resistant to the any influx from experience and is rigidly non-modifiable in its pattern and devoid of any relevant individual variations. This move also showed the English ethologists' intention to find a kind of reconciliation with the behaviorists, with whom they had been having a long-standing lively quarrel considered excessive, by many.

Nevertheless, Lorenz underlined that the denial of any distinction between innate and learned components of behaviour would be met with preventive censorship of any relevant scientific research, mainly dictated by "diplomatic" concerns, and opposed to the criticisms of his concept of "innate", advanced from several fronts, a series of countercriticisms.

According to Lorenz, the hypothesis that every form of behaviour of any organism, independently from its evolutionary status, is produced "in its elementary constitutive units" by an interaction between inborn components and learning processes can appear formally correct, because also the forms of individual modification of behaviour found among the simplest existing unicellular organism are already classifiable as learning. In fact, ethologists define them as forms of "*non-associative learning*" (Mainardi 1992: 49–50 and 66–68). But he underlined that these basic forms of learning, or "habituation and its counterpart, namely sensitization, are the only forms of adaptive modification of behaviour ever found in protozoa and in organisms with a diffuse nervous system" (Lorenz 1965: 30). That is to say that in unicellular organisms, characterized by a diffused sensitivity, it is possible to observe *only* non-associative forms of learning. From an ontogenetic standpoint this means that these organisms perform only hereditarily fixed action patterns throughout their entire life cycles. In other words, they can learn to vary the intensity of their reaction according to the frequency of an external stimulus, but cannot associate their reaction different stimuli (associative learning). Finally, from a *phylogenetic* standpoint, this means that, if life has existed for about three and a half billion years, as calculated on the basis of the most ancient organic fossils (stromatolites and microfossils), for most of this time (2830–2930 million years, or, from the appearance of the earliest organisms to the advent of the first metazoans) the survival of living beings has been *exclusively due to genetically fixed reactive and interactive motor patterns and habituation and sensitization processes.*

But his deep knowledge of animal behaviour allowed Lorenz to demonstrate that even among mesozoans, which are capable of very complex forms of learning,[21] it is possible to observe species-specific motor modules, the execution of which is poorly

[19]The divergences between Lorenz and the English ethologists, on the nature-nurture theme, touched many other points that are not considered here.

[20]On the earliest "irrelational" phase of Lorenz's theory of instinct, see De Crescenzo (1975, Chapter I–IV), Brigandt (2003), Burkhardt (2005, Chapter III).

[21]Among multi-cellular animals, already "in the phylum of the *plathelmynts, there is evidence of associative learning*" (Mainardi 1992; entry "apprendimento, capacità di": 53. My translation).

dependent on learned information. "There is not and there cannot be any argument", wrote the author in *Evolution and Modification of Behavior*, "about the fact [...] that a stickleback responds to the key stimulus 'red below' by performing the motor patterns of rival fighting and that a male stickleback is indeed red on the ventral side" (Lorenz 1965: 32). Indeed, even among mammals and birds, i.e. in those classes of animals in which individual and social learning play, beyond doubt, a central role, ethological studies have revealed the presence of rigid motor sequences (FAP: fixed action patterns) which, although presenting little individual variation, are clearly recognizable as species-specific, and are also performed by young individuals that have not yet been able to observe them in others.[22] The motor patterns of bone burial observable in dogs, those of hiding nuts carried out by squirrels, the catching and killing movements in various carnivores, the escape reactions in chicks and the alarm signals in many bird species fall into this typology.

According to Lorenz, the substantial independence of this kind of behavioural sequence from acquired information is demonstrable through an accurate use of the deprivation or "Kaspar Hauser experiment"[23] and through some observable processes which he had already described in the early 1930s. In fact, in many animal species, the prolonged absence of environmental stimuli capable of triggering some species-specific motor patterns engenders phenomena of the lowering of reaction thresholds, active search for stimuli or "vacuum activity" as in the case of segregated songbirds performing courting motor sequences while facing a corner of their cage (Lorenz 1978: Sects. 5.9, 5.10, and 5.11).

In short, although the English ethological school had undoubtedly made an important move in overcoming the early Lorenz's rigid instinctivism, the Austrian ethologist's counter-critiques showed that the obtained result—the affirmation that all behaviour is a mixture of innate and learned—was more of an academic compromise to extinguish the controversy between psychologists and ethologists within university departments and scientific meetings than a way to account for the knowledge acquired in the ethological field up to that time.[24]

However, Lorenz in turn had to admit that the first generation of ethologists, including himself, had simplistically opposed "the innate" to "the learned" and, in

[22] In humans, for instance, long rigid motor sequences, classifiable as hereditary are found in the behaviour of infants (search for the breast suction, reflex of clinging to the maternal breast, swimming) but are almost absent in adults.

[23] The protocols of these deprivation experiments have undergone severe criticism. Lorenz himself in his 1965 essay raised several critiques of their unsound use (Lorenz 1965: 83–100). Criticisms focussed on the difficulty of rigorous application and possible traumatic consequences for subjects, particularly if experiments are invasively and inappropriately performed. Lorenz's remarks were mainly aimed to restrain their use to cases of real need and reduce the procedures, thus sparing the animal cognitive and emotive damage.

[24] The two major critiques examined in the book are the following: (1) innate and learned are only defined by reciprocal exclusion (Hebb 1953); (2) one cannot establish the innate character of a behavioural module beyond doubt as "it is never possible to exclude learning *in ovo* or in utero" (Lehrman 1953). Chapters 3 and 4 of the work are dedicated to the rebuttal of these critiques.

doing so, they "were neglecting a most important function of the majority of the phylogenetically adapted behaviour mechanisms: the function of teaching" (Lorenz 1965: 81). Or, as he specified later, the function of *teaching to learn, to encourage learning, to instruct behavioural programs made to receive further training from interaction between the intra- and inter-specific environment.* A very important discovery on which also the great biologist E. Mayr would have later reflected (Mayr 1974: 650–659) and that Lorenz himself would have valued as an important integration to his "Natural History of Human Knowledge" in the work *Behind the mirror* (Lorenz 1973).

1.6 Eibesfeldt and the Deterministic Approach of the First Human Ethology

Between the sixties and the seventies, human ethology began to establish itself as a specific research domain; its main spokesman was Eibl Eibesfeldt,[25] one of Lorenz's earliest and later best known collaborators.

Undoubtedly, Eibesfeldt gave a great impulse and contribution to both field research and the compilation of data related to human behaviour. He also had the merit of having taken (at least in part) his distance from some metaphysical and apologetic interpretations of aggression and biologistic justifications of war as those voiced since the 1960s by authors such as Desmond Morris, Robert Ardrey, Robet Fox, Lionel Tiger, Alain de Benoist.

However, despite clearly defining the field of ethology as the study of both the hereditary bases of behaviour and the processes of learning, Eibesfeldt *preserved and strengthened his master's innatist outlook*. This is clearly reflected in the theoretical framework of his wide-ranging research, mainly focused on the identification of the hereditary components of human social behaviour as well as in some of his public outings on problems related to peaceful coexistence between different human communities.

According to Eibesfeldt, "biological inheritance determines human behavior […] in precisely definable parameters" (Eibesfeldt 1984 [1989]: 3), including a spontaneous appetence to aggression, the tendency of appropriating a territory and erecting barriers against intruders, the intolerance of coexistence with populations of different habits or "physical-anthropological" tracts, the "predisposition to submission" and "rank aspiration", a division of labour with assigned roles within the group respectively to females and males.

[25]The birth of the International Society of Human Ethology and the first publication of a journal specialized in this field dates back to 1978.

He re-proposed the hypothesis, already sustained by Lorenz and Ardrey, that there is in man, as in other animals, a *specific appetence (Appetenz) for aggressive behaviour*, and that if repressed it accumulates to the point of uncontrollably exploding.[26] In this perspective, innate human aggressive tendencies cannot be completely inhibited or eliminated through education. However, it is possible, through *ethos*, recreational activities, sports and engagement in social usefulness to channel them towards non-destructive forms of expression. Eibesfeldt argues that our aggressive propensities, originally depending upon self-preservative functions, are not to be assimilated with Ardrey's "murdering disposition" or "killer instinct" (Ardrey 1961), and nor Freud's self-destructive "death drive" devoid of any function useful to life, as already observed by Lorenz (Lorenz in Evans 1975: 53–55). According to these ethologists, in natural conditions aggressive behaviour plays an important role in the survival of human and animal species. Its "pathological" degeneration and exponential growth shown in modern human societies are to be understood, according to Lorenz and Eibesfeldt, as deleterious effects of inborn programs originally performing a positive function in the preservation of species. Effects due to the discrepancy between rapid cultural and technological development and slow phylogenetic evolution, or to the living conditions created by industrialization and to the manipulation of some inborn tendencies typical of our species *implemented by* economic potentates, demagogues, and political parties.

Distancing himself from those who had tried to present war phenomena as something "inevitable" and deeply rooted in man's "animal" nature, Eibesfeldt proposed a distinction between "aggression", which he considers "innate", and "war", instead representing *a product of human social history arising at a social level instead of an individual one* (Eibesfeldt 1984 [1989]: 402–422). In this perspective, if aggression is an inborn trait, war is certainly not: it is the "product of cultural evolution. Therefore, it can be overcome culturally" (Eibesfeldt 1984 [1989]: 421). Eibesfeldt agrees with Lorenz on the presence in man of an inborn inhibition against aggression and killing of conspecifics, and on its possible role as a factor in maintaining social equilibriums, where aggression is not enhanced by cultural factors. To his mind, therefore, the paradox of war is that "over the biological norm filter that inhibits destructive aggression in man as in other creatures; a cultural norm filter is superimposed that commands us to kill" (Eibesfeldt 1975 [1979]: 123). Peace among men, according to Eibesfeldt, is therefore possible, provided some of their drives are taken into account and balanced by adequate measures.[27] But the means he

---

[26] As pointed out by Alexander Alland jr. (Alland 1972: 41–42), the experiments reported by Ardrey (in which electrical stimulation of some cerebral areas elicits aggressive reactions) demonstrates that animals undergoing such treatment are capable of aggressive reactions, *not that these necessarily have a spontaneous, endogenous, and cyclical character*, as maintained by the author of *African Genesis* and *The Territorial Imperative*. Such aggressive reactions in this kind of experiments are, in fact, obtained through the very use of external stimuli, albeit directly applied to the brain, without the mediation of the sensory organs.

[27] Although opportunely marking the difference between his own view and Ardrey's thesis of the "asocial and homicidal essence of man", Eibesfeldt seems to re-introduce some elements of a

proposes in order to secure peaceful relations among individuals and communities appear rather traditional, almost reactionary: hierarchy and rank, within the social group, barriers and safety distances among different ethnic groups (Eibesfeldt 1984 [1989]: 314–320). We are here taking into consideration one of the points on which Eibesfeldt's approach was more exposed to accusations of "racist implications": from his point of view, "cultural and religious diversity" conjures with different "physical and anthropological characters", making an integration and pacific coexistence between different human communities difficult, if not outright impossible. Consequently, Eibesfeldt thinks that immigration, in these cases, is always a cause of social tension and conflicts, inevitably triggering processes of marginalization and self-marginalization, because it is perceived as an authentic "invasion".

To both Eibesfeldt and Lorenz the hierarchic order also has an innate foundation in our species: "The human disposition to form rank orders is based on our primate heritage [...]. Obedience and readiness to become subordinated are as innate in humans as striving for rank" (Eibesfeldt 1984 [1989]: 314). The Austrian ethologist also dedicates a similar analysis to "territorial" behaviour, which he considers, in man as in other species, to have a hereditary basis: "Human groups occupy territories and demarcate themselves territorially [...]. Within its own region a group claims privileges above those of others and therefore is dominant. Within group territories there are subgroups (such as families) and individuals laying claim on their small zones, marking them accordingly" (Eibesfeldt 1984 [1989]: 339). Eibesfeldt concludes that territoriality may be assumed to be "a phylogenetically acquired trait", however, he "basically" agrees with Rada Dyson-Hudson and Eric Alden Smith stating that it is something much less binding than a "genetically fixed trait", in the sense of "a fixed action pattern" (Eibesfeldt 1984 [1989]: 334). In fact, in marking his distance from Ardrey, Eibesfeldt remarks how inappropriate it would be to refer to a "territorial imperative", and how "ethology has never equated territoriality and fixed-action patterns!" (Eibesfeldt 1984 [1989]: 334).

Contemporary ethology has in many respects abandoned the substantially deterministic and genocentric approach, characteristic of the conceptual framework of classical and first human ethology, and later took to the extremes by two fathers of sociobiology as Edward O. Wilson and Richard Dawkins.

But as it will be shown in the next chapter, in the seventies, Konrad Lorenz himself was to become the main promoter of a profound theoretical renewal of Darwinism and neo-Darwinism, centered on the belief that every adaptation is a *cognitive process*, that every organism plays in this process an "extremely active" role, and that the whole process of phylogenesis can be understood as a cognitive increase, a process of accumulation, differentiation and transmission of information useful for survival which integrates embodied and acquired information, genetic and social components.

---

biologistic justification of war phenomena when he includes among the "innate tendencies" of our species, not only defensive aggressiveness and exploratory trends, but also a "typical motivation to fight and dominate, especially in the male", or when he suggests a parallel between "exploratory aggression, with which children and young people test their social behavioral liberties" and some kinds of aggressive or expansionistic state policies: "Newly formed states use the same exploratory strategy in international relations" (Eibesfeldt 1984 [1989]: 396).

1.7 Darwin's Darwinism Reconsidered Today

In 160 years of probing, Darwinian theory has reached an unquestionable scientific maturity. Especially after its confluence with Mendelian Genetics which occurred during the first half of the twentieth century, it has allowed considerable developments in all the biological domains, and in every field related to the understanding of living beings, revealing unsuspected horizons even to the reconstruction of human proto- and pre-history. From a theoretical and explanatory point of view, Darwinism to this day constitutes a fertile conceptual nucleus that in the last fifty years has inspired ever-new formulations and issues, without appearing atrophied or obsolete. Moreover, the need for a critical revision of the theory that allowed to overcome the most mechanistic and deterministic aspects of *Neo*-Darwinian Synthesis has been emerging for several decades: ongoing research which today takes shape in several attempts to reach a "post-modern" or "extended" synthesis of the theory of descent with modifications (Jablonka and Lamb 2005; Pigliucci and Müller 2010).

As mentioned, this reworking of the Darwinian and Neo-darwinist theories, still in progress both in the fields of behavioural sciences and evolutionary studies, is today leading towards a both post-genocentric and post-anthropocentric approach. The potential of Darwinian Darwinism as a scientific and cultural revolution, capable of shattering the disciplinary boundaries between natural sciences and humanities and of opening new roads toward the understanding of ourselves and other animals, finds, in my opinion, its best enhancement in this theoretic and methodological developments.[28] Yet in some areas of behavioural sciences such as evolutionary psychology and sociobiology, mechanistic and genocentric explanatory models still prevail despite the results of contemporary molecular, evolutionary and developmental biology having made them obsolete.

These ultimately determinist models aim at supporting the idea that there is a "human nature", understood as a set of psychic and behavioural predispositions fixed in a modular way (divided into behavioural modules or patterns independent from one another), that is the result of natural selection, is transmitted through genetic inheritance and remains therefore substantially impermeable to or hardly influenced by environmental, family, cultural, and social influences.

Leading exponents of this school of thought such as Steven Pinker, Marc Hauser, Jonathan Haidt believe that the development of personality and even of individual moral, political and religious tendencies are substantially guided by innate characteristics and that the living or training environment has little or no influence on them (Boyer 2001; Pinker 2004; Alford et al. 2005; Hauser 2006; Haidt 2012). As we have already seen, although having been proposed by scholars who like to call themselves Darwinists or "ultra-Darwinists", this thesis is the exact opposite of that advocated by Darwin himself in *The Descent of Man*. Above all, this thesis is now rendered anachronistic by developments in areas of research such as behavioural, social and cultural epigenetics that are based on statistical, historical, empirical and

[28] With them I will deal a little more extensively in other two essays of this volume.

experimental data. The latter in fact show that not only environmental stimuli and personal experiences, but also the habits and experiences of immediately preceding generations trigger processes of methylation and demethylation, or deactivation and reactivation of the genes which influence the development of important attitudes such as the propensity to explore and establish social bonds, the degree of exposure to stress, or the mental ductility (Jablonka 2013, 2014, 2017).

Unsettling developments for every form of genetic determinism, which also attest to their distance from those who, like Weissmann and Dawkins, Pinker and Hauser, imagined hereditary material as something largely impermeable to external influences, and the natural selection of genetic variants as the *only* law regulating biological evolution, and the critical foresight of Darwin.

The latter is wisely collected in the few sentences with which Darwin concluded the introduction of *The Origin of Species*: "I am convinced that natural selection was the main cause, but not the only one, of modifications" (Darwin 1859: 5), and with the same pluralistic spirit, in the Chapter V dedicated to the "laws of variation", he admitted that some "Lamarckian" factors (climate, nutrition, habits, use and disuse of organs) had their weight in the differentiation of varieties and species (Darwin 1871: 211).

References

Alford, John, Funk Carolyn, and Hibbing John. 2005. Are Political Orientations Genetically Transmitted? *American Political Science Review* 99 (2): 153–157.

Alland Jr., Alexander. 1972. *The Humane Imperative*. New York: Columbia University Press.

Ardrey, Robert. 1961. *African Genesis*. London: Collins.

Boyer, Pascal. 2001. *Religion Explained: The Evolutionary Origins of Religious Thought*. New York: Basic Books.

Brigandt, Ingo. 2003. Gestalt Experiments and Inductive Observations: Konrad Lorenz's Early Epistemological Writings and the Method of Classical Ethology. *Evolution and Cognition* 9 (2): 157–170.

Burkhardt, Richard W., Jr. 2005. *Patterns of Behavior. Konrad Lorenz, Niko Tinbergen, and The Founding of Ethology*. Chicago and London: University of Chicago Press.

Burkhardt, Frederick, and James A. Secord (eds.). 2005 (1867). *The Correspondence of Charles Darwin*, vol. 15. Cambridge: Cambridge University Press.

Cavazzini, Andrea. 2009. Uomini e animali, tra natura e cultura. *L'Ateo* II: 5, 6.

Choi, Jung-Kyoo, and Samuel Bowles. 2007. The Coevolution of Parochial Altruism and War. *Science* 318: 636–640.

Darwin, Charles. 1859. *The Origin of Species by Means of Natural Selection*. London: John Murray.

Darwin, Charles. 1871. *The Descent of Man*. London: John Murray.

Darwin, Charles. 1872. *The Expression of Emotions in Man and Animals*. London: John Murray.

Darwin, Charles. Posthumous: 2009. *Charles Darwin's Notebooks, 1836–1844: Geology, Transmutation of Species, Metaphysical Enquiries*. Cambridge: Cambridge University Press.

De Crescenzo, Giovanni. 1975. *L'etologia e l'uomo*. Firenze: La Nuova Italia.

Eibesfeldt, I. Eibl. 1975. *Krieg und Frieden aus der Sicht Verhaltensforschung*. Engl. Transl. 1979. *The Biology of Peace and War*. New York: Viking Press.

Eibesfeldt, I. Eibl. 1984. *Die Biologie Des Menschlichen Verhalten*. Engl. Transl. 1989. *Human Ethology*. New York: Aldine de Gruyter.

Evans, Richard I. (ed.). 1975. *Konrad Lorenz. The Man and His Ideas*. New York: Harcourt Brace Jovanovich.
Gould, Steven Jay. 1977. *Ever Since Darwin*. New York: W. W. Norton.
Haeckel, Ernst. 1868. *Natürliche Schöpfungeschichte*. Berlin: Georg Heimer.
Haidt, Jonathan. 2012. *The Righteous Mind: Why Good People Are Divided by Politics and Religion*. New York: Pantheon.
Haslam, Michael, et al. 2016. Pre-colombian Monkey Tools. *Current Biology* 26 (13): R521–R522.
Hauser, Marc. 2006. *Moral Mind*. New York: HarperCollins.
Hebb, Donald. 1953. Heredity and Environment in Mammalian Behaviour. *Journal of Animal Behabiour* 195 (1): 43–47.
Himmelfarb, Gertrude. 1959. *Darwin and the Darwinian Revolution*. New York: Chatto and Windus.
Hobaiter, Catherine, et al. 2014. Social Network Analysis Shows Direct Evidence for Social Transmission of tool use in wild Chimpanzees. *PLoS Biology* 12 (9): e1001960. https://doi.org/10.1371/journal.pbio.1001960.
Jablonka, Eva. 2013. Behavioral Epigenetics in Ecological Context. *Behavioral Ecology* 24 (2): 325–326.
Jablonka, Eva. 2014. Epigenetics and Behavior. In *Oxford Bibliographies in Evolutionary Biology*, ed. Jonathan Losos. New York: Oxford University Press.
Jablonka, Eva. 2017. Cultural Epigenetics. *Sage Journal* 64 (1): 42–60.
Jablonka, Eva, and Marion Lamb. 2005. *Evolution in Four Dimension*. Cambridge, MA: MIT Press.
Journal of the History of Biology. 2005. 38 (1).
Kant, Immanuel. 1790. *Kritik der Urteilskraft*. Engl. Transl. 2005. *Critique of Judgment*. Mineola and New York: Dover.
La Vergata, Antonello. 1999. *Positivismo e darwinismo sociale*. In *Il pensiero politico. Idee, teorie, dottrine*, III (1), ed. Pasquino Gianfranco. Torino: Utet.
La Vergata, Antonello. 2001. *Il darwinismo sociale*. In *Giovanni Canestrini. Zoologist and Darwinist*, ed. Alessandro Minelli and Sandra Casellato. Venezia: Istituto Veneto di Scienze, Lettere ed Arti.
La Vergata, Antonello. 2005. *Guerra e darwinismo sociale*. Soveria Mannelli (Catanzaro): Rubbettino.
Lehrman, Daniel. 1953. A Critique of Konrad Lorenz's Theory of Instinctive Behavior. *Quarterly Rewiew Biology* 28: 337–363.
Lorenz, Konrad. 1963. *Das sogenannte Böse. Zur Naturgeschichte der Aggression*. Engl. Transl. 1966. *On Aggression*. New York: Bantam Books.
Lorenz, Konrad. 1965. *Evolution and Modification of Behavior*. Chicago: University Press.
Lorenz, Konrad. 1973. *Die Rückseite des Spiegels*. Engl. Transl. 1977. *Behind the Mirror: A Search for a Natural History of Human Knowledge*. London: Methuen & Co.
Lorenz, Konrad. 1978. *Vergleichende Verhaltensforschung: Grundlagen der Ethologie*. Engl. transl. 1981. *Foundations of Ethology*. New York: Springer Verlag.
Lorenz, Konrad, and Nikolaas Tinbergen. 1938. Taxis und Instinkthandlung in der Eirollbewegung der Graugans. *Zeitschrift für Tierpsychologie* 2: 1–29.
Mainardi, Danilo. (diretto da). 1992. *Dizionario di Etologia*. Torino: Einaudi.
Malthus, Thomas. 1798. *An Essay on the Principle of Population*. London: Johnson.
Marx, Karl, and Friedrich Engels. 1975–2004. *Marx/Engels Collected Works*, vol. 41. London: Lawrence and Wishart.
Mayr, Ernst. 1974. Behavior Programs and Evolutionary Strategies: Natural Selection Sometimes Favors a Genetically "Closed" Behavior Program, Sometimes an "Open" One. *American Scientist* 62 (6): 650–659.
Mercader, Julio, et al. 2007. 4300-Year Hold Chimpanzee Sites and the Origins of Percussive Stone Technology. *Proceedings of the National Academy of Sciences* 104: 3043–3048.
Monod, Jacques. 1970. *Le hasard et la nécessité*. Paris: Edtions du Seuil.
Nietzsche, Friedrich. Posthumous 1868. *Philosophisce Notizen*. Engl. Transl. 2009. *Writings from the Early Notebooks*. Cambridge: Cambridge University Press.

Nietzsche, Friedrich. 1874. *Vom Nutzen und Nachtheil der Historie für das Leben*. Engl. Transl. 1980. *On the Advantage and Disadvantage of History for Life*. Indianapolis and Cambridge: Hackett.
Nietzsche, Friedrich. 1887. Zur *Genealogie der Moral*. Engl. Transl. 2006. *On the Genealogy of Morality*. Cambridge: Cambridge University Press.
Nisbett, Alec. 1976. *Konrad Lorenz*. New York and London: Harcourt Brace.
Pigliucci, Massimo and Müller, Gerd. 2010. *Evolution—The Extended Synthesis*. Cambridge: MIT Press.
Pinker, Steven. 2004. Why Nature & Nurture Won't Go Away. *Daedalus* 133 (4): 5–17.
Ruse, Michael. 1979. *The Darwinian Revolution: Science Red in Tooth and Claw*. Chicago: University of Chicago Press.
Spencer, Herbert. 1857. Progress: Its Law and Causes. *The Westminster Review* 67: 445–465.
Spencer, Herbert. 1862. *A System of Synthetic Philosophy: First Principles*. London: Williams and Norgate.
Spencer, Herbert. 1887. *The Factors of Organic Evolution*. New York: Appleton.
Tinbergen, Nikolaas. 1955. Some Aspects of Ethology. *Advancing Science* XII: 17–19.
Tinbergen, Nikolaas. 1963. On Aims and Methods of Ethology. *Tierpsychologie* XX: 410–433.
Tinbergen, Nikolaas. 1965. *Animal Behavior*. New York: McGraw-Hill.
Tinbergen, Nikolaas. 1968. On War and Peace in Animals and Man. *Science* 160: 1411–1418.
Tort, Patrick. 1983. *La pensée hiérarchique et l'évolution*. Paris: Aubier.
Tort, Patrick. 1992. La seconde révolution darwinienne. In *Darwinisme et societé*, ed. Patrick Tort. Paris: Puf.
Tort, Patrick. 1999. *L'anthropologie inattendue de Charles Darwin*. In Charles Darwin. *The Descent of Man*. Engl. Transl. 1999. *La Filiation de l'homme*. Paris: Syllepse.
Tort, Patrick. 2000. *L'antropologia di Darwin*. Roma: Manifesto Libri.
Wallace, Alfred Russel. 1870. *Contributions to the Theory of Natural Selection*. London and New York: Macmillan.
Weismann, August. 1896. Germinal Selektion. *The Monist* 6: 250–293.
Wilson, David Sloan. 2019. *This View of Life: Completing the Darwinian Revolution*. New York: Pantheon.

Chapter 2
From Evolutionary Epistemology to an Extended Evolutionary Synthesis

Marco Celentano

Abstract The first part of this essay reconstructs the scientific and philosophical paths that led, between the 1940s and the late 1990s, to the formulation and development of Evolutionary Epistemology (EE). This was an "open theory" and a research program aimed at studying analogies and differences between biological and social and scientific human evolution focused on the hypothesis that both can be considered as effects of a "cognitive increase", or a process of knowledge. Konrad Lorenz, Karl Popper, and Donald Campbell were the promoters of this approach and the Altenberg Circle (*Altenberger Kreis*), animated by scholars such as Rupert Riedl, Erhard Oeser, Franz M. Wuketits, was its forge. The second part of the essay shows how the EE project continued to develop under different forms over the last decade, despite being partly shipwrecked. In fact, thanks to its critical reworking and scientific update promoted by members of the Konrad Lorenz Institute For Evolution and Cognition Research such as Werner Callebaut, Karola Stotz, Massimo Pigliucci, Gerd Müller, it acted as a basis for the formulation of a new scientific project: that of formulating an "extended synthesis" of the Darwinian theory of descent with modifications able to take into account the most important discoveries of the last decades pertinent to developmental processes and epigenetic forms of inheritance.

2.1 Introduction

The two initial sections of this chapter examine the short but intense collaboration between the young ethologist Konrad Lorenz and the philosopher Eduard Baumgarten from which arose, between 1940 and 1941, a fruitful attempt to rework the Kantian doctrine of knowledge in the light of the Darwinian theory of natural selection.

The third section focuses on the "natural history of human knowledge" which Lorenz proposed in his book *Behind the Mirror* (Lorenz 1973), developing the hypothesis that every adaptation of organisms to their environment is the result of a "cognitive process" (Lorenz 1973, 1983). The following three parts analyze three main developmental stages of Evolutionary Epistemology: its foundation, by Konrad Lorenz, Karl Popper and Donald Campbell, in the 1970s; the birth of the

R. Marchesini and M. Celentano, *Critical Ethology and Post-Anthropocentric Ethics*,
Numanities - Arts and Humanities in Progress 16,
https://doi.org/10.1007/978-3-030-74203-4_2

Altenberg Circle, and later of the Konrad Lorenz Institute; the "constructivist turn" of EE promoted in the 1990s by Rupert Riedl.

The seventh section focuses on the impact the reformulations by Riedl of the "body plan" concept from pre-Darwinian morphology in an evolutionary and systemic key in the context of contemporary evolutionary biology.

The following section illustrates the contributions some of the members of the Konrad Lorenz Institute gave to a critical re-examination of EE and shows that, since 2007, this project has been merging with the attempt to reach an "extended synthesis" of the Darwinian theory, capable of integrating the most important discoveries emerging in contemporary eco-evo-devo (ecological, evolutionary, developmental) biology.

The tenth section attempts to clarify, in a nutshell, how the anti-mechanistic and *etho-centric* model that emerged in the field of evolutionary and ethological studies presented in the previous sections allows both an anti-deterministic and anti-biologistic approach to the study of human behaviour and social evolution and a critical reflection on their effects.

Finally, in the brief open conclusions, I propose some reflections oriented to the elaboration of a post-anthropocentric and a post-genocentric, inter and trans-specific concept of knowledge.

2.2 The Meeting Between Lorenz and Baumgarten and the Project of a "Darwinian" Re-formulation of the Kantian Apriorism

On September 2nd of 1940, Konrad Lorenz arrived in Königsberg, Immanuel Kant's hometown. He had been offered to chair Comparative Psychology at the Albertus Universität, where the great Illuminist philosopher had taught for almost fifty years. Lorenz's major sponsor there was Eduard Baumgarten, full professor of Kantian Philosophy and pragmatically oriented thinker, follower of John Dewey and expert in the work of Ralph Waldo Emerson, Nietzsche, and Max Weber. He was interested in a critical reworking of Kantian philosophy and went looking for "a second teacher with gnoseological interests, but at the same time endowed with a solid biological background" (Wuketits 1990: 60).[1] In Königsberg, he "brought together very bright scientists and men of letters for evening discussions, with the ambition of paving the way to a theoretical and methodological synthesis of the two fields" (ibid.). In the Institute he co-directed with Lorenz, "philosophical anthropology was combined with comparative behavioural research" (Lorenz 1992: 75). At that time, they both were active members of the *Kant Gesellschaft Königsberg* where they promoted heated debates.

[1] This and all the other quotations taken from essays that do not have an English translation, contained in this chapter, are my translations.

Lorenz remained in Königsberg only thirteen months, after which he enlisted in the army.[2] In his brief time there, however, together with Baumgarten he sketched a phylogenetic and "non-transcendental" revision of Kant's theory of knowledge set to become, in the second half of the twentieth century, the first pillar of Evolutionary Epistemology. Lorenz summarized this theoretical position in the essay "Kants Lehre vom Apriorischen im Lichte gegenwärtiger Biologie" (Lorenz 1941), originally published in the *Blätter für deutsche Philosophie*.[3]

As Donald Campbell remarked later, in writing "Kant's Doctrine of the A Priori in the Light of Contemporary Biology", "the young Lorenz creatively solved a major epistemological puzzle" (Campbell 1974: 96) and, as Lorenz remarked in turn, this puzzle had been evidenced by Kant himself:

> «If one were to entertain the slightest doubt that space and time did not relate to the Ding-an-sich but merely to its relationship to sensuous reality, I cannot see how one can possibly affect to know, a priori and in advance of any empirical knowledge of things, i.e. before they are set before us, how we shall have to visualize them as we do in the case of space and time» (section 11 of *Prolegomena to the critique of pure reason*)
>
> Kant was obviously convinced that finding an answer to this question in terms of natural science was impossible. He found clear proof that our forms of ideation and categories of thought, in contrast with what Hume and other empiricists claimed, are not the products of individual experience; they are logically necessary a priori, and therefore cannot have 'evolved' (Lorenz 1973: 9).[4]

But building on Kantian premises, i.e. on a pluralistic realism acknowledging the reciprocally independent existence of "external" entities and of a "subjectivity" that experienced them, and on a theory of knowledge founded on the acknowledgment of "a priori forms of sensibility" as conditions of possibility of experience, Lorenz intended to demonstrate that a consistent response to Kant's problem was made possible by the Darwinian theory, or more precisely, by a specific interpretation of it. The ethologist in fact wrote:

> The system of sense organs and nerves that enables living things to survive and orientate themselves in the outer world has evolved phylogenetically through confrontation with an adaptation to that form of reality which we experience in phenomenal space. This system thus exists a priori to the extent that it is present before the individual experiences anything and must be present if experience is to be possible. (ibid.)

In other words, organisms are pre-adapted, already from birth, to the interaction with a given environment, and the human mind itself is pre-adapted to this interaction, but this condition is an a priori only for the individual, not for the species.

As noted by Franz M. Wuketits, Lorenz was attempting, with this first strategic move, "a synthesis between Kant's theory of knowledge and Darwin's theory of

---

[2]Lorenz went to Königsberg on September 2, 1940 and joined the army on October 10, 1941.

[3]The essay was first translated in von Bertalanffy and Rapoport, A. (Eds.) General Systems, Yearbook of the Society for General Systems Research, vol. III, 1962: 23–35, and later reprinted in Evans (1975: 181–217).

[4]The quotation by Kant, contained in the Chapter 11 of Prolegomena to the Critique of Pure Reason (1783) is extracted from the 1973 English edition of Lorenz mentioned in the References.

evolution" (Wuketits 1990: 83). A deep understanding of anatomy and behaviour of organisms allowed him to provide new empirical foundations for a concept already brought to philosophical clarity by Nietzsche: the true possibility of experience is not to be found in a (non-existent) transcendental structure of reason, immutable and free from historical and empirical influences, as maintained by the Kantians. The real subject of experience is the human body, as a product of natural, social, and individual history (Nietzsche 1882).

According to the philosopher Aldo Masullo this conception was already present in the confutation of Idealism attempted by Kant in the second edition of the *Critique of Pure Reason* (1787). Kant's transcendental idealism was primarily aimed at undermining the "immaterialism" and "spiritualism" upheld by Berkeley and Descartes. It was meant to overcome that form of idealism which frames the knowing subject in an incorporeal dimension and defines the known object as "mere subjective representation". In this sense, "if never explicitly [...] the theme of corporeality of subjects constitutes the unifying focus of Kantian theoretical thought" (Masullo 1986: 34).

However, Kant's failure to make this theme completely explicit had important theoretical consequences, highlighted by Lorenz in his 1941 essay:

> The only thing we can assert about the thing-in-itself, according to Kant, is the reality of its existence. The relationship which exists between it and the form, in which it affects our senses and appears in our world of experience [...] is determined by the ideal forms and categories of intuition. (Lorenz in Evans 1975: 182)

But these forms, in a Kantian theoretical horizon, cannot be related "to the laws inherent in the «thing-in-itself» by abstraction or any other means" (ibid.).

What derives from this approach is a radical dualism, according to which the value of a priori forms of reason is considered "in principle independent from the laws of real nature, based only on the faculties of the subject, while the thing in itself appears in principle unknowable" (1975: 183). For Lorenz, this approach generates some questions that biologists "have to ask" Kant:

> Is not human reason with all its categories and forms of intuition something that has organically evolved in a continuous cause-effect relationship with the laws of the immediate nature? Can an organ that has evolved in the process of continuous coping with the laws of nature have remained so uninfluenced that the theory of appearances can be pursued independently of the thing in itself, as the two were totally independent from each other? (ibid.)

Lorenz's second step was to translate the Kantian concept of "a priori form of sensible intuition" in that of a historical state of pre-adaptation of the organs, produced by selection and heredity (the "experience of the species"), which is a necessary condition for both existence and experience for each and every individual. This means that, to Lorenz, the conception of the "a priori" as an organic function "means the destruction of the concept: something that has evolved in evolutionary adaptation to the laws of the natural external world has evolved a posteriori in a certain sense" (ibid.).[5]

[5]Lorenz's formula "what is a priori for the individual is a posteriori for the species" was not an absolute novelty in the post-Darwinian debate. In a 1876 anthology of earlier journal articles, St.

The third theoretical step, in the final section of the essay, consists in an attempt to demonstrate that some categories, or central conceptual nexuses such as causal relation, may have also originated from genetically and physiologically based learning programs, i.e. the conditioned reflex. In fact, for Lorenz, human understanding "does not prescribe the laws of nature" (1975: 186) because, quite like a horse's hoof, it continually stumbles over "unforeseen changes in its task, highlighting the inadequacy of its hypotheses" (ibid.):

> [...] the fundamentals of pure reason are just as imperfect and down to earth as the band saw, but also just as real. Our working hypothesis should read as follows: everything is a working hypothesis. This holds true not only for natural laws which we gain through individual abstraction a posteriori from the facts of our experience, but also for the laws of pure reason. The faculty of understanding does not per se constitute an explanation of phenomena, but the fact that it projects phenomena for us in a practically usable form onto the projection screen of our experiencing is due to the formulation of working hypotheses; developed in evolution and tested through millions of years! (ibid. 199)

According to Lorenz, precisely the fact that human beings, since they exist have had to interact with beings and phenomena that do not passively submit to their efforts to "shape" them, precisely the experience of these "resistances" from the external environment, accumulated in the biological and cultural patrimony of our species ensure that we can indeed rely on the capabilities we possess, albeit within our species-specific and historical limits.

Despite the unbridgeable gap introduced by these theoretical shifts, Lorenz did not ignore many points of a substantial convergence between Kant's transcendental idealism and his own genealogical materialism. Like Kant, he opted for a "critical realism", while distancing himself from any form of "naïve realism":

> [...] we are perfectly aware that what exists in itself will never be completely at hand, except within the limits imposed even to theoretically higher living forms by the categorical forms of our thought, [...and] even if we as natural scientists are in a certain sense naïve realists, we still do not take the appearance for the thing in itself, nor the experienced reality for the absolutely existent! (ibid. 191)

Compared to the Kantian model, Lorenz's approach both strengthened and weakened human pretensions in the cognitive field. Whereas Kant's transcendental idealism stated the impossibility of a positive knowledge of real aspects or features of things, according to Lorenz's hypothetical-critical realism "evolutionary success does not entail that all our innate hypotheses be true, but only that they cannot be completely false" (Vollmer 1983: 49). As Riedl later remarked, no organism could survive if its sensory organs and relational modalities did not put it in the condition of detecting any real feature of the elements it really deals with in its own environment (Riedl 1980:

George Mivart wrote, criticizing Scottish sensism: "in this way Mr. Spencer conceives that what is a priori to the individual is but a posteriori to the race and he thus claims to have reconciled the two schools of thought, namely, those who assert and those who deny the derivation of our ideas exclusively from sensation and experience". He went on: "As it is manifest, however, he gives the substantial victory entirely to the sensists, and denies to all ideas any higher origin than mere incipient sentiency" (Mivart 1876: 425).

56). In this perspective, the very fact that organisms can interact with other entities in a way that is functional to their survival demonstrates a positive know-ability of real entities or processes and an actual capacity of knowing, present in different forms and degrees in every living being. About human knowledge, it shows that the relation "between the real within and the real outside ourselves" is "explorable in principle", but always and only indirectly, through an understanding that is constantly putting to test *vis à vis* every day acting and living, and which is not absolutely true or false, but rather more or less useful to face the needs, the circumstances and perils of life.

2.3 Toward a "Natural History of Human Knowledge"

It was only a quarter of century later, with *Behind the mirror* (Lorenz 1973), that Lorenz tackled again the question of the genesis of animal and human forms of knowledge. In this work, the genealogical approach to the theory of knowledge was recast in a model comprising, besides the phylogenetic version of apriorism, a critical interpretation of Darwinism and neo-Darwinism: a re-formulation of the problem of understanding, but also of the theory on the origin and transformation of the species. In the "Epistemological Prolegomena" which opened the work, Lorenz subtly criticized the position expressed a few years earlier by Jacques Monod in *Chance and Necessity* (Monod 1970) by observing how "it is undeniably true, yet at the same time misleading, to say that living organisms are at the mercy of purely random changes and that evolution only takes place through the elimination of the unfit" (Lorenz 1973: 27). In his polite rebuttal of Monod's claims, Lorenz was marking his own distance from a specific interpretation of the neo-Darwinian canon, linked to the then dominant approaches in molecular biology. In a nutshell, the interpretation consists in the idea that evolution is essentially based on the interaction between two factors: "chance", embodied by favourable genetic mutations, and "necessity", embodied by external selection. Lorenz, instead, stressed a third factor, namely, the "extremely active" care of their living conditions which all the organisms manifested through their behaviour and physiology.

In other words, his interpretation implied that living beings search in an "eminently active" way and tend to accumulate "both a fund of energy and a stock of knowledge, the possession of the one being instrumental to the acquisition of the other" (ibid.). According to Lorenz the lack of appreciation for this "exploratory" aspect of behaviour makes it impossible to account for two fundamental features of the evolutionary process, its "speed" and its "directness", without resorting to metaphysical and finalistic hypotheses. If evolution "depended simply on the random elimination of the unfit, then the period of a few thousand million years which has been calculated by physicists as a few thousand million years based on the rate of decay of radioactive substances would hardly be enough for man to have evolved from the most primitive organisms" (1973: 28). By the same token, the appearance of beings endowed with a growing degree of organic complexity and behavioural capacities, which Lorenz considered an established fact, can be explained, without resorting to teleology and ultimately to theology only by acknowledging that life is also a process of acquisition of "information".

The history of living systems can be described as a process of knowledge acquisition, meaning that in order to survive and reproduce, organisms have had to learn to distinguish the things that have an impact on their physiological condition and their likelihood of survival so as to exploit or avoid them. This means organisms have turned their own living conditions and the factors influencing them into objects of knowledge, however indirect, without this process implying any form of predestination. Organic evolution "does not follow a predetermined plan" but derives its direction from the reciprocal selection among organisms, from their attitude to explore both the external environment and their behavioural capabilities, from their active search and construction of specific internal and external conditions.

Lorenz's organic history can thus be understood without reference to any kind of determinism, be it finalistic, genetic or environmental. In *Behind the Mirror*, the ethologist was therefore proposing a general reinterpretation of Darwinism in which differentiation and preservation of living species are conceived as effects of a "process of acquisition of knowledge" in the sense of an increase, selection, and differentiation of "information" embodied and potentially embeddable in organisms themselves, and usable to the survival of individuals and species.

From this perspective, whatever its level of internal complexity may be, a living organism can never be considered as an entity simply undergoing an external selection: it must at the same time be considered a selecting agent.

The behaviour of organisms must be therefore analyzed as both a product of phylogenetic, social and individual history and as one of the main selecting factors orienting phylogeny itself and the history of the species with it.

2.4 The Project of an "Integrated Theory" of Evolution

The earliest version of Evolutionary Epistemology sprang from the integration of three independently developed approaches, those of K. Lorenz, K. Popper and D. Campbell. The latter, back then less known than the others, was a psychologist interested both in the theoretical aspects of Lorenz's approach and in the evolutionary reinterpretation of the falsificationism propounded by Karl Popper since the 1960s. It is to him that we owe the invention of the formula "Evolutionary Epistemology" (Campbell 1974). Campbell conceived EE as a research program targeted to an "integrated theory", whose was to identify analogies and differences between biological and socio-cultural human evolution, biological adaptation and scientific progress. In the perspective of its founders, EE implied first of all "the hypothesis that biological evolution in itself represents a cognitive process, independent from the appearance of the human species" (Somenzi 1996: 238) and the conviction that the human condition is a "product of biological and social evolution" (Campbell 1974: 413). The common denominator among processes of such diverging order, complexity, and origin is to be found, according to EE, in a process based on "trials" and selective preservation of efficient solutions, which underlies both natural selection and individual associative learning. This process would have its most meaningful precedent and functional analogue (not teleologically oriented and not responding to any conscious immanent or transcendent design) in the "positive" interaction between genetic variance and environmental selection, which neo-Darwinism took as the moving force of biological evolution. The process of natural selection, favoring in terms of differential reproduction the organisms best fitted to their environments, produces effects analogous to a learning process, unrolling through trials and errors. In other words, natural selection and descent, without being pre-oriented in any direction, have de facto triggered a process leading to the elimination of errors, seen as inefficient solutions to the problems of survival and reproduction, and to their replacement with more efficient forms of behaviour and internal organization. According to Lorenz, Campbell and Popper, in the course of phylogeny, all individual learning configurations, from the simplest to the most complex, have developed and differentiated starting from this first form of "learning of the species".

But despite these important points of convergence, and Campbell's attempt to mediate them, Lorenz and Popper's theoretical positions presented irreducible differences (Stanzione 1984; Celentano 2000, 2011). Popper's approach to Evolutionary Epistemology was founded on "genetic dualism", a theoretical formulation admittedly very close to "a mind-body dualism". It presupposes the possibility of identifying, even in "very simple organisms" and a fortiori in more complex ones, an organization based on "two distinct parts: roughly speaking a behaviour-controlling part like the central nervous system of higher animals, and an executive part, like the sense organs and the limbs, together with their sustaining structures" (Popper 1972: 273). Each organism, then, would be divided into an "aim-structure" and a "skill-structure" and, according to Popper, in the course of phylogeny, the development of teleological structures has preceded and favoured that of performative structures, thus endowing evolution with a course ever less subject to happenstance and progressively characterized by orthogenetic developments.

> Once a new aim or tendency or disposition, or a new skill, or a new way of behaving, has evolved in the central propensity structure, this fact will influence the effects of natural selection" and this, to Popper, meant "that the evolution of the executive organs will become directed by that tendency or aim, and thus 'goal-directed'. (1972: 278)

Lorenz's mastery of comparative anatomy made him aware that this hypothesis was untenable in the case of "lower" organisms, devoid of a centralized nervous system. Popper's dualism arbitrarily extends to all or almost all living organisms a model, derived from the neurophysiologic organization of 'higher' animals endowed with a central nervous system (a level of organization arising only in a very advanced phase of phylogeny). Cognitive performances of some complexity are, instead, observable in the protozoans or lower metazoans, whose physiological organization shows no trace of the division between two different mechanisms devoted to central coordination and executive performances, adumbrated by Popper.[6] But, according to Lorenz, even for "higher" organisms Popper's dualistic model was valid only in part. In almost

[6] In coelenterates (medusa) we find groups of cells with specialized perception, sensitive to light and to position equilibrium. A type of cephalic specialization, still very far from a centralized nervous

all his works, Lorenz remarked how nervous structures originated from the integration of "already functioning" parts, with a certain degree of reciprocal autonomy, and that the integration is only partial at each stage of evolution and never devoid of dysfunctions.

Popper instead opted for the inclusion of all animals in the genetic-dualism model, to the point of denying the distinction, at the time common among biologists, between the capacity to react to stimuli, or "excitability"[7] considered present in all living cells and "sensation", traditionally restricted to the animal world. One must admit, considering present developments in plant ethology, that Popper was ahead of his time on this point when he observed that plants "do have something like sensations or perceptions" (Popper 1990: 35).[8] However, the point at stake here is mainly philosophical. Even though he was not arguing for "conscious" knowledge processes in other organisms, Popper, following Kant, maintained that interpreting the behaviours of living beings "as if" they acted according to patterns of finalistic reasoning, analogous to that of humans, was the only way to underscore the active and selective character of those behaviours. On the contrary, in all his works, Lorenz tried to interpret the sequences that, in many different animal species, lead from appetitive behaviours to the execution of a "consummatory act",[9] taking into account the fact that, to the animal, each and every phase of the sequence is "self-compensatory" and acts as a sort of "present aim" (Lorenz 1937: 298).

2.5 The EE and the Debate on Similarities and Differences Between Natural and Social Selection

Even sharper were the differences between Lorenz and Popper's approaches in the interpretation of organic and human social evolution, and of the effects of modern science and capitalistic economy. Lorenz was highly critical of the idea of an evolutionary process generally following "the direction of a greater completeness of adaptation" (Lorenz 1983: 40) and offered a lot of empirical evidence against it. Despite

system, appears in the phylogenetic line with anellids. In these animals it is possible to observe "a metameric system, with groups of nerve cells (ganglia) organized in pairs, in every ring" anterior to the sense organs. In insects "besides metameric groupings, made more numerous by the fusion of metameres [...] it emerges a very advanced specialization of the system of cephalic ganglia, anticipating the future development of a brain" (Fancello 1985: 110–111).

[7]"The capacity to react to stimuli (excitability) is a basic property of all living organisms, including plants" (*Nuovo Atlante Biologico*, Milano, Garzanti, 1989: 339).

[8]For an updated overview of the sensory and perceptual plant systems see Baluška et al. (2007), Mancuso (2018).

[9]A "consummatory act" is defined as the final sequence of a hereditary motor co-ordination, as distinct from "appetitive behaviour" (active search of triggering stimuli) preceding it and "in natural conditions, it leads to the disappearance of the pulsion" (Craig 1918). In complex motor sequences, however, "an act may represent at once an 'appetitive behaviour' for what follows and a 'consummatory act' for what precedes" (Heimer 1977: 32).

his intention to correct some "mistakes" of neo-Darwinism, Popper instead found himself in a general agreement with the idea that organic evolution was explainable as a gradual and progressive emergence of "the fittest", and extended this model to the interpretation of human social, scientific and political evolution. For Popper (and Campbell's position was similar on the subject), the "evolution of scientific knowledge is, in the main, the evolution of better and better theories" (Popper 1984: 395) and this is in every aspect "a Darwinian process. The theories become better adapted through natural selection: they give us better and better information about reality. They get nearer and nearer to the truth" (Popper 1984: 396). With these passages, Popper let his original falsificationist position "slip", as he put it, from the methodological to the theoretical domain, making it a model for the interpretation of the whole history of Western science as a gradual progress towards better theories. Unfortunately, history does not seem to confirm this hypothesis: for centuries competition among scientific theories led to outcomes far different from those imagined by Popper. The fact that some scientific theories hindered an adequately critical study of empirical phenomena, instead favouring superstition and social privileges, has in many instances been the very reason of their success. On the contrary, the fact that certain theories offer tools for validating truths towards which mass media controllers are hostile lead even today to their boycott, as shown by numerous sources. Popper's model, therefore, seems highly simplified and idealized, inasmuch as it arbitrarily removes the processes of conscious and unconscious manipulation of information and processes of social selection of knowledge not aimed at the critical development of knowledge itself, but rather subordinated to other individual or collective goals, such as social control or profit.

After all, Popper extended his optimistic model of organic evolution and modern scientific progress to the political sphere: taking the US system as a model, he maintained that we are actually living in the best possible world, and that "democracies are always open to ideas, especially those coming from the opposition. Far from being masked dictatorships, democracies are always open to self-doubting" (Popper in Arrigoni 1991: 226).

Lorenz dissented from this idyllic approach, which in the last years of his life the same Popper had doubted (Popper and Condry 1994), arguing that human socio-cultural evolution, especially in the age of advanced capitalism and triumphant technocracy, was led by selective processes different from those regulating organic evolution. In *The Waning of Humaneness*, he elaborated a perspective in which the "creative selection" underlying organic evolution "has ceased to influence humans. Creative selection has been replaced by intra-specific selection" (Lorenz 1983: 12). It is intra-specific, social selection, Lorenz argued, namely the selection of man by man, that now determines the "direction of development" of human evolution, and it is "our present technocratic world order" that sets this social selection (1983: 13). For these reasons, to Lorenz, not just dictatorial regimes, but also the present democratic systems were taking on "more and more totalitarian aspects" (1983: 187) and the increasingly pervasive power to manipulate individual and mass behavior offered to the powers in force by today's technologies represents one of the most serious dangers that humanity has to face.

2.6 Riedl's "Constructivist Turn"

The developments of EE can be divided into three main phases.

The first lasted from the mid-1970s to the late 1980s and includes the constitution and developments of the Altenberg Circle (*Altenberger Kreis*), around the ethologist K. Lorenz, the biologist R. Riedl and the philosopher O. Oeser. This circle was an ever-growing interdisciplinary group of scholars, who regularly met at Lorenz's house in Altenberg, to discuss the theoretical implications and possible developments of the "evolutionary and cognitive" approach.

The second phase began after the death of Lorenz (1989), covered the 1990s and marked the transition to the foundation of the Konrad Lorenz Institute for Evolution and Cognition Research (1990–1991), culminating in the turning point toward a "constructivist extension of EE" propounded by its first director, Rupert Riedl (Riedl 1995). These developments contributed to introduce EE in the contemporary international debate about the "naturalistic" approach in epistemology, and about the emergent eco/evo/devo (ecological, evolutionary and developmental) approach in biology. On the other hand, they opened the theory to many different explanatory models and presumptive domains of application causing a considerable weakening of its internal consistency between basic assumptions and developments.

The third period started with the new millennium but was anticipated by the new research project expressed in the programmatic paper *Lean Evolutionary Epistemology* (1998) in which philosophers Werner Callebaut and Karola Stotz attempt a radical reform of EE and a more rigorous formulation of its basic assumptions. This project in the following years converged into an even more ambitious one: that of an "extended synthesis" of the contemporary theory of evolution, aimed at going beyond the Modern Synthesism and integrating the recent results of disciplines as molecular archaeology, genetics, epigenetics, neurophysiology, cognitive and cultural ethology in a systemic eco/evo/devo approach (Pigliucci and Müller 2010). An attempt which is still in progress. After having analyzed the first phase of this process in the previous sections, we will now describe and discuss the other two.

Towards the mid-1970s, the time was ripe for the convergence program, of widely diverging perspectives, interests and theoretical orientations in the EE research program. In 1975, only a year after Campbell's programmatic essay *Evolutionary Epistemology*, German physicist Gerhard Vollmer,[10] in his *Evolutionäre Erkenntnistheorie* (Vollmer 1975), tried to "design the structure of an «Evolutionary Epistemology» with a view to the whole" (Riedl 1980: 3).

According to EE, "evolutionary success does not entail that all our innate hypotheses be true, but only that they cannot be completely false" (Vollmer 1983: 49), because neither the human being nor any organism could survive if their sensorial organs and their relational modalities did not grasp any real aspect of the elements with which they have to deal with in their own environment. Both the EE founding

---

[10] Born in 1943 in Speier, Rheinland, Vollmer studied mathematics, physics and chemistry in Munich and Berlin. After an early career in theoretical physics, he moved on to linguistics and philosophy, focusing on logical and gnoseological issues.

fathers and the members of the Altenberg Circle shared this hypothesis, but Vollmer reworked it in a way that aroused some perplexity among the Circle. According to Vollmer, the cognitive modalities that we have inherited from our phylogenetic past prove to be valid in the "world of medium dimensions" (*Welt der mittleren Dimensionen*), or "mesocosm", that would correspond to the environment with which man has had to deal with during his prehistory. Developing this approach, Vollmer ended up supporting an adaptationist justification of both human common sense and scientific knowledge, on the basis of the assumption that our hereditary cognitive apparatuses are well adapted to the world of average conditions, against which they were refined during phylogeny, and only fail when they are taken away from such a world. In other words, they prove inadequate only when our experiences look beyond the threshold of the mesocosm, as happens, for example, in sub-atomic physics experiments. Science, however, through its means of research, verification and control, may overcome these inborn limitations of the cognitive apparatuses, allowing us to "know something not only about ourselves, but also about the world (the thing in itself). This is the reason why objective knowledge is actually possible" (Vollmer 1975: 189).

This stance delved into a subtle, but theoretically not irrelevant, divergence between Vollmer's approach and that of the Altenberg Circle. In fact, it risked nullifying the differences between a traditional "objective realism" and the "critical-hypothetical realism" that Lorenz, Popper, and Campbell had considered a theoretical pillar of EE, according to which to every representation of reality that we enact, from the perceptive to the theoretical one, from an epistemic point of view we can only assign, the value of a "working hypothesis". Therefore, Vollmer's position exposed itself to the already mentioned criticisms that Lorenz had addressed, in 1941, in perfect agreement with Kant, to every "naïve" realism: "even if we, as natural scientists, are in a certain sense naïve realists, we still do not take the appearance for the thing in itself, nor the experienced reality for the absolutely existent" (ibid.). In fact, the EE approach, as originally conceived, implies the idea that *the relationship* between our phenomenal representation and the external reality is *a real interaction between real entities* (human beings and their environments), and that it is "by principle investigable", but does not allow the belief that the scientific study can be resolved, to a certain degree of its development, in an objective mirroring of the studied entities, process or events.

The philosopher Erhard Oeser and the biologist Rupert Riedl, two scholars at that time very close to Lorenz, would have contributed in the following years to relaunch, in a constructivist key, this approach, the development of which continued, after the death of the ethologist (1989), through the foundation of the Konrad Lorenz Institute for Evolution and Cognition Research, currently based in Klosterneuburg, near Vienna.

In that same year, a criticism of the adaptationist approach, prevalent in the first formulations of the EE and particularly in Vollmer's version of the same, was exposed by the philosopher Eve Marie Engels in the book *Erkenntnis als Anpassung? Eine Studie zur Evolutionären Erkenntnistheorie* (Engels 1989). Engels argued for the impossibility of explaining the complexity and variety of the mental, social and cultural human evolution through the biological concept of "adaption". This criticism led the promoters of the Altenberg Circle to clarify the non-exclusively adaptationist assumptions implicit in the EE that some years later would have found a programmatic expression in Riedl's paper "Deficiencies of Adaptation in Human Reason" (Riedl 1995). In that essay the author explicitly called for a "constructivist extension of Evolutionary Epistemology", aimed at:

a. integrating the concept of external selection with that of self-organization, and the adaptationist approach with the systemic-constructive one;
b. offering an evolutionary explanation, not only of the resources, but also of the limitations and "deficiencies" of our cognitive apparatuses, clarifying that those defections occur because we live in an environment which is now profoundly different from the one in which our species evolved, that is to say, in an almost totally anthropized world.

Not by chance, Riedl had been the first, within the Altenberg Circle, to try a *rapprochement* between selective and constructivist models and, more precisely, between the concept of "natural selection" introduced by Darwin and the hypothesis of an "internal selection" (*innere Selektion*), a kind of selection which is intra-specific and intra-organismic, relatively independent from the inter-specific one, already introduced by heterodox scholars such as Lancelot Whyte and Sewall Wright (Riedl 1975). Species are at first transformed in response to changes in the environmental contexts in which they live, and therefore under the pressure of "natural selection", understood as a set of selective processes that take place in a given ecological context (of which each species, group, individual is both an object and an active participant). Instead, according to Riedl, the selective pressures that lead to the *stabilization and conservation* of species-specific characteristics, or of those of larger taxonomic groups, although conditioned by a wider ecosystem context, depend to a large extent on processes of *intra-specific trans-generational self-regulation*. Processes in which each species draws information from itself and becomes a selective environment for itself. In other words, in Riedl's model, if the external environmental selection is the main cause of the transformations that the species have undergone in the course of their history, "internal selection" is, on the contrary, the main cause of the stability of their anatomical and morphological basic characters, of their preservation over time, and of the resistance to changes that phylum, classes, genera and species manifest. According to his hypothesis, in fact, the basic characteristics of taxonomic groups are fixed "more by the internal systemic conditions of the organism than by the external environment" (Riedl 1975: 297) and are protected by constraints which make drastic changes of their load-bearing structures during embryonic development extremely improbable. "The order of the living", that is the stability of the taxonomic groups that organic evolution has produced and the irreversibility of the process that led to their differentiation, depends on these constraints.

To clarify the nature and functions of internal selection, Riedl took up and reworked, in a systemic-evolutionary key, a concept strongly rooted in the tradition of pre- and post-Darwinian German biology from Goethe, to Haeckel, to Uexküll: that of a body's "structural", "anatomical", or "development plan" (*Bauplan*) which guides the development of each individual organism. As Günter Paul Wagner was later to write, according to Riedl's approach the structural plan is something like a "spectrum of adaptive degrees of freedom within the plan itself" (Wagner 1996: 20), the genesis of which must be explained in a historical phylogenetic, anatomical-morphological and probabilistic perspective. This means that the stability of the basic structures that characterize each clade depends on the role of conditions of possibility of all the further stages of development they played during the evolution and still play in every process of ontogenesis. In other words, during the course of evolution, the development of these structures has become a *conditio sine qua non* for the formation of an ever-greater set of characteristics and functions, in turn indispensable for the further development and operation of organisms. This has made any modification of these first, crucial phases of cell differentiation more and more unlikely, because, given the enormous complexity of the inputs and processes that depend on them, every slight variation from their stabilized pathway could cascade over all others, compromising performance.

As an instance of Riedl's evolutionary and systemic concept of *Bauplan*, Günter Paul Wagner refers to the structural organization of vertebrates being completely disposed (*gruppiert*) around the spinal cord. This conformation has a specific counterpart in the process of embryonic development: all the signals "that are necessary to the development of axial organs" and thus for the rest of the body, in fact depart from the construction of the spine. As Wagner writes the spinal column "is a structural plan, in that nearly all the other characteristics of a vertebrate depend on its presence" (ibid.).

As Waddington had already shown (Waddington 1975) the crucial development phases of these load-bearing structures are encoded allowing them to be restored in their course even after disturbances or anomalies, provided that these are not too drastic. As Riedl argued, these resetting mechanisms have the important function of protecting species, populations, and individuals from genetic mutations or changes in the epigenetic regulation of genetic expression that would affect a large number of the functions of an organism, with likely drastic or lethal consequences. Or, as Gerd Müller wrote in the early 1990s: "In the case of vertebrate limbs, where most of these mechanisms were studied, it is difficult to imagine any kind of novelty that could not be readily and automatically integrated by the epigenetic cascade" (Müller 1990: 119).

2.7 The Concept of "Body Plan" Today

Riedl's conception of "internal selection" as a function that presides over the stabilization of body plans and regulates the process of development, limiting "the ability of the phenotype to evolve" and binding it "to follow a determined path", was taken up and reworked by many researchers in the ensuing years. Only four years after the first edition of *Die Ordnung des Lebendigen*, the two eminent scholars Stephen Jay Gould and Richard Lewontin, in the well-known essay *The Spandrels of San Marco and the Panglossian Paradigm: A Critique of the Adaptationist Program* (Gould and Lewontin 1979), praised Riedl's "integrative" model as a useful attempt to overcome the unilateral adaptationism of modern synthesis and widen the horizons of evolutionary theory. Later, numerous scholars engaged in different research areas contributed to test and develop concepts such as "body plan" and "internal selection". The geneticist Wallace Arthur, the physicist-philosopher-engineer William C. Wimsatt, the social psychologist Jeffrey C. Schank, the cell biologist Stuart Newman, the evolutionary geneticist Günter Paul P. Wagner, the expert of vertebrate limb development and evolution Gerd Müller, the philosophers of science Werner Callebaut and Massimo Pigliucci are just some of the best known.

Wimsatt and Schank, already since the 1980s, had reworked Riedl's notion of *Bauplan* by developing the concept of "generative entrenchment" and defining it as follows:

> The generative entrenchment of an entity is a measure of how much of the generated structure or activity of a complex system depends upon the presence or activity of that entity. It is argued that entities with higher degrees of generative entrenchment are more conservative in evolutionary changes of such systems. (Schank and Wimsatt 1986: 33)

In the 1990s Arthur proposed a model in which internal selection is considered a "selection for co-adaptation" which "presides over the co-evolution of genes and their products, in order to select genes «downstream» according to their ability to adapt to those «upstream» in the morphogenetic process" (Caianiello 2013: 115). In his picture, "the reason that some character combinations are fitter than others may sometimes be determined primarily by the prevailing environmental" but "often the «internal selection» that drives co-adaptation is also important" (Arthur 2004: 285).

Newman, professor of cell biology and anatomy at the *New York Medical College* and editor-in-chief of *Biological Theory*, the Konrad Lorenz Institute for Evolution and Cognition Research scientific journal, has taken up once again the problems faced by Riedl, studying development and evolution in systemic terms. That is, as multi-level phenomena that must be analyzed with a multi-scalar approach. Particularly interesting it is the fact that Newman, documenting the relevance of environmental and epigenetic factors in evolutionary processes, uses the anti-genocentric perspective, which Riedl introduced to explain *the stability* of morphological and anatomical structures, to explain *evolutionary changes* instead. In fact, the biologist, starting from an analysis of the physical-chemical constituents of organisms, and of the constraints that their characteristics impose on both individual and species-specific morphogenesis, attempted to reconstruct the processes that led from the explosion of biological forms, peculiar of the Cambrian, to that bottleneck narrowing which later led to the circumscribed number of body plans existing now (Newman 2014). According to Newman, this process was influenced more by changes in the epigenetic structure (phenomena concerning the regulation of gene expression that in no way modify DNA sequences) than by genetic mutations, as the Synthetic Theory argued. In fact, as Newman writes and as it is today demonstrated, "heritable morphological changes were seen to be capable of occurring abruptly with little or no genetic change" (Newman 2014: 2403). We know that they require a "significant involvement of the external environment and, in several documented cases, appear to be not purely happenstance as the neo-Darwinian approach predicted, but oriented in preferred directions" (ibid.).

Gerd Müller, who along with Newman has produced important and innovative essays (see, among others, Müller and Newman 2003, 2005; Newman and Müller 2010), observes in this regard: "the majority of novelties arise as secondary by-products of epigenesis that appear when quantitative modifications of developmental processes reach a threshold of the affected system" (Müller 1990: 124).[11] Some of the consequences of this approach appear relevant for a consistent overcoming of the genocentric one:

> If morphological novelties are initially epigenetic by-products, which arise as a consequence of threshold properties in development, it follows further that it is not necessary to evoke new genes for their origin, as had been proposed on previous occasions [...]. Rather, we may find at the genome level an epigenetically induced, modified activation of existing genes. This does not exclude the possibility of later genetic assimilation of the new character [...] and its exposure to natural selection, but genetic mechanisms will not of necessity have to be held responsible as initiating causal agents. In addition, it is noteworthy that the three properties of development discussed - threshold phenomena, intermediate structures, and sequential transition of mechanisms - share the capacity to produce discontinuity within brief periods of time. (Müller 1990: 123)

Müller, who had already anticipated these ideas in the early 1990s when epigenetics and the evo-devo approach were in their infancy, in the following decades, together with scientists-philosophers such as Werner Callebaut and Massimo Pigliucci, gave important contributions to the most ambitious tasks that contemporary life science had set itself: to insert the new concepts and discoveries emerging from fields such as "molecular biology and evolutionary developmental biology, the recognition of ecological development, niche construction and multiple inheritance systems, the '-omics' revolution and the science of systems biology" in a "renewed and extended theoretical synthesis" (Müller 2017: 1) of the theory of descent with modifications.

2.8 The Project of an Extended Synthesis

Werner Callebaut was born on October 7, 1952 in Mechelen, Belgium, and in 1983 obtained his degree in philosophy at Ghent University. Three years later, he became one of the protagonists of the international debate on Evolutionary Epistemology by organizing an important workshop in Mechelen connected to the one promoted two years earlier in New York by Donald Campbell and Alex Rosen. Together with Rik Pinxten, he was later become editor of the collective volume which contained the proceedings of that meeting: *Evolutionary Epistemology: A Multi-paradigmatic Program* (Callebaut and Pinxten 1987).

---

[11] G. Müller, *Le origini della novità morfologica*, cit. p. 270.

In the following years, Callebaut worked at the universities of Brussels, Limburg, and Ghent where he became professor of philosophy in 1995. Then, after two periods as "visiting professor" at the Konrad Lorenz Institute for Evolution and Cognition Research, in 1999 he became, its scientific manager and later, its scientific director.

What sort of contribution was the philosopher giving to the EE debate in those years?

During the 1990s, as a diversity of interpretative models proliferated under the "umbrella" of Evolutionary Epistemology, the conceptual vagueness of its theoretical core (the idea of an indissoluble relationship between "life" and "knowledge" and between evolution and cognitive increase) became ever more apparent with a lack of a rigorous set of procedures for the control and falsification of assumptions and the inadequacy of the EE approach to the study of human social, cultural and scientific evolution. This situation convinced Werner Callebaut that EE needed to be radically re-founded. He worked on the project with the philosopher Karola Stotz and in 1998 they published "Lean Evolutionary Epistemology" (Callebaut and Stotz 1998): a sort of manifesto-essay conceived as a comprehensive review of the potentialities and limits of EE and a revision of its main theses.

Presenting EE as a *descriptive*, rather than normative, theory, and as an *open research program* based on a "multi-paradigmatic" approach and a methodological pluralism, the authors declared the dual intent to both go back to its roots and to renew it. Summarizing these goals in the introductive section of the essay, they wrote:

> What we recommend instead is a critical reflection on the naturalistic roots of EE (the quest for a scientific, that is, anti-transcendent and anti-transcendental epistemology for limited beings) which we hope may inspire a version of EE apt to face the future. (Callebaut and Stotz 1998: 11)

In fact, Callebaut and Stotz deemed it necessary to clarify and re-assert the cornerstones of the earliest versions of EE, proposed by the founding fathers Popper, Lorenz and Campbell, but also to overcome them towards a naturalistic but not genocentric, modular but not only adaptationist, selectionist and at the same time constructivist model of the evolutionary processes. Therefore, they proposed the rejection of every dichotomy between internal and external causes in favour of an integration between "interactionism" and "constructivism", or selectionism and self-organization-based models, already attempted by Riedl, enriched with the new discoveries about epigenetic inheritance systems acquired in the 1990s. In a far-sighted way, they emphasized the importance of the then very recent discovery that what distinguishes the morphology of a fish from that of an insect or amphibian is not the presence of class-specific or species-specific genes, but rather the way in which the expression of some genes which are common to all these classes and species is regulated trough processes of methylation and de-methylation (Callebaut and Stotz 1998: 19). A key point in their approach was the acceptance of a fundamental, although never absolute, *autonomy of the epigenetic and social evolutions, changes and developments* from the genetic mutations.

The authors' orientation towards an overcoming of the genocentric approach clearly emerged also in their critique of the apodictic and popularized version of the "central dogma" of molecular biology, according to which the phenotype is a faithful execution of a program already entirely written in the genotype. Referring, back then, to important studies by Jablonka and Lamb (1995), Müller and Newman (1999, then in press) and others, Callebaut and Stotz explicitly affirmed the centrality of the interaction between the cellular system and the external, environmental stimula for the modulation of gene expression during embryonic development and in the later stages of the life cycle.

This way they offered a model of both evolutionary and developmental processes which "treats non-genetic developmental resources as equally important to the course of evolution as genetic resources" (Callebaut and Stotz 1998: 19), and that is grounded in a "«constructivist interactionist vision of ontogeny and phylogeny»" (Oyama 1999 quoted in Callebaut and Stotz 1998: 20).

In short, the critical review of EE proposed by Callebaut and Stotz gathered several important questions that would become crucial to the bio-evolutionary research of the following two decades. However, at first their invitation to a collective re-elaboration of the EE program obtained a scarce echo and in the following years the project seemed destined to fail. With the death of Riedl in 2005, and the coeval closure of the six-monthly scientific journal *Evolution and Cognition*, official organ of the Konrad Lorenz Institute, a cycle seemed to be closing and the reworking of EE seemed to stall. However, the most active members of the Institute did not give up and produced an important breakthrough to their scientific activities. They began to converge towards a complete reworking of the theory of descent with modifications capable of taking into account what, in those years, was being discovered on the interactions between development and evolution, internal regulatory factors and environmental influences, selective and constructive processes, both in the biological and social sphere. In a few years, in fact, important steps forward had been made in this field.

In 2005, Eva Jablonka and Marion Lamb had published their revolutionary book *Evolution in four dimensions*, prompting a wide debate (Jablonka and Lamb 2005). The book appeared aimed at enhancing, in all their theoretical scope, the discoveries and "conceptual changes" which had deeply renewed almost all the branches of biology in the previous decades, in view of an overcoming of the "genocentric" approach. Discoveries and changes that the two authors summarized, in the prologue of the work, in four points:

- there is more to heredity than genes;
- some hereditary variations are non-random in origin;
- some acquired information is inherited;
- evolutionary change can result from instruction as well as from selection. (Jablonka and Lamb 2005: 1)

The new outlook that they introduced was relevant and able to solve the aforementioned classic "Darwinian dilemma" (Celentano 2013) already formulated in 1867 by Fleeming Jenkin (also known as Lord Kelwin): the hypothesis advanced by Darwin in *The Origin of Species* (1859) that the slow accumulation of random hereditary favourable variations produced by natural selection is the "main" spring of evolutionary changes can hardly explain the origin of complex organs such as the eye or the brain, at least because the time since the emergence of life on earth to today would be too short (Jenkin 1867; Lorenz 1973).

Jablonka and Lamb documented the fact that, in the course of phylogeny, alongside the slow processes of genetic variation, three other kinds of selection, heredity and variation, respectively defined epigenetic, behavioural and cultural, cooperated with the first reciprocally producing phenotypic adaptations independently of genetic or genomic mutations. In Chap. 4, they described four different kinds of Epigenetic Inheritance Systems ("Self-sustaining loops" or "memories of gene activity", "structural inheritance" or "architectural memories" and "cell structures"; "Chromatin marking systems" or "chromosomal memories"; "RNA interference" or "Silencing of the Genes") which have in common the ability to transmit from mother to child cells information "that is not related to DNA" (Jablonka and Lamb 2005: 402). Already present in the protozoa, fundamental to the evolution of multi-cellular organisms, EIS are indispensable to every kind of organisms in dealing with rapid changes and contiguous variations of their living and social environments. They are triggered by behavioural habits and/or environmental stimuli, and can preserve or modify, within very few generations, food preferences, immune systems, cognitive abilities, psychophysical and emotional attitudes. The book reported a rich documentation on cases of transmissions of food preferences taking place before and independently of any form of induction or imitation learning, in animals as rabbits, rats and humans (Jablonka and Lamb 2005: 203–207), also illustrating cases of epigenetic transmission of the effects of stress and traumatic experiences and immune deficiencies through cellular memory. It also describes cases in which new phenotypes are produced in the absence of any DNA modification (Jablonka and Lamb 2005: 339) and cases of no random genetic mutations whether induced by stress or changes in the environment (Jablonka and Lamb 2005: 97, 99, 109, 115–116).

The rapidity with which the process of differentiation and stabilization of the body plans took place, still a mystery within the theoretical framework of the "new synthesis", according to which evolutionary changes depend almost exclusively on slow accumulation of random favourable genetic mutations then "rewarded" by natural selection is in this way finally explicable.

Not surprisingly, in the concluding dialogue of the work, Mistabra, the imaginary interlocutor with whom the authors discuss at the end of each chapter, observes:

> But of more general interest are the implications your version of Darwinism has for the dynamics of evolutionary change. It implies that evolution can be very rapid, because often an induced change will occur repeatedly and in many individuals simultaneously; there is also a good chance that such a change will be of adaptive significance, since it stems from already-evolved plasticity. Even without selection, evolved plasticity will bias the direction of evolution, simply because induced variations are non-random. However, as I see it, one of the most important implications of the version of Darwinism that you have espoused is probably that when the conditions of life change drastically, it may induce large amounts of all sorts of heritable variations. The genome, the epigenome, and the cultural system (when present) may all be restructured, with the result that there can be rapid evolutionary changes in many aspects of the phenotype. (Jablonka and Lamb 2005: 350, 351)

As of 2006, the research project carried out by the Konrad Lorenz Institute for Evolution and Cognition Research, previously focused on a critical redefinition of

Evolutionary Epistemology, has increasingly come to converge with this project of a post-genocentric synthesis of Darwinian genealogical theory, and several members of the Institute have given significant contributions to this collective effort. In addition to the aforementioned biologists Gerd Müller and Stuart Newman, the protagonists of this phase were two philosophers of science such as Werner Callebaut, scientific director of the institute, who unfortunately died in 2014, and Massimo Pigliucci, professor of philosophy at the CUNY- City College of New York. In 2006, together with Müller, Callebaut founded the scientific journal *Biological Theory* and became "an early supporter of the extended version of evolutionary theory currently in the making, having himself contributed to it with his conceptualizations of biological modularity (Callebaut and Rasskin-Gutman 2005) and the organismic Systems Approach (Callebaut et al. 2007)" (Müller 2015: 2). In July 2008, Pigliucci, Müller and Callebaut organized an important international workshop in Altenberg, involving "a group of 16 prominent evolutionary biologists" (Pigliucci and Müller 2010: VII) in the discussion of the ambitious project. The proceedings of the meeting would be published a couple of years later, in the collective volume edited by Pigliucci and Müller, *Evolution—The Extended Synthesis* (Pigliucci and Müller 2010).

The evolutionary model they proposed presents some notable differences if compared to the classical Neo-Darwinian one.

To highlight only a couple of the qualifying points:

- it envisages a both functional and evolutionary autonomy of each level of the biological organization and evolution, particularly of the epigenetic, behavioural, social and ecological level from the genetic one. Therefore, this model imposes the need to study changes that occur, or have occurred, in each level (from the molecular to the social) using the spatial and temporal scales appropriate to it (Callebaut 2009, 2012);
- it does no longer place genes and genetic mutations in the role of unique or principal driving forces of evolution, but in the role of "followers" of evolutionary divergences that start from the epigenetic and ethological spheres. That is, from the differentiation of environmental contexts, uses, and habits between individuals and between populations of the same or related species, in biologically relevant contexts such as diet, explored and frequented niches, mating rituals, communicative traditions (Jablonka 2006; Pigliucci and Müller 2010).

In this regard, in the book Pigliucci wrote that today "genes could come to be seen as «followers» rather than leaders in the evolutionary process" (Pigliucci in Pigliucci and Müller 2010: 370) and "the pre-eminent role that behaviour plays in directing evolutionary changes" (2010: 371) is now undeniable.

In short, if the mechanistic model of development proposed in the late nineteenth century by Haeckel and re-proposed in the following century with updated languages by different versions of genetic determinism affirmed that "phylogenesis is the mechanical cause of ontogeny",[12] the approach of contemporary scholars such

[12]E. Haeckel, *Anthropogenie, oder Entwickelungsgeschichte des Menschen. Engelmann*, Leipzig 1877, ed. or. 1874, p. 7, my translation.

as Callebaut, Müller, Pigliucci and Newman, as noted by Salvatore Tedesco, comes rather to the opposite hypothesis: "ontogenesis creates phylogenesis".[13]

Meanwhile the work aimed at offering a "postmodern" synthesis of the Darwinian theory of descent with modifications has become a *collective international enterprise* in which the efforts of many scholars specialized in different disciplines, and living in different countries, are converging.

2.9 Towards a Post-Genocentric Conception of Human Behaviour and Social Evolution

The post-genocentric approach nowadays prevailing in both ethological and evolutionary studies offers innovative answers to another *vexata quaestio* that accompanied the history of Darwinism and evolutionism:

To what extent are human mind and behaviour the product of inherited adaptations fixed in an irreversible or scarcely modifiable way in all members of our species?

To what extent, are they instead influenced by economic, social and cultural selection, by tradition, institutions and communication, and can therefore develop significant, useful or harmful, changes in the rapid times in which these kinds of social factors operate?

The old question now seems through the developments of genetics and epigenetics, molecular and developmental biology, cognitive and cultural ethology, *to the detriment of the genetic determinism that dominated the theories of inheritance and evolution in the twentieth century*. The reasons for this change are easily understandable even to non-experts. The theoretical framework of the "modern synthesis" suggested that social environment and cultural influences had not been able to make significant modifications in the human phenotype and in its physiological, psychic and behavioural propensities. The phenotypic conformation was, in fact, considered the result of the faithful execution (except for random errors) of a program entirely written in the DNA and hardly influenced by the environment. Therefore, significant phenotypic modifications were considered possible only in the long times required by relevant genome mutations, that is, over millions of years. The empirical and experimental evidence discussed in this essay instead allows the corroboration of opposite approaches and conclusions: as Darwin and Lorenz had already guessed, in human social evolution, the influence of selective pressures exerted by social environment on individual mental and behavioural orientations has been growing over the years, to detriment of the influence of the selective pressures exerted by the natural environment and the inter-specific context. The social selection of behaviours and propensities has increasingly replaced that interaction "of everything with everything" through which natural selection in a Darwinian sense is regulated. The environment in which the selection of human behaviour takes place has already for

[13]S. Tedesco, *Introduzione*, in A. Pinotti, S. Tedesco, *Estetica e scienze della vita*, Parte seconda, *Evo-Devo e morfologia*, Raffaello Cortina Editore, Milano 2013, p. 188.

millennia been and become an increasingly *anthropic* environment. An environment in which the conditions for survival and success are socially produced by our species. Today the developments of epigenetics enable us to demonstrate that, generation after generation, this social context in which experiential and selective dynamics take place selects orientations and contents of emotions, sensations and propensities, cognitive and health resources and deficits, paths of mental and behavioural development, not only in the individual, but also in their immediate descendants, thus triggering the inhibition and/or reactivation of the expression of certain genes (Jablonka 2006, 2013, 2017). Thus, we can now affirm that:

- for millennia, the most important biological modifications taking place in the human organism, those that channel and guide the development of main aspects of the emotional, cognitive and social attitudes, are shaped by inter-specific selection, by the social environment, and by the rules and living conditions that they impose, more than by "natural selection" understood in the Darwinian sense;
- the effects of these ethogenetic and epigenetic inheritances are transmitted not with the slowness of genetic modifications that require millions of years to set new characters, but with extreme rapidity. In fact, as we shall see in greater detail in the final essay of the volume, it is today possible to demonstrate that the experiences and lifestyles of every single generation have an immediate impact on morphogenesis and organization (micro-geography and functionalities) of the brain's and organic apparatuses in immediately ensuing generations.

In other words, today biology and ethology themselves show that our "inside", understood both as organic profile and subjective experience, is to a large extent a *social product*, a product of the human, social, economic and cultural system in which it takes shape. As Jablonka and Lamb pointed out in their book, this makes *anachronistic* every claim to explain the "social-behavioural status quo" of contemporary humanity as an expression of innate tendencies irreversibly fixed in our genetic code millions of years ago (Jablonka and Lamb 2005: 424–427).[14]

After all, in the last fifteen years, the developments of this field of research made increasingly evident the close correlation between EIS (Epigenetic Inheritance Systems) and BIS (Behavioral Inheritance Systems), leading to the birth of a new scientific discipline or sub-discipline: behavioural and cultural epigenetics (Jablonka 2013, 2017; McGowan and Szyf 2010; Champagne and Rissman 2011). Its task is "the investigation of the role of behaviour in shaping developmental-epigenetic states and the reciprocal role of epigenetic factors and mechanisms in shaping behaviour" (Jablonka 2017: 42). This development of behavioural and cultural epigenetics open the possibility of a social history of human biological evolution and condition based on an anti-separatist but at the same time anti-deterministic approach.

They invigorate and raise the need of that two-stage evolutionary model that had already been introduced in 1987 by the philosopher Erhard Oeser (1987) in the context of EE, according to which behavioural and perceptive habits, mental categories and expectations, interactive modes and evaluative criteria develop in

---

[14] *Ivi*, p. 473.

the human individual, over the course of life, thanks to so-called "open programs", that is to say innate learning programs, of which Lorenz and Mayr had already spoken.[15] In other words, through a series of sensitive phases of biological and cognitive maturation, marked by species-specific forms of *active research of the stimulus* leading individuals to assimilate mental and behavioural models from society, from the behaviour and teachings of con-specifics, from the parental and socio-cultural figures of reference through some key phases of psychic development. According to this perspective, cerebral and cognitive formation of every human being starts, inevitably, from an assimilation and a re-elaboration of inputs, models, behavioural patterns, emotional reaction norms, drawn from the social environment, which in the first years of life take place in a largely unaware form. The brain and psychic development is thus configured as a "biographical path" that leads to the structuring of "a hierarchy of dynamically interdependent layers, in which the component of mere adaptation progressively decreases, giving way to constructive elements that take over in an ever more incisive manner" (Oeser 2000: 338). Callebaut and Stotz also referred to this model in their aforementioned 1998 essay, framing human thought and behaviour as socially produced and individually learned and re-elaborated expressions, starting from an interactive cultural and linguistic environment that exercises training functions on them. As we have seen, today studies aimed at the formulation of an "extended" or "post-modern" synthesis of Darwinian theory, reaching a full recognition of the autonomy of socio-cultural evolution and individual behaviours from direct and binding genetic influences. Moreover, the research of contemporary ethologists such as Frans de Waal, who for over twenty years investigated the phylogenetic and historical origins of "morality", also led to a similar recognition. In fact, the long de Waal study confirms the decisive influence of social factors and early experiences on the development of moral codes and ethical choices (de Waal 1996, 2006; Celentano 2013) and clarifies an important concept: in our species, the innate components which guide the development of ethical evaluations and choices manifest themselves not as innate rules or principles as some evolutionary psychologists have recently argued (Hauser 2006; Pinker 2008), but as learning programs. Not by chance, on this matter the ethologist writes:

> We are born not with any specific social norm, but with a learning agenda that tells us which information to imbibe and how to organise it. […] In a sense, we are imprinted upon a particular moral system through a process that, though hundreds of times more complicated than the imprinting of birds, maybe just as effective and lasting. (de Waal 1996: 36)

So, in a nutshell, today it is exactly the same sciences that every genetic determinism has traditionally used to argue the (still unproven) hypothesis that human behaviours are prescribed in a binding way by our genes that contribute decisively to proving the opposite. They thus prepare us for a task from which contemporary culture can no longer escape, if not at the risk of compromising its own ability to produce critical

---

[15]See K. Lorenz, *Evoluzione e modificazione del comportamento*, tr. it. Boringhieri, Torino, 1971; E. Mayr, *Behavior Programs and Evolutionary Strategies: Natural selection sometimes favors a genetically "closed" behavior program, sometimes an "open" one*, *American Scientist*, 62, 6, 1974, pp. 650–659.

self-reflection: understanding what consequences have derived and are deriving from the fact that the society, and no longer nature, in its inter-specific wealth, produces the environment in which the "human" is selected and cultivated, in an increasingly pervasive way.

This target implies a triple commitment that, in my opinion, should today concern transversely all the scientific disciplines:

- to contribute to understanding the ways in which intra-specific selection and social context impose developmental constraints and directions on the biological, behavioural and cognitive expressiveness of individuals, populations, and species, i.e., to develop a comparative, synchronic and diachronic study of the "canalizations" (Waddington 1975) that social evolution imposes on biological evolution;
- to contribute to explain how human biological history has evolved and is evolving *within* human social history and under the dominance of intra-specific human selection. Trying to understand which transformations social life has inscribed and is inscribing in physiological, sensory and cognitive activities, as well as in the "appetences" and propensities of humans. To develop a synchronic and diachronic comparative study of the structural constraints and evolutionary orientations that currently dominant development models in human social organization are imposing on humans themselves, on almost all the other organisms and on our and their living environment (Jablonka and Bronfman 2014; Jablonka 2017);
- to contribute to the scientific identification and concrete social removal of all those forms of discrimination, exploitation and social manipulation of human beings, and of other sentient beings, which, as we know today thanks to cultural epigenetics, leave scars, wounds and traumatic effects not only in those that suffer them in a direct way, but also in their descendants.

2.10 Genealogical Perspectivism: Towards an Ethological Conception of Knowledge

Like fish capable of exploiting undulating motions of water so as to move and live in it, or birds, able to fly, hunt and sometimes sleep gliding over air currents, we only possess *knowledge without "truth"*. We can elaborate complex forms of schematization of the "entities", "elements", "processes" or "events" with which we interact. These allow us to survive in the environments we colonize, as the fish does in water and the bird in air, but we hold no "truths", if we intend them in the traditional, strong and emphatic sense of the term, that of metaphysical laws, or laws of nature that are universally valid in space and time. The "knowledge" we incorporate through inheritance and learning never translates into statements that can be considered unquestionably true or exact from any point of view and that remain so over time. Our cognitive activity, not unlike that of any other body, is never resolved into an objective and neutral reflection of what we come into contact with. Rather, it manifests itself always and exclusively as a production of behaviours, of forms of assimilative, exploratory, transformative and self-regulating interaction with what coexists with us.

As some prominent contemporary scholars have reiterated, this limitation concerns not only the ability of discernment of other living beings and our common sense, but also our *scientific* knowledge. In fact, even the latter cannot overcome these limits. This finding led contemporary scholars like Ronald Giere and Werner Callebaut to assume an epistemological position they called *scientific perspectivism* (Giere 2006; Callebaut 2009, 2012; Celentano 2018). Its basic theses can be summarized as follows:

- Each detection system, whether biological, technological, or hybrid, responds only to a particular and limited set of signals. Like our organs, scientific "instruments are sensitive only to a particular input and blind to everything else; and their output is a function of both their input and their internal constitution" (Callebaut 2012: 76).
- No system, whether organic or artificial, offers a mere neutral mirroring of data, because all the detection systems process the collected data according to criteria intrinsic to their own structure, and not (or not only) to the entities or processes they detect.
- Different sensor organs and instruments receive and represent the same phenomena in different ways.
- The scientific hypotheses and theories are "models of aspects of the world", and the so-called natural laws are highly generalized models "that characterize a scientific perspective" (ibid.).
- "As a consequence, we cannot transcend our human perspective" (Brandon 2007).

Starting from them, I would like to propose, in these concluding pages, some reflections on the possibility to elaborate an ethological conception of knowledge and to clarify the basic concepts of a perspectivism that is not only scientific, but also *genealogical*. Or rather, of an ethological and genealogical approach to the problem of knowledge which affirms, not only that every human and not human form of knowledge is based on a limited perspective and makes only certain aspects of reality perceptible, or conceivable, but also that these limits are essential to make this knowledge useful for the survival and for the modification of the living conditions of the organisms. This approach radically modifies the concept of "knowing", which was intended since the age of Aristotle until the contemporary one. In fact, it sets the evaluation parameters and criteria for putting to test our knowledge, no longer with the pretense of progressively approaching the formulation of alleged immutable "natural laws", or to chase an impossible exact representation of the "in itself", but solely in its concrete, practical effects. That is, in the qualitative and quantitative modifications, in the protection or in the worsening, of the living conditions of acquaintances and known subjects that this knowing produces.

Cognitive activities are always simultaneously self-regulative interactions, forms of energetic and informative exchange with the species-specific and inter-specific environment, and exploratory behaviours or forms of testing and implementation of the practical skills of an organisms.

The forms of the perceptive, mental and symbolic representation of the reality we elaborate, may be more or less functional to the resolution of practical, individual and collective problems, more or less useful to develop certain systems of representation of reality that we use in our forecasting and calculation operations, but not absolutely true.

In fact, as Nietzsche pointed out, the existent does not, in itself, have an aspect. It can take on an aspect, it can present itself under certain forms and characteristics, only in relation to a living being that perceives it and to the different situations in which this happens. Therefore, we don't have to establish a difference between the way we perceive something and the way that something would appear "in itself". Rather, we must take note that every existing "x" must necessarily appear in as many different ways as they are the organs, apparatuses or instruments engaged in understanding and interacting with it and the different circumstances in which this interaction takes place.

Every existing entity, or ongoing process will thus necessarily appear differently to the many species of organisms that will interact with it because each species (and to a lesser extent each population and individual) has developed different perceptive and cognitive apparatuses in relation to particular vital needs and evolutionary contingencies.

A similar argument also must be applied to human abilities: our cognitive organs developed their skills over the course of phylogeny and of social, cultural and individual history. This means that they acquired and modified their shape not only through a mental representation of the external environment, but also through *a material interaction with it.*

Consequently, we need a complete overcoming of the traditional concept of knowledge as a form of compliance of human ideas to external things, or *adaequatio rei et intellectus*, and of the idealistic conception of the cognitive apparatus as a set of functions delegated to perform *a mere mental representation* of the external environment.

The brain, or eye, as the fin of a fish, rather than merely representing an external environment, serve to move in it, to interact with it, to implement vital and social functions in a determined (but changing) ecological context. In this sense, we can affirm that a cognitive activity primarily manifests itself in all living beings as the ability to *produce body plans and behavioural forms* suitable enough to guarantee their survival and reproduction as an individual, populations or species. This is the core of an ethological conception of knowledge. We can today recognize that not only human beings, but all organisms, in order to stay alive and reproduce need to carry out and exercise cognitive activities within the limits imposed to them by their own evolutionary and ontogenetic history. These cognitive activities manifest themselves as a production of physiological and behavioural forms and activities which place crucial self-regulating functions and are capable of implementing qualitative changes in the physiological states and life conditions of an organisms.

But to what extent is it possible to affirm that organisms are capable of acquiring and transmitting, in ways that go from the genetic to the cultural heritage, information that is useful to the survival or to the qualitative improvement of their living conditions?

From a genealogical point of view, we could answer that the cognitive endowments of the organisms have proved to be useful enough to ensure that life, for more than three billion years (that is from the moment of its appearance until today), has not died out. We could not allow more than this, with regard to their degree of reliability, but it would be difficult to deny them at least this degree of "utility". The organisms "know" some factors of the real world just enough to survive and reproduce in some of its areas. This is what we can say about the "validity" of the systems of codification and decoding of reality developed both by other living beings and by ourselves.

This also means that, even though we cannot reach absolute certainties, each of our cognitive resources, from sensation to emotion, from perception to imagination and theories, just like the skills of other organisms, can be put to the test. In fact, we could test their real effectiveness on the condition that we modify as a yardstick of our knowledge, by placing it no longer on the horizon of an alleged progressive approach to the perfect correspondence between things as they actually are and their scientific representation, but rather exclusively in terms of its practical effects. That is, in terms of the damages or benefits it entails. Knowing serves to live, to face life. We can't demand more.

The first consequence of this approach is that claiming a cognitive methodology is more valid than any other, and therefore preferable under all circumstances, is as vain as expecting a microscope to be equally useful for observing microorganisms as to avoid the dangers of a car-filled street and studying the stars. In fact, the utility of a presumed knowledge or methodology is detectable only in relation to the specific area of problems that they are called upon to face and resolve.

In conclusion, genealogical perspectivism is based on an ethological conception of knowledge according to which cognitive processes are not reducible to mere processes of assimilation of information or mere activities of perceptive, mental or symbolic representation of known "objects". In fact, they always imply an effective interaction between the knowing subject and the known "object" (which of course may also be one or more other living beings) that transforms them both and that is indispensable for the environmental monitoring, self-monitoring and self-regulation that every organism must perform to keep itself alive.

Moreover, according to an ethological conception of knowledge, as the fathers of the Evolutionary Epistemology already guessed, we can include in the cognitive activities not only processes that allow individual organisms to monitor the environment and interact with it, but also those that regulate morphogenesis, ontogenesis, and phylogeny. Living being learn while taking shape, both physically and mentally in the first place. Organisms acquire knowledge by transforming themselves and the context in which they live, and this is true both for the homeorhetic processes (Waddington 1975), or the vital parabola of individuals as well as for trans-generational processes that allow the species to preserve and/or change. Therefore, development and evolution are both cognitive processes, inextricably intertwined, in which living beings play an eminently active role, contributing with their intelligence to shape themselves shaping and to the evolution of the environments in which they live.

References

Arrigoni, Ermanno. 1991. *"L'uomo a una dimensione" di Marcuse e l'alienazione dell'individuo nella società contemporanea secondo gli autori della Scuola di Francoforte*. Torino: Paravia.

Arthur, Wallace. 2004. The Effects of Development on the Direction of Evolution. *Evolution and Development* 6 (4): 282–288.

Baluška, Frantisek, Stefano Mancuso, and Dieter Volkmann. 2007. *Communication in Plants: Neuronal Aspects of Plant Life*. New York: Springer.

Brandon, Ed. 2007. Ronald N. Giere: Scientific Perspectivism. *Metapsychology: Online Reviews* 11 (41).

Caianiello, Silvia. 2013. L'interno della selezione. In *Confini Aperti: Il rapporto esterno/interno in biologia*, ed. Continenza Barbara, Elena Gagliasso, and Sterpetti Fabio. Milano: Franco Angeli.

Callebaut, Werner. 2009. Multi-scale Phenomena in Biology and Scientific Perspectivism. In *Multi-scale Phenomena in Biology. Melville: American Institute of Physics Conference Proceedings*, ed. R. M. Sinclair and K. M. Stiefel, 57–69.

Callebaut, Werner. 2012. Scientific Perspectivism: A Philosopher of Science Response to the Challenge of Big Data Biology. *Studies in History and Philosophy of Biological and Biomedical Sciences* 43 (1): 69–80.

Callebaut, Werner, and Rik Pinxten. 1987. *Evolutionary Epistemology: A Multi-paradigmatic Program*. Dordrecht: Reider.

Callebaut, Werner, and Karola Stotz. 1998. Lines Evolutionary Epistemology. *Evolution & Cognition* 4–1: 1–36.

Callebaut, Werner, and Diego Rasskin-Gutman. 2005. *Modularity: Understanding the Development and Evolution of Complex Natural Systems*. Cambridge: MIT Press.

Callebaut, W., Müller Gerd, and Newman Stuart. 2007. The Organismic Systems Approach: Evodevo and the Streamlining of the Naturalistic Agenda. In *Integrating Evolution and Development: From Theory to Practice*, ed. Sansom Roger and Brandon Robert, 25–92. Cambridge: MIT Press.

Campbell, Donald T. 1974. Evolutionary Epistemology. In *The Philosophy of Karl Popper*, vol. I, ed. Paul A. Schilpp, 413–463. La Salle: Open Court.

Celentano, Marco. 2000. *Etologia della conoscenza*. Napoli: Città del Sole.

Celentano, Marco. 2011. *Konrad Lorenz e l'etologia contemporanea*. Milano: Franco Angeli.

Celentano, Marco. 2013. Etologia e antropologia. Utilità e limiti del principio di continuità. In *Una prospettiva evolutiva sulle emozioni. Da Charles Darwin alle neuroscienze*, ed. Bernardino Fantini, Giovann Destro Bisol, Fabrizio Rufo (a cura di), 47–66. Pisa: Edizioni ETS.

Celentano, Marco. 2018. Prospettivismo genealogico. Proposte integrative per un programma di ricerca sull'etologia del conoscere. In *Pluriversi cognitivi. Questioni di filosofia ed etologia*, ed. Marco Celentano and Roberto Marchesini. Milano-Undine: Mimesis.

Champagne, Frances and Rissman, Emilie. 2011. Behavioral Epigenetics: A New Frontier in the Study of Hormones and Behavior. *Hormones and Behavior* 59 (3): 277–278.

Craig, Wallace. 1918. Appetites and Aversions as Constituents of Instincts. *Biological Bulletin* 34 (2): 91–107.

de Waal, Frans. 1996. *Good Natured: The Origins of Right and Wrong in Humans and Other Animals*. Cambridge, MA: Harvard University.

de Waal, Frans. 2006. *Primates and philosophers*. Princeton, NJ: Princeton University Press.

Evans, Richard J. 1975. *Konrad Lorenz: The Man and His Ideas*. New York: Harcourt Brace Jovanovich.

Fancello, Omiti. 1985. *Dalle molecole all'uomo* [From molecules to man]. Roma: Editori Riuniti.
Giere, Ronald. 2006. *Scientific Perspectivism*. Chicago: University of Chicago Press.
Hauser, Mark. 2006. *Moral Minds*. New York: Ecco/Harper Collins Publisher.
Heimer, Armin. 1977. *Ethologishes Wörterbuch*. Berlin: Paul Parey.
Jablonka, Eva. 2006. Genes as Followers in Evolution—A Post-synthesis Synthesis. *Biologyand Philosophy* 21: 143–154.
Jablonka, Eva. 2013. Behavioral Epigenetics in Ecological Context. *Behavioral Ecology* 24 (2): 325–326. http://scholar.google.com/scholar_lookup?publication_year=2013&pages=325-326&issue=2&author=E.+Jablonka&title=%E2%80%98Behavioral+epigenetics+in+ecological+context%E2%80%99&http://scholar.google.com/scholar_lookup?publication_year=2014&issue=2&author=E.+Jablonka&author=Z.+Z.+Bronfmanauthor=J.+Losos&title=Oxford+Bibliographies+in+Evolutionary+Biology&.
Jablonka, Eva. 2017. Cultural Epigenetics. *Sage Journal* 64 (1): 42–60.
Jablonka, Eva, and Marion Lamb. 2005 (revisited edition 2014). *Evolution in Four Dimensions*. Cambridge, MA: MIT Press.
Jablonka, Eva, and Zohar Bronfman. 2014. Epigenetics and Behavior. In *Oxford Bibliographies in Evolutionary Biology*, ed. Losos Jonathan. New York: Oxford University Press.
Jablonka, Eva, and Marion Lamb. 1995. *Epigenetic Inheritance and Evolution*. Oxford and New York: Oxford University Press.
Jenkin, Fleeming. 1867. [Review of] The Origin of Species. *The North British Review* 46 (June): 277–318.
Kant, Immanuel. 1783. *Prolegomenazu einer jeden künftigen Metaphysik*. Engl. transl. 1994. *Prolegomena to any future metaphysics*. Indianapolis: Pearson.
Kant, Immanuel. 1781 (1787). *Kritikderreinen Vernunft*. Engl. transl. 1999. *Critique of Pure Reason*. Cambridge: Cambridge University Press.
Lorenz, Konrad. 1937. Über die Bildung des Instinktbegriffes. *Die Naturwissenschaften* 25 (19): 289–300.
Lorenz, Konrad. 1941. *Kant's Lehrevom Apriorischen im Lichte gegenwärtiger Biologie*. Engl. transl. 1975. *Kant's Doctrine of the a Priori in Light of Contemporary Biology*. In R. I. Evans 1975: 181–217.
Lorenz, Konrad. 1973. *Die Rückseite des Spiegels*. Engl. transl. 1977. *Behind the Mirror: A Search for a Natural History of Human Knowledge*. London: Meuthen & Co.
Lorenz, Konrad. 1983. *Der Abbau des Menschlichen*. Engl. transl. 1987. *The Waning of Humaneness*. Boston: Little, Brown.
Lorenz, Konrad. Posthumous 1992. *Die Naturwissenschaft vom Menschen*. Engl. transl. 1995. *The Natural Science of the Human Species*. Cambridge, MA and London: MIT Press.
Mancuso, Stefano. 2018. *The Revolutionary Genius of Plants: A New Understanding of Plant Intelligence and Behavior*. New York: Atria Books.
Masullo, Aldo. 1986. *Fichte. L'intersoggettività e l'originario*. Guida: Napoli.
McGowan, Patrick and Szyf Moshe. 2010. The Epigenetics of Social Adversity in Early Life. *Neurobiology of Disease* 39 (1): 66–72.
Monod, Jacques. 1970. *Le Hasard et la Necessité*. Paris: Editions du Seuil.
Mivart, St. George Jackson. 1876. *Lessons from Nature*. London: John Murray.
Müller, Gerd. 1990. Developmental Mechanisms at the Origin of Morphological Novelty: A Side-Effect Hypothesis. In *Evolutionary Innovations*, ed. Nitecki Matthew, 99–130. Chicago: University of Chicago Press.
Müller, Gerd. 2015. Liberté, Egalité, Modularité: In Memory of Werner Callebaut (1952–2014). *Biological Theory* 10 (March 1): 1–4.
Müller, Gerd. 2017. Why an Extended Evolutionary Synthesis is Necessary. *Interface Focus* 7: 1–11.
Müller, Gerd, and Stuart Newman. 1999. Generation, Integration, Autonomy: Three Steps in the Evolution of Homology. *Novartis Found Symposium* 222: 65–73; discussion: 73–9.

Müller, Gerd, and Stuart Newman. 2003. Origination of Organismal form: The Forgotten Cause in Evolutionary Theory. In *Origination of Organismal Form: Beyond the Gene in Developmental and Evolutionary Biology*, ed. G. B. Müller. Cambridge, MA and London: MIT Press.
Müller, Gerd, and Stuart Newman (eds.). 2005. The Innovation Triad: An Evo-Devo Agenda. *Journal of Experimental Zoology* 304: 487–503.
Newman, Stuart. 2014. Form and Function Remixed: Developmental Physiology in the Evolution of Vertebrate Body Plans. *Journal of Physiology* 592 (11): 2403–2412.
Newman, Stuart, and Gerd Müller. 2010. *Morphological Evolution: Epigenetic Mechanisms*. New York: Wiley.
Nietzsche, Friedrich. 1882. *Die fröhliche Wissenschaft*. Engl. transl. 2001. *The Gay Science*. Cambridge: Cambridge University Press.
Oeser, Erhard. 1987. *Psichzoikum. Evolutions und Mechanismus der menschlichchen Erkenntnisfähigkeit*. Berlin und Hamburg: P. Parey.
Oeser, Erhard. 2000. Anatomia della coscienza: il metodo trascendentale kantiano come neuroepistemologia. In *Natura e cultura*, ed. Giuseppe Cantillo and Rossella Bonito Oliva (a cura di), 325–341. Napoli: Guida.
Pigliucci, Massimo, and Gerd Müller. 2010. *Evolution—The Extended Synthesis*. Cambridge: MIT Press.
Pinker, Steven. 2008. *The Moral Instinct*. lumerinnovato. blogspot.com/.../listinto-morale-di-steven-pinker.html.
Popper, Karl. 1972. *Objective Knowledge an Evolutionary Approach*. London: Oxford University Press.
Popper, Karl. 1984. *Evolutionary Epistemology*. In *Open Questions in Quantum Physics*, ed. Gino Tarozzi and Alwyn van de Merwe. Dordrecht, Boston, Lancaster: D. Reidel Publishing Company.
Popper, Karl. 1990. *A World of Propensities*. Bristol: Thoemmes.
Popper, Karl, and John Condry. 1994. *Cattiva maestra televisione*. Venezia: Marsilio.
Riedl, Rupert. 1975. *Die Ordung des Lebendigen*. Hamburg and Berlin: Paul Parey.
Riedl, Rupert. 1980. *Biologie der Erkenntnis*. Berlin: P. Parey.
Riedl, Rupert. 1995. Deficiences of Adaptation in Human Reason: A Constructivistic Extension of Evolutionary Epistemology. *Evolution and Cognition* 1 (1): 27–37.
Schank, Wallace, and William Wimsatt. 1986. Generative Entrenchment and Evolution. *PSA: Proceedings of the Biennial Meeting of the Philosophy of Science Association* 2: 33–60.
Somenzi, Vittorio. 1996. *L'epistemologia evoluzionistica*, in *Un'introduzione all'epistemologia contemporanea*, ed. Antiseri Dario, Gava Giacomo, Petroni Angelo, Somenzi Vittorio, Stanzione Massimo. Padova: Cleup.
Stanzione, Massimo. 1984. *Epistemologia evoluzionistica: confronti e critiche*. In *Evoluzione e modelli*, ed. Continenza Barbara, Cordeschi Roberto, Gagliasso Elena, Ludovico Anna, and Stanzione Massimo. Roma: Editori Riuniti.
Vollmer, Gerhard. 1975. *Evolutionäre Erkenntnistheorie*. Stuttgart: Hirzel.
Vollmer, Gerard. 1983. *Mesokosmos und objektive Erkenntnis*. In *Die Evolution des Denkens*, ed. Konrad Lorenz and Frans Wuketits. München: Piper.
Waddington, Conrad Hal. 1975. *The Evolution of an Evolutionist*. Edinburgh: Edinburgh University Press.
Wagner, Gunt P. 1996. Die Größte Herausforderung der EE. In *Die Evolutionäre Erkenntnistheorie im Spiegel der Wisssenschaften*, ed. Riedl Rupert, Delpos Manuela. Wien: WUV Universitätverlag.
Wuketits, Franz M. 1990. *K. Lorenz. Leben und Werk eines großen Naturforschers*. Italian translation. 1991. *Konrad Lorenz. Una vita*. Milano: Mondadori.

Chapter 3
Interspecific Cultural Studies and Numanities: The Comparative Study of Animal Traditions Beyond the Separation Between Humanities and Life Sciences

Marco Celentano

Abstract The introductory part of the chapter will display its whole purpose and topic. First of all, this section of the book aims to clarify the situation of both humanities and behavioural sciences after the discovery of animal thought and cultures. In other words, after the fundamental theoretical assumptions of these two scientific traditions were empirically refuted: the idea of man as the only thinking and cultural animal and other animals as mechanically explainable entities. From this the need for a critical and self-critical re-foundation of both humanities and behavioural sciences arises. A process which is in fact already underway but is still not reflecting enough on the level of theoretical elaboration and on the practical level of a reform of scientific training and research. This need for a new organization of university and post-university education, transversal to the bipartition between human and life sciences, and of meta-disciplinary forms of organization of the basic and applied research, on which the chapter aims to focus, is in various respects close to the goal of a radical self-reform of humanities proposed in Martinelli's *Manifesto of Numanities* (Martinelli 2016). In this specific case, the pivot or pillar of this "revolution" of humanities is identifiable in the attempt to reorganize the humanistic field as *Interspecific Cultural Studies*. That is, as a meta-disciplinary area able to assume a post-anthropocentric approach towards its topics and collaborate, each sector starting from its own specificity, on an enterprise that we are attempting in our age for the first time: to insert the study of past and present human cultures into the broader context of a comparative study of *all animal cultures, existing and existed*. This enterprise would imply, as its indissoluble condition, the commitment to protect the survival of these animal cultures and of the natural environments in which they have evolved. The following section presents, in extreme synthesis, the state of the art of cultural ethology. The third section introduces, in an equally concise way, the emerging etho-centric approach to the explanation of evolutionary processes in contrast to the geno-centric one, recognizing not genes, but explorative and cognitive behaviours, experiences and cultural traditions as the main driving forces of animal evolution. The fourth section illustrates the basic characteristics of the meta-disciplinary area indicated in the chapter as *Interspecific Cultural Studies*

R. Marchesini and M. Celentano, *Critical Ethology and Post-Anthropocentric Ethics*, Numanities - Arts and Humanities in Progress 16,
https://doi.org/10.1007/978-3-030-74203-4_3

(ICS) and their close affinities with the program of Numanities. The fifth section focuses, within the ICS framework, on a particular project and object of research: the study of the cases of *Interspecific Cultural Convergences* (ICC). These are cases in which a technique, an invention, a discovery, an expressive form or use have been independently developed not only by different populations of the same species, but also by *societies and traditions of different animal species*. The last part illustrates one of the best-known ICC cases: *singing*. A widespread expressive form in all human cultures and in primates genetically and phylogenetically quite distant from us such as Hylobatidae, Tarsius, Indri and Callicebus yet not among our sister species (chimpanzee, bonobo, gorilla). An expressive form developed by animal species as diverse and from a genetic, phylogenetic, morphological and ecological point of view as different as birds, mice and whales. The diffusion of singing in so different clades and environments obviously cannot be explained as a case of homology (similar characteristic inherited by common ancestor), because the ancestors common to birds and mammals did not sing, just as as those common to insects and birds did not have wings. It is instead the result of mutually independent, but in some aspects similar, evolutionary processes and social or ecological selective pressures. It can be adequately understood only by identifying and comparing the *biological and social functions* that this kind of expression plays, and the *forms* it has assumed, in all these animal communities, just as is commonly done by comparing human singing traditions and performances. In the ICS perspective, this approach can be extended to the study of *all aspects of animal cultures* and of all cases of ICC found.

3.1 Introduction

From Aristotle's time to the first half of the twentieth century, studying past or existing cultures, reconstructing their history, analyzing and comparing their languages, productions and traditions signified dealing exclusively with *human* culture, as Man was considered the only "cultural animal".

Today, just over half a century since the beginning of that scientific revolution which, in the 1960s led to the discovery of different traditions in various species of primates, cetaceans and birds, cultural ethology is still a very young discipline, but we can affirm that the existence of non-human cultures is widely proven and the hypothesis that *all* existing species of mammals and social birds have, over millions of years, developed their own uses and dialects is highly probable.[1]

[1] How are the concepts of "culture" and "tradition" in the ethological field defined today? Ethologists are in broad agreement on a trans-specific notion of "culture" that implies, as its necessary and sufficient conditions, the existence of systems of transmission of experiences and uses to other individuals and generations, through learning/teaching processes (de Waal 2001: 11; Martinelli 2011: 230). They "refer to as 'cultural traits' those behavior patterns shared by members of animal populations that are to some degree reliant on socially learned differences between individuals,

It is also demonstrated that this differentiation of uses and languages, handed down through social learning in different populations of the same species, is not an exclusive prerogative of mammals and birds. Research performed over the past forty years has highlighted the existence of intra-specific local traditions in various species of teleost fish, particularly among species that live in the coral reefs (Helfman and Schultz 1984; Bshary et al. 2002; Laland et al. 2011), and this fact reinforces the hypothesis that analogous phenomena can be found in all aquatic and terrestrial vertebrates that derive from bony fish.

Therefore, the world of animal cultures appears to us, today, as an immense, yet almost completely unknown universe, because so far, humanity has lived next to it without recognizing it as such. At the same time, it presents itself to our eyes as a world whose survival is increasingly threatened by anthropic impact, or rather by our models and processes of economic and social development.

In our age the greatest mass extinction of animal and plant species ever recorded is currently underway and consequently, almost all existing animal cultures are, in fact, *hunted cultures*, from which larger and larger portions of living environment are subtracted on a daily basis for purposes related to human profit.

In my opinion, it is exactly this interweaving of ethological, ecological, socio-cultural, philosophical and ethical problems, that anyone who studies animal cultures is used to facing, that today requires a full development in the field of research indicated in this essay as Interspecific Cultural Studies. An area to be understood not just as something completely new that the author of this paper would like to propose, but rather as the set of heterogeneous skills, transversal to traditional disciplinary blocks, necessary to investigate a new object of study, about which we began to learn just sixty years ago, such as animal cultures. The problems inherent in their study, in fact, can be adequately addressed, *to a theoretical level*, only by attempting a critical overcoming of both the anthropocentrism and anthropo-denial rooted in the humanist tradition and mechanistic approaches to the study of the evolutionary processes and animal behaviour, rooted in the biology and behavioural sciences of the twentieth century. They can be adequately addressed, *on a practical level*, only through a radical reform of paths of scientific training and organization of basic and applied research. Reform that must aim at overcoming the bipartition between humanities and life sciences and educating generations of students, scholars, environmental, social and cultural operators to manage skills that are transversal to those offered by these two traditional disciplinary blocks. This overcoming undoubtedly implies a not easy, individual and collective effort of self-renewal and dialogue, a willingness of all the 'actors' involved to explore fields other than those of their traditional sector, to exercise a self-critical reflection on one's disciplinary, methodological, theoretical and conceptual traditions. It appears, however, indispensable for safeguarding and re-launching the critical vocation, the social relevance and reliability of the sciences and of their forms of application.

---

observed within or between populations, that are to some degree attributable to differences in what they learned socially. We treat tradition' and 'culture' as synonyms" (Laland et al. 2009: 178, 179).

The objective of the ICS is to outline graduate and post-graduate training programs and research methods aimed at providing and applying the intertwining of ethological, ecological and socio-cultural competences that is today necessary to critically connect the comparative study of human cultures and the comparative study of the cultures of other animal species.

For humanities, first of all, this task requires the effort to insert, acknowledging the facts, the history of human cultures in the much older and broader history of animal cultures, the comparative study of human cultural productions and of their forms of diversification and contamination, within the immensely wider, but still debutant, field of the comparative study of animal cultures and of the processes that lead to their birth, differentiation, dissemination and contamination.

The project to overcome traditional didactic and scientific paths, from which these pages are inspired, presents evident points of convergence with that of Numanities, outlined by Dario Martinelli in the volume *Arts and Humanities in Progress. A manifesto of Numanities* (Martinelli 2016). The need to "defend and promote critical thinking" proposing a reflection on "the position of the humanities in modern society" and on their "crisis" (Martinelli 2016: 11) and the effort for "working on relocating and redefining the humanities" (Martinelli 2016: 9), which the author indicates as the global goals of Numanities, are also fundamental to the ICS program, as provisionally outlined in this chapter. The method focused on discussing "the current crisis of the humanities and its possible solutions, in a spirit that should be both critical and self-critical", based on very "Multi/Inter/Cross/Trans-disciplinary dialogues between humanities and social and/or natural sciences". The attention to "the context, dynamics and problems of current societies" (Martinelli 2016: 9), that the "Manifesto" proposes also characterizes the ICS approach.

The effort to promote a transition of humanities beyond the anthropocentric approach and rethink them as Interspecific Cultural Studies, or as a set of disciplines devoted to the comparative study of all existed and existing cultures (not only human), should be part of the programmatic objectives of Numanities and could constitute the specific contribution of the ICS to their development.

To give a pair of non-random examples, I think that today, in any university course and whatever its disciplinary field may be, it should not be considered tolerable that students can finish their studies without having taken courses and topics that allow them to develop an awareness of the current global environmental crisis and of the global and local ecological problems linked to their field of study, or of the environmental impact of the production and distribution cycles that concern them. In fact, it is unthinkable that there are now areas of knowledge that can be called out of the problems linked to the devastating effects of the anthropic impact on the environment, on millions of animal and plant species, on all the human beings. Social ecology, as Murray Bookchin called it (Bookchin 1980, 1990), represents, therefore, an area of study that our age must consider transversal, and necessary to all the forms of knowledge and formative paths.

Similarly, in my opinion, no disciplinary field should today endorse that traditional kind of human mythology according to which "man and culture originated simultaneously, by definition" (White 1959: 5), because it is now adequately demonstrated that this is not the case.

Contributing to the de-mythologizing of this atavistic anthropocentric arrogance, and to a transition of the humanistic culture beyond it, also means contributing to the search for models of society and forms of social and scientific development alternative to those in force (descendants of that haughtiness) and capable of reversing their course showing respect towards the environmental, social and cultural disasters they have provoked.

In fact, the ability of human and natural sciences to preserve their social relevance, exercise their critical role rise to the challenges that await them will also depend on the ability of the present generations of scholars and students to question themselves and to overcome secular prejudices and obsolete paradigms, trying to boost a collective critical and self-critical re-establishment of human knowledge.

3.2 From the Discovery of Animal Cultures to Contemporary Cultural Ethology

In the second half of the twentieth century developments in ethology led to one of the most revolutionary discoveries of contemporary science: the existence of animal cultures.

This falsified (in the Popperian sense), or empirically refuted one of the fundamental assumptions of our philosophical and scientific tradition, that of man as the only "cultural animal". Therefore, it questioned the same *founding partition of western sciences: the division between humanities*, conceived as sciences of culture, *and natural sciences.*

Two field studies, introduced to the scientific community in the mid-1960s, allowed this amazing discovery. The first, directed by Junichiro Itani, Shoji Kawamura and Masao Kawai, disciples of the Japanese ethologist Kinji Imanishi, began in 1948 on the island of Koshima, where a community of macaques (*Macaca fuscata*) lived then and still does today. The second, promoted by Louis Leakey, the most authoritative anthropologist of the time, began in 1960 and was carried on by Jane Goodall, who was the first scholar to study the behaviour of chimpanzees in their natural environment, in the Gombe Stream Chimpanzee Reserve, in Tanzania.

In 1953, the observation of macaques made Satsue Mito, an inhabitant of Koshima aide to the three ethologists, the first human witness to the birth of a tradition within a community of non-human animals (de Waal 2001). Western scientific community became aware of this discovery in 1965, when Kawai published a paper about it in the scientific magazine *Primates* (Kawai 1965).

Meanwhile, in 1960, Jane Goodall had begun studying the chimpanzees at the Gombe Stream Chimpanzee Reserve, a site that, thanks to her efforts, would become a protected area from 1968. She was the first scholar to discover several important aspects of the social, emotional and cognitive life and material culture of chimpanzees: their ability to build wooden tools and exploit them to obtain food, the techniques used to open coconuts by choosing, carrying and using different stones in the form of anvils and hammers, the existence of cultural differences between groups, the complexity of their social structures and the differences in sensitivity, intelligence, character and preferences in every single individual.

But, in the 1960s, also another sub-field of ethological research began to contribute to the birth of cultural ethology: the study of the communicative systems of singing birds. It led to the discovery of "dialects", which are regional and macro-regional differentiations of songs within a species. In fact, Peter Marler and Miwako Tamura, pioneers in this development, had in the early 1960s already discussed "Song «Dialects»" (Marler and Tamura 1962) and "culturally transmitted patterns of vocal behavior" by sparrows (Marler and Tamura 1964).

The debate on the philosophical and scientific consequences of these discoveries began to develop in the 1970s and intertwined with the discussions on animal minds arisen by some comparative psychologists who studied the ability of "higher primates" to learn man-made languages as the ASL, or *American Sign Language* (Miles 1994; Fouts 1997; Patterson 1999), and other techniques of interactive use of human lemmas or symbols (Premack 1986; Savage Rumbaugh 1977), to recognize themselves in the mirror (Gallup 1970; Povinelli 1987) and solve several cognitive problems (de Waal 2016).

Despite their methodology, based on observations in captivity and initially set up assuming the anthropocentric presupposition of the Cartesian matrix which equated the intelligence of other animals to their ability to acquire and use human language or tools, these experiments opened up a window on an unpublished scenario: *the translation into human languages of the thoughts, moods and experiences of other animals* like chimpanzees, bonobos, gorillas, orangutans and parrots, *made by the animals themselves* (Warren et al. 1996; Patterson 1999; Pepperberg 2002).

Between the 1990s and the next decade, a new *philosophy of ethology* oriented both in a post-anthropocentric and post-genocentric direction began to emerge, intersecting with Animal Studies and the rising Critical Animal Studies. Books like *Visions of Caliban* (Peterson and Goodall 1993) and *Species of Mind* (Allen and Bekoff 1997) gave a first significant boost in this direction. Then, ethologists, philosophers of ethology and zoo-anthropologists such as Marchesini (Marchesini 1999), Lestel (2001), de Waal (2001), Despret (2004), Martinelli (2007), Best (2007) and Nocella et al. (2014)—to name but best known—contributed to set the comparative study of animal minds, cultures and societies on new both post-mechanistic and post-idealistic bases.[2]

Cultural ethology has since then gained increasing media attention to the extent that no adequately informed scholar now denies the existence of animal cultures.

[2]Though it circulated mostly in Italy and Austria, I would include my *Etologia della conoscenza. Per una teoria critica del comportamento umano* (Celentano 2000) in the list of texts indicative of the birth of contemporary Philosophy of Ethology.

However, the legacy of anthropocentric traditions and ontological separatisms is still crawling in studies that defend the thesis according to which cultures of non-human animals lack relevant characteristics such as active teaching, cooperation, imitation in the strict sense of the word, syntactically organized languages, which would be exclusive to human cultures (Boyd and Richerson 1985; Heyes 1993; Tomasello 1994, 2014).

Other scholars, such as Laland, Kendal and Kendal, while contrasting these theses and highlighting data that suggest their fallacy, underline that it is very difficult to document on the field, in wild species, processes of active learning correction, forms of cooperation, or even more the birth of new traditions, and believe that, for now, only imitation can be considered fully proven (Laland et al. 2009). However, a conspicuous number of studies has provided empirical and experimental evidence of these characteristics in different animal cultures (for a first approach see: de Waal 2001, 2016).

We know, for example, that forms of active orientation of learning through discouragement or encouragement and even more complex educational processes are documented in many species of mammals. Each mother "cat (*Felis catus*), through a complex procedure that requires the succession of many different phases, teaches her cubs to hunt" (Mainardi 1992: 63),[3] and behaviours with similar instructive value have been studied in other felines as tigers, cheetahs, and desert lynxes. Chimpanzees and bonobos dissuade puppies from manipulating dangerous objects such as attractive but inedible fruits. Scenes in which chimpanzee mothers correct their offspring's attempts to break coconuts with a wooden stick, taking the branch away from them and re-placing it in their hands in a functional position, were filmed by Christophe Boesch (AA.VV. 2006). Experiences of multi-decade studies on chimpanzees living in semi-captivity conditions have provided surprising evidence of their capacity for cooperation (Fouts 1997; de Waal 2016). If scholars like Tomasello still doubt that chimpanzees can spontaneously develop or adopt cooperative behaviours (Tomasello 2009, 2014), Roger Fouts, in his *Next of kin* (Fouts 1997), already over twenty years ago, described the attempted escape of a group of chimpanzees placed on an island surrounded by a fence, in terms that left little doubt about any cooperation. The observation of this event, gained by spying on the chimpanzees from a closed window, still represents a precious testimony for at least four good reasons: it documents a case of cooperation not induced by man; the chimpanzees *took turns* in carrying out the laborious task of twisting the final piece of the heavy net that surrounded the colony: when one got tired, another took its place; the cooperation concerned both the performance of this task and its *dissimulation*, or the immediate suspension of activities when the chimpanzees were observed by men; the attempt lasted for several days, making thus further clear its character as a planned and agreed action (Fouts 1997: 180–181).

More recently, several experiments have confirmed the remarkable cooperative capacity of our close relatives. In a study published in 2016 in *Proceedings of the*

---

[3]This and all the other quotations from essays that have not been translated into English, contained in this chapter, are my translations.

National Academy of Sciences, Frans de Waal and his collaborators demonstrated both the cooperative abilities of chimpanzees and a series of social constraints or dynamics that reinforce them. They wrote on the subject:

> Our species is routinely depicted as unique in its ability to achieve cooperation, whereas our closest relative, the chimpanzee (*Pan troglodytes*), is often characterized as overly competitive. Human cooperation is assisted by the cost attached to competitive tendencies through enforcement mechanisms, such as punishment and partner choice. To examine if chimpanzees possess the same ability to mitigate competition, we set up a cooperative task in the presence of the entire group of 11 adults, which required two or three individuals to pull jointly to receive rewards. This open-group set-up provided ample opportunity for competition (e.g., freeloading, displacements) and aggression. Despite this unique set-up and initial competitiveness, cooperation prevailed in the end, being at least five times as common as competition. The chimpanzees performed 3,565 cooperative acts while using a variety of enforcement mechanisms to overcome competition and freeloading, as measured by (attempted) thefts of rewards. These mechanisms included direct protest by the target, third-party punishment in which dominant individuals intervened against freeloaders, and partner choice. (Suchak et al. 2016: 10215)

Other studies have clarified that among chimpanzees cooperation is enacted not only between related individuals but that adults who have known each other for a short time also cooperate to obtain advantages or alleviate discomforts (Langergraber et al. 2007; de Waal 2016).

Moreover, sharing and cooperation have been documented, both in captivity and in nature, even among non-anthropomorphic monkeys: "Cooperation is common, for example, among capuchin monkeys. These monkeys are not only willing to help others obtain resources, but are more likely to share with individuals who help them" (Brosnan 2010: 11).

As for the birth and spread of new traditions in animal societies, if in the 1960s, the aforementioned birth of a new tradition among the macaques of the island of Koshima caused a sensation, further empirical evidence has been accumulated over the following decades. To limit ourselves just to a few examples, in 2007, in Côte d'Ivoire, a coconut crushing site that had been used by local chimpanzee populations for no less than 4300 years was discovered (Mercader et al. 2007). The use of stone tools has also been observed in some anthropoid monkeys and, in 2016, a site for crushing cashew nuts, used by local communities of striped cebi (*Sapajus libidinosus*) for over 700 years was found in Brazil, in the National Park of Serra da Capivara (Haslam et al. 2016). Moreover, in 2014, an article by Catherine Hobaiter and her collaborators, published in PLoS Biology, for the first time documented a phenomenon of transmission of a cultural innovation consisting in the invention of a sponge made with leaves and mosses among a group of wild chimpanzees (Hobaiter et al. 2014).

Regarding the structure of the languages used by different cultural species, and their differentiation through habits handed down for social learning, studies conducted on cetaceans, on hundreds of species of songbirds and on some singing primates show a tendency that is receiving continuous confirmation: the more our technical abilities to record and analyze these languages increase, the more they reveal a structural complexity, expressive variety, a network of relationships and local and regional divergences comparable in various aspects to those of human verbal and singing languages (Campan and Scapini 2002; Naguib and Riebel 2014).

As for the capacity of social learning through imitation, its spread among not only so-called "higher animals" but also in different species of invertebrates is now established. As early as 1992, an essay published in *Science* by Graziano Fiorito and Pietro Scotto spread the discovery that octopuses are capable of learning by imitation (Fiorito and Scotto 1992). More recently, Lars Chittka and other researchers from the Queen Mary University in London have published, in the same scientific journal, a study in which they demonstrate that bees are able of solving complex problems by imitating and *improving* on the behaviour of others that they have observed (Alem et al. 2016).

3.3 Behaviour and Cultural Innovations as Driving Forces of Animal Evolution

As I tried to clarify in the previous chapters, the developments of cognitive and cultural ethology and those of evolutionary studies today seem to converge towards a both post-genocentric and post-anthropocentric interpretation of the behaviour and history of organisms.

In the perspective of contemporary ethology, behaviour is framed as *a self-regulative and cognitive interaction* of organisms with their inter- and intra-specific environment.

"Self-regulative activity and cognitive interaction" means that all organisms, of every species at any time need to maintain or restore internal processes and physiological states which allow them to stay alive and perform this function through *explorative and energy trading activities*, absorbing and transforming matter and energy present in the external environment and *modifying both the latter and themselves.*

What does "cognitive" mean here? In the perspective of contemporary ethology, we can call "cognitive" all activities through which organisms explore their survival chances and test their ability to actively change or regulate their physiological and/or perceptual states. Each "cognitive" activity is in this sense, *a production of behavioural forms*, or of *self-regulative internal and external interactions*, enabling the performance of an organism's life cycle. In this perspective, cognitive activities are notable not only in animals, but in all organisms, because the simple fact that organisms are able to survive constitutes evidence of their ability *to make an object of knowledge out of their own living conditions* (Lorenz 1973; Riedl 1980; Celentano 2000, 2017).

These self-regulating and cognitive activities are obviously channeled and limited through constraints imposed by the anatomy and morphology of the species, intra-specific and interspecific context, individual characteristics and contingencies. However, all this allows us to understand both the history of each existed and existing species and group, and the history of each body as an active and selective exploration of their environment, and an active construction of their ecological and social niche.

This post-mechanical conception of behaviour has since the 1990s assumed, a relevant role in both field and laboratory ethology, as well as in new models of evolutionary biology derived from the developments of epigenetics and the introduction of the evo-devo perspective.

In particular two notions, previously introduced by two eminent scholars of the twentieth century like Conrad Hal Waddington and Jean Piaget, began to find consensus and gain relevance through experimental findings such as "behavior as motor of evolution" (Piaget 1976), and the existence of *non-genetic hereditary systems* capable of producing phenotypic modifications much faster than genetic mutations (Waddington 1975; Piaget 1976), which we now call *Epigenetic Inheritance Systems* (Jablonka and Lamb 2005).

What does it mean, in this new context, that behaviour is a "motor of evolution"?

To conceive behaviour as a driving force of the differentiation of organisms means that when encountering environmental changes that endanger their survival or offer them new growth opportunities, individuals, populations and species do not passively wait for a favorable genetic mutation that will allow some of them to overcome those obstacles or exploit those resources. Organisms, in the face of any change, engage all the innate and/or learned cognitive and physiologic, social and (possibly) cultural resources they possess to take advantage of it. This means, in turn, that except for rare cases in which they derive from significant genetic mutations, evolutionary divergences start from the sphere of behaviours, from changes in ethological attitudes, which are active responses to changes in an environmental, social or individual context.

This approach to the comparative study of behaviors and evolutionary processes, which in contrast to the genocentric one could be defined as etho-centric, also allows us to recognize the role of input that cultural differentiations can play in the processes of animal speciation and phylogeny. Scholars such as Jablonka, Lamb, Whiten, van Schaik, Dugatkin, Beans have recently suggested that animal cultures, by passing on and differentiating behaviour, can influence the evolution of species in various ways: cultural innovations as tools or tactics to avoid predators and social processes that stimulate the ability to learn, preserve and transmit useful information can increase the chances of survival and reproduction of certain populations compared to others (Whiten and van Schaik 2007). The development of complex languages and social interactions can stimulate, as many scholars believe happened to our species, the evolution of higher dimensions and performance of the brain (Dugatkin, 2001).

I personally think that it is fully correct to hypothesize that cultural differentiations, modifying niches, diets and habits have also contributed to the differentiation of somatic and physiological characteristics, to the processes of speciation, to the differential reproduction. But, in my opinion, the possibility that this process will continue to occur *today* and in the future is drastically limited by the effects of *anthropic* impact. At least one million non-domesticated plant and animal species and, among these, most cultural animal species are currently at high risk of extinction due to human intrusiveness, that is, the processes of destruction/anthropization of the environments in which they lived (Ceballos and Erlich 2018).

At the time being, we can therefore consider highly probable, and partially proven, that an incalculable number of animal traditions, such as many human cultures, has disappeared, is disappearing or is living in conditions of regression, dispersion and homologation, due to the human impact. Most existing animal cultures are now living a *hunted existence*, seeing their traditional habitats eroded day after day. Of course, this trend is not absolute. Some social species that manage to effectively parasitize humans, such as rats, are probably experiencing phases of increasing cultural differences and culturally transmitted information. But, from a global point of view, these few processes of differentiation are in no way comparable to the rhythms and vastness of the processes of destruction of animal societies and impoverishment of animal cultures induced by the anthropic impact in the last century, and currently underway. This is also attested by a study conducted over ten years on 144 groups of chimpanzees of central and southern Africa by a team of more than seventy primatologists whose results were then compared with those collected in other 100 communities of chimpanzees in the *Pan African Programme: The Cultured Chimpanzee*. The study, coordinated by primatologist Ammie Kalan, led to the cataloguing of 31 cultural behaviours and also showed that in groups of chimpanzees that live in closer contact with human settlements the probability of encountering cultural behaviour is 88% lower than in groups established in "low-impact areas" (Kühl et al. 2019: 1453). *Practically, the chimpanzee groups that live closer to our species have preserved 3 behavioural patterns handed down through social learning at most, while the communities located far from human settlements show between 15 and 20 cultural behaviours.*

3.4 Interspecific Cultural Studies (ICS) and Numanities

The discovery of animal thought and cultures has forced the humanities to begin a self-critical review of the anthropocentric assumptions on which their tradition has been based, for millennia (man as the only cultural, thinking, linguistic animal, the one capable of feelings, cooperation, inventions, innovations etc.). The goal of an adequate development of the comparative study of animal cultures, languages and forms of thought required however much more effort: a comprehensive reform of both scientific education and research organization oriented towards a full overcoming of the division between life sciences and humanities. This means towards the development of a meta-disciplinary area capable of combining biological, ethological and ecological skills with the cognitions and methods of modern anthropological, social, linguistic, aesthetic, and more generally humanistic studies in the contents of a comparative study, not only of the human uses and traditions, but of *all known and knowable animal cultures*. It is this meta-disciplinary which, as a first approximation, I indicate as being the field of *Interspecific Cultural Studies*. Its development is, in many ways, a process already underway. In fact, as I explained, the attempt to develop this area should not be understood as the proposal of something completely new that the author of these pages intends to introduce to the scientific field. Rather, it is a project aimed at gathering, in both the fields of education and research, the set of heterogeneous skills, transversal to the traditional disciplinary blocks, required by a new and very complex subject of study which Western sciences began to address only sixty years ago: *animal minds and cultures, their history, their expressive forms and productions.*

For the humanities, a sector with which the study of cultural processes and phenomena has traditionally coincided, this change implies an extension of their field of study, a critical renewal of their theoretical and methodological tools and an integration of their formative methods. That is: a (self) critical re-founding of all their sectors aimed at inserting, in a coherent and competent way, the comparative study of human cultural activities in the virtually much bigger but from a scientific point of view, still nascent field of the comparative study of animal cultures.

In some areas of the human sciences this transition has now been underway for quite some time; in others, the resistances to these transformations and extensions are much wider. For example, not surprisingly, the post-anthropocentric turn we discussed is clearly perceptible in a branch of the human sciences that since the eighteenth century has been the closest to life sciences: anthropology. Today, in fact, by anthropology we no longer intend only the study of man, as it traditionally took place, because the anthropological field has increasingly merged with those of primatology and cultural ethology (Rodman 1999). One of the skills currently required for a good anthropologist is the ability to frame the comparative study of past and present human cultures within the broader horizon of the comparative study of the societies and cultures of our pre-human ancestors and of our sister species: the great apes traditionally referred to as "anthropomorphic". In turn, the comparative study of anthropomorphic cultures finds its historical and genealogical placement within the bigger horizon of a study of all the societies and cultures belonging to the great suborder of *Anthropoidea*. Finally, the latter goes into the wider horizon of a comparative study of all the animal cultures. Anthropology and human ethology are thus closely related to cognitive and cultural ethology and evolutionary studies. It is not by chance that, within the anthropological sector, the greatest resistance to these contaminations is recorded, albeit with some exceptions, in the specific sub-field of philosophical anthropology, particularly in the continental area and, within the latter, by the Italian tradition. That is, in the disciplinary areas and in places that have been cradles and emblems of the humanistic tradition.

Even the psychological sciences today include among their competences the ability to compare the human mind with that of other animals and, even between traditional resistance and cyclical counter-offensive, manifesting a tendency to overcome both the anthropocentric and deterministic-biologistic approaches, which are the two major paradigms that have competed in the past for the supremacy in this field.

A similar situation, which sees at the same time the development of openings to the interspecific sphere and the persistence of resistances and closures towards it, can also be recorded in the context of contemporary linguistic and semiotic studies, in which zoo-semiotics have gained their niche but only partially have managed to trigger a collective work of re-foundation of the basic concepts and methods of the sector.

A third example of a humanistic discipline that has opened itself up, in recent decades, to fruitful though still insufficient contaminations with cultural ethology can be identified in the musicological field. In the last decades, with the birth of zoo-musicology, the comparative study of human musical traditions has begun to be rethought, re-elaborated and integrated by starting from a comparative study of the traditions and productions of all animal species that practice an "aesthetic use of communication sound" (Martinelli 2011). A study that implies the analysis of the social and biological functions that these expressive traditions perform in the animal societies that exhibit them, the reconstruction of the processes that lead to their genesis and the comparison of the structural and formal aspects of their products. This contamination, or rather integration, offers to ethno-musicology, understood as the comparative study of the human musical traditions, the possibility to develop solid zoo-musicological and ethological bases and, at the same time, allows cultural ethology and zoo-musicology to take advantage of the immense patrimony of research and reflection developed by ethno-musicology and other branches of musicological studies.

This integrative or contaminative approach can be extended to many disciplinary sectors that deal with cultural activities and productions, therefore, to many fields related to humanities.

From the comparative study of the learning processes, or of the techniques for finding, using and transforming raw materials and energy sources, to the analysis of languages, dances, songs, and processes of ritualization of behaviors, from the study of the phenomena of dissimulation, deceit and espionage between members of a group to that of the forms of resolution of social conflicts, or of dissemination of innovations, there is almost no sphere of social and cultural action in which the comparison between human societies and the community of other animal species is not revealing instructive and surprising discoveries.

This comparison, however, will be adequately developed only if our university systems are equipped to offer training courses and research structures capable of integrating biological, ethological and ecological skills with the knowledge and study methods developed by the humanistic tradition.

In other words, a research area such as the Interspecific Cultural Studies, and the training courses suitable for its implementation, can be adequately developed only when they become the object of a collective effort of the scientific community and a structural pivot of university organization. Its implementation requires profound and both practical and theoretical upheavals, which can be consolidated only through a reallocation of several sectors of humanities within an *interspecific comparative perspective and a meta-disciplinary operational context*. Moreover, they require, in the field of the behavioural sciences, the overcoming not only of the traditional dualistic Cartesian models, but also of the "psycho-hydraulic" and mechanistic model of classical and early cognitive ethology (Marchesini 2016), of the genocentric one of "classical sociobiology" (de Waal 2001), as well as of the substantially deterministic approach still today dominant in evolutionary psychology (Lieberman 2013).

Listed in an extremely synthetic way, the main goals of Interspecific Cultural Studies can be summarized in the following ten points:

- To promote a full overcoming of the division between life sciences and humanities in the organization of both scientific training and research, aimed at training generations of citizens and workers equipped with skills transversal to these two traditional blocks and useful to tackle some of the new important scientific and social challenges posed by our era;
- to revolutionize the traditional forms of human self-representation, paving the way for post-anthropocentric forms of self-understanding in which man is *just one* of the cultural animals, and to refound the methodologies, epistemological references and narrative background of the humanities with a post-anthropocentric and interspecific setting.
- to refocus ethology and behavioural science on a post-genocentric, post-deterministic and post-mechanist approach, which considers all organisms as selective agents capable of cognitive and explorative activities and cognitive and selective behaviour as one of the main driving forces of evolution.
- to learn to compare human and non-human cultures and societies without falling into the traditional opposition between anthropomorphism and anthropodenial;
- to collectively construct, through research, comparisons and debates, a meta-disciplinary lexicon capable of attributing to concepts such as "culture", "traditions", "invention", or "singing", "dance", "ritual", meanings usable in reference not only to humans, but also in non-human contexts;
- to critically insert the (chronologically) short history of human cultures into the greater history of animal experiences, traditions and cultures spanning hundreds of millions of years;
- to increasingly correlate research on animal cultures with a commitment to protect them and the natural environments in which they are rooted;
- to increasingly implement the extension in the fields of cognitive ethology and animal psychology of ethical rules that guarantee the non-invasiveness of study methods and respect for the freedom of the subjects studied as in in the fields of human ethology and psychology;
- to contribute to form new generations of teachers and scholars, students and socio-cultural or environmental operators equipped with theoretical and practical skills transversal to the traditional bipartition between humanities and life sciences and capable of using them to adequately understand the environmental and social impact of human activities;
- to contribute, as far as possible, starting each from their own specific field of study or work, to the search for models of social and scientific development that are alternative to those now dominant and capable of reversing the line promoted by them that led to the environmental and social disasters now underway.

The program outlined in these ten points to me seems to converge, on several points, with the intent of Numanities "to discuss the current crisis of the humanities and its possible solutions, in a spirit that should be both critical and self-critical", trough very "Multi/Inter/Cross/Trans-disciplinary dialogues between humanities and social and/or natural sciences", and "in the context, dynamics and problems of current societies" (Martinelli 2016: 9).

ICSs follow in the wake of the proposal of Numanities to enhance humanistic skills in new or broader research fields (in this case, the comparative study of animal cultures), and at the same time promote a radical renovation of their methods, theoretical assumptions and empirical contents.

The objective of contributing to the transition of humanities beyond anthropocentric prejudice (and related mechanistic way of seeing all other organisms), which characterizes ICS, in my opinion should in fact constitute the distinctive and characteristic factors of Numanities, to the extent that I believe it should be made even more explicit in a possible revision of their programmatic points, presented in the *Manifesto of Numanities* (Martinelli 2016: 11–83).

In favour of the idea that this orientation could aim to, imagining a under construction Numanities building of, the metaphor of a master wall on which to graft its various articulations would represent, in my opinion, the same intellectual and professional path of its main promoter. Indeed, the meta-disciplinary attitude that guided Martinelli's studies led him to make important contributions to the development of areas such as zoo-semiotics (Martinelli 2007, 2010, 2017) and zoo-musicology (Martinelli 2009, 2011) and to an updated criticism of anthropocentrism. His anti-hierarchical sensibility very soon pushed him towards anti-speciesist ethics. In short, his whole intellectual path attests to the centrality that the father of Numanities recognizes to the anti-anthropocentric commitment, with all that it entails in terms of criticism of modern and contemporary human societies and their models of development.

Finally, although this aim is not particularly emphasized in the programmatic points, the centrality that the post-anthropocentric approach plays in the Numanities project is especially signaled by the fact that the case studies 2A and 2B, proposed in the second part of *Arts and Humanities in progress*, are focused precisely on it. They respectively regard the alleged "special specificity" of man, or human uniqueness (Martinelli 2016: 144–160), and the relationship between "language and interspecific communication" (Martinelli 2016: 161–201).

The ICS approach aims at overcoming the traditional man-animal dichotomies on the direction of a post-anthropocentric and at the same time post-genocentric approach to the study of animal (and therefore also human) behaviour and evolution. They represent an attempt to contribute to promoting forms of self-understanding of human beings and societies emancipated from developmental myths and based on a global ecological-ethological perspective that takes into account the interests of the community of living beings in which we are immersed and the repercussions that human activities have on it.

Indeed, contributing to the overcoming of the anthropocentric approach means, today, also contributing to a radical criticism of the goal of an ever more *total domination, and of an ever more indiscriminate exploitation, of all human and non-human natural resources* that the modern and contemporary societies continue to pursue, even though it has proved patently unsustainable. For the same reasons, to contribute to the research, testing and spreading of ethical parameters and models of social and scientific development in opposition to those that predominate today seems to me the most important and ultimate objective which the attempt to promote a transition from Humanities to Numanities should aim for.

3.5 Interspecific Cultural Convergences (ICC): A New Object and Project of Study

Among the objects of Interspecific Cultural Studies, a particular group of them should occupy a privileged place for its relevance to both evolutionary and ethological perspectives: the cases of *cultural convergent evolution among different species*, or *Interspecific Cultural Convergences* (ICC).

In ethology, as in morphology and in the anatomical area, cases in which, during phylogeny, different species have developed similar structural and/or functional traits that are not inherited from common ancestors are called *convergent evolutions*, *evolutionary convergences*, or simply *convergences* (Heymer 1977: 74; Mainardi 1992: 221–222). The wings in flying insects, bats and birds are a typical example.

I here suggest extending the concept of *evolutionary convergence* to the phenomena inherent in *cultural evolution*, defining:

- as *cultural convergences* or *Cultural Convergent Evolutions* (CCE) all (and exclusively) cases in which it is historically proven that a technique, an invention, a discovery, an expressive form or a use have been developed by different cultures and populations in reciprocal independence[4];
- as *Interspecific Cultural Convergences* (ICC) all (and only) cases in which cultural convergences occur not only between populations of the same species, but also between *societies and traditions of different species.*

It should however be clarified that the concept of ICC so intended presents some differences from that of "convergent evolution" normally adopted in evolutionary studies: "Convergent evolution is typically defined as the repeated evolution of similar traits in independent evolutionary lineages inhabiting similar environments" (Harmon 2013). However, cases of ICC can also occur among species *living in very different environments.* The case of singing is emblematic in this regard: from a taxonomic point of view, singing is a phenomenon appearing in the animal world in a miscellaneous way. It appears in species that are genetically, phylogenetically and ecologically different from one another as cetaceans, monkeys as Hylobatidae, Tarsius, Indri and Callicebus, as all human cultures spread over the planet, as the mice and thousands of species of singing birds (Celentano 2016, 2018).

The fact that singing is developed in species so distant from each other means that this convergence cannot be explained on the basis of "homology", intended as characteristics inherited through a common ancestor.

[4]This concept of CCE should not be confused with that of "Convergence Culture", recently introduced by H. Jenkins (2006), which refers to the effects of interactions between traditional and new digital media.

The diffusion of singing in so different clades and environments is therefore the result of mutually independent, but in some aspects similar, evolutionary processes and selective pressures. For this reason, it can be adequately understood only by identifying and comparing the *functions* that this kind of expression plays, and the *forms* it has assumed, in all these animal societies, just as is normally done in the comparison of human singing traditions and performances. This approach can be extended, in the perspective of ICS, to all the cases of ICC.

The cataloging of ICC cases and the research on the causes of these evolutionary convergences are still at their first steps. To became able to deepen our knowledge on these phenomena we will need to integrate the methodologies of the comparative study of uses and customs, communication systems and expressive forms, social regulation devices and material techniques, developed by the humanities, with the observation and intra- and interspecific comparison methods of contemporary ethology. We will also need to construct open databases to determine a methodical comparison between products, forms and intra-specific differentiations of all the animal cultures.

I would like to conclude this section by proposing, in the following figure, a first provisional mapping of the most common cases of ICC and of some of the factors that may have contributed to their genesis. It is of course only a first sketch that can be widened and further articulated with the contributions of other scholars.

3.6 ICC: The Case of Singing

In this final section I intend to illustrate the biological and social functions singing performs in different animal species and some converging aspects they present. I chose to proceed in a way that may seem strange: comparing the functions that ethologists have attributed to animal songs with those the traveler, ethnologist and writer Bruce Chatwin attributed to the "songs of the ancestors" of the Australian Aborigines (Chatwin 1987). The reasons for that are the following: the communities of Australian natives are those that longer than any other were preserved from exchanges with others (practically, until Cook's expedition in 1770). Singing played a very important social role in their traditions. Finally, although some biologically and socially important features of the songs, such as courtship, are not reflected in his descriptions,[5] Chatwin's analysis in a surprising way illuminates some of the features and uses of songs which can also be found in other animal communities.

Chatwin attributes three different functions to the songs of the ancestors:

- totemic memories of their clan and individual recognition documents. Indices of the familiar and mythical roots from which an individual comes, the songs allow the identification of each member of the group through his affiliation with his "totemic" ancestors (Chatwin 1987: 4, 12–13).[6]
- Melodic and vocal maps of a territory, travel guides for migration or occasional shifts, information vehicle about territorial features and boundaries that cannot be crossed without risk (Chatwin 1987: 13, 14, 69, 134–135).
- "Pass": sound attestations that allow to recognize a person as "the owner of that path"; documents transmitted by cultural inheritance, in order to identify who has the right of transit in a given territory and the right to give or deny to others the transit permission on it (Chatwin 1987: 14, 70).

Is it possible to find equivalents of these three functions in the songs of other species? Here is a brief analysis of these three points.

Songs as individual recognition 'documents', indices of the geographical and family roots from which an individual comes, allowing mutual recognition among members of a group or colony.

At the end of the 1950s, Weeden and Falls interpreted some duets between male birds in neighbouring territories as exchanges destined to get to know each other (Weeden and Falls 1959), and Marler suggested that the melodies of birds could provide information for individual identification (Marler 1960). A decade later, two studies conducted in different areas (Thompson and O'Hara Rice 1970; Emlen 1971), documented this feature in the song of the male of the *Passerina Cyanea*: in case of a sound intrusion of new neighbors, males modified their singing adding to the specific

[5]Chatwin's notes privilege, within a rich set of local songs, only a few. He did not aim at an exhaustive cataloging of native songs, but the existence of courtship serenades in aboriginal traditions is attested by other authors (Lockwood 1962; Englaro 1998; Gioia 2015).

[6]Similar cases of anthroponymic functions of songs are reported by other scholars concerning Australian populations such as Warramunga (Bosi 1994: 116) and Aranda (*Ibidem*: 95–96).

sequence of their species some individually differentiated final parts. The songs of all the members of the group were so marked by a different end. Further studies have shown that there are intermediate layers between the songs of a species and its individual variations. According to Feekes (1977), the *Cacicus cela* emits colony-specific songs that have the function of a colony password and similar functions are found (Bailey and Baker 1982) in the Virginia quail (*Colinus virginianus*). Starting in the 1960s, Peter Marler and Miwako Tamura (1962, 1964), William Thorpe (1961, 1972), Wolfgang Wickler (1986) and many others contributed to the early stages of the study of local and regional dialects. The existence of "micro geographic (or local dialects) and macro geographic differences (regional dialects)" (Martinelli 2011: 238) was in the meantime discovered in the communicative systems of other animals, like cetaceans, and it is now regarded as a phenomenon widespread in mammals and birds. For example, studying the songs of the *Batis molitor* in nine different regions of East Africa, ethologists and zoo-musicologists found dialectal variations that concerned two aspects in each of them: the presence of a sequence of three descending sounds or of longer sequential sequences, and differences in the order of the three base sounds, in which the middle height may be in the second or third position (Wickler 1986: 76–77).

In many cases, the development of local song traditions is a prerequisite for the invention of personal songs and their use for identifying individuals and reinforcing parental or couple ties. We find an interesting example of this in the African *Lanarius aethiopicus major*: "here, the members of a pair learn to perform duets with one another and, while adopting certain phrases and rhythms which are characteristic of the locality, work out between themselves the duets which are sufficiently individualistic to enable a bird to distinguish and keep contact with its mate by singing duets with it or, to be more exact, singing antiphonally with it in the dense vegetation in which they usually live" (Thorpe 1972: 160–161).

These performances of the Ethiopian shrike revealed, in later studies, even more complex interactions which include a dozen of different pair duets and many duets between competing males and/or neighbors, both divisible, from a formal point of view, into two subgroups: unisons and antiphonal duets. The latter, in the case of male territorial duets, are, in turn, divided into exchanges of identical notes and varied exchanges. There are also cases in which an individual sings by issuing two different voices at the same time and cases where individuals who lost their companion, using this technique, alone run the sequence they used to perform as a couple (Harris 2000). Finally, there are songs that mix different dialects and cases of simultaneous running of two different types of duet, one of courting or strengthening the couple tie, the other as a sort of duet/duel with a rival (Wickler 1986).

Songs like melodic and sung maps of the territory, guides for migrations and occasional or cyclical displacements, which transmit information about resources and dangers and on "borders" that cannot be overcome without risk.

Well known in this category, are the cases of the Lira bird (*Menura novaehollandiae*), that includes in its own sounds environmental and animal sounds collected from the surrounding territory, thus offering an acoustic mapping of it (Dalziell and Magrath 2012), and that of the Australian magpie which exchanges information on food sources and migratory routes with con-specifics through songs (Rogers and Kaplan 1998: 86). Also well-documented is the use of vocalizations with information and referential functions in birds such as the northern royal gull (*Larus argentatus*), or the *Indicator indicator* that use song to locate food resources.

With regard to whales, Roger Payne was the first to hypothesize, even if in a doubting form, that "the humpbacks use their songs a bit like Australian aborigines, whose songs contain descriptions of the road and the points where you are and tell about the characteristics of the scenery you are in" (Payne 1995: 165). Martinelli observed, in turn, that "migratory species of cetaceans use songs as geographic maps, in a way that cannot help but recall Chatwin's songs" (Martinelli 2011: 163). Stimpert, Peavey, Friedlaender and Nowacek (Stimpert et al. 2012), conducting a study on ten male individuals of humpback whale to which they applied multi-sensors that allow deep recordings, have reinforced this hypothesis. Their research led them to conclude that the choir repertory of *Megaptera novaeangliae* males does not exclusively include courtship songs and does not only appear in the breeding season. In the vicinity of the migratory season, the individuals they watched were leaving for food, and using songs that were significantly different from those of courtship to communicate remotely.

Songs as a "pass" that allow to recognize an individual as "path owner": a person who has the "right" of transit on that path that may enjoy the resources that it offers and grant or deny to others transit.

The words "right" and "owner" that Chatwin chooses to describe this use of songs, and the reference to the bargaining practices that take place through songs exchanges, at a first glance would seem to preclude a comparison with non-human cultures. However, we are here facing notions of "right" and "property" that are very different from those used by the Western traditions. In fact, they do not sanction the fixed property of a territory and do not permanently interrupt the other's right of transit in it or of usufruct of its products. They only attest that someone has the privilege of crossing it, practicing hunting, gathering or exchanging without being attacked and receiving help when needed during this crossing. Looking at this profile, this type of use of songs presents remarkable analogies with the "territorial" delimitation function that ethologists have found in the song of the adult males of many of the singing birds.

But to determine whether we can detect analogies or convergences with this use of songs in other animal species, one should ascertain whether other animals can, through intense gradients or formal differentiations of their singing, *not only signal the presence of an x male in a y territory*, and also not only send a generic message of transit prohibition or allowance. We should check as well:

- whether the resident male responses depend or not on the ability/inability of the intruder to be individually recognized through their song;
- whether or not we can find any differences in the songs that the resident male performs in the presence of intruders depending on their being unknown individuals, new entries or long-term frequentations;

- whether adjustments and "bargaining" actually occur between these animals through singing or not.

Recent observations indicate that generally both the bird that enters the territory of another and the one already allocated in it may adopt various communication strategies that reduce or intensify aggressive reactions. These choices are manifested by variations in the form and intensity of the song, or through options which consist in overlapping or not on the other's emissions. Recently, a study titled *The Social Interaction of Song in Song Sparrows* introduced the concept of "acoustic ownership marker" (Burt and Beecher 2008). The study shows that the songs of the resident males can perform the function of deterrent to many varying degrees of intensity, some of which seem to include the possibility of resolving small boundary controversies between neighbors without any physical clashes, only through the exchange of songs. According to the authors, these interactions appear like a continuous strategic game of escalation and/or de-escalation of aggressive elements, driven by executive stench and other parameters such as the overlapping or not of the songs and the repetition or variation of the verse performed on the other. Different dynamics reveal the exchange of songs between "first-year neighbors" and "neighbors of long time" (*Ibidem*). The authors suggest that the use of a certain kind of "conventional matching", which arises an agreement on repertoires that can be paired or alternated (*Ibidem*, p. 89), can be attested between long time neighbors.

Territorial defense and recognition of the con-specifics that pass through the exchange of songs are closely linked to the last group of the biosocial functions of singing we have mentioned: those related to courtship, mating and strengthening of the couple's bond. Two are the main groups of non-human protagonists of the studies about this kind of song: singing birds (for an introduction, Marler and Slabbekoorn 2004: 39–78; for an update, Naguib and Riebel 2014) on Mysticeti (for an introduction, Payne 1995; for an update, Suzuki et al. 2006). More recently, these model species have been added to others. For example, individual differences and local dialects have been identified in the male-female couple duets of the crested gibbon (Geissmann and Nijman 2006; Thinh et al. 2011). The courtship vocalization of mice, brought to frequencies which are audible and distinguishable by human ears, revealed melodic qualities comparable, for beauty and complexity, to those of the birds (Chabout et al. 2015).

Among Mysticeti, the species whose song has been most studied is *Megaptera novaeangliae*. Payne and McVay (1971) were the first to decode the structures and functions of their songs. Thanks to their efforts, it is today well-known that the male humpback whale produces melodies which are differentiated by geographical area and are renewed year after year or, more drastically, in multi-year cycles. They have complex structures, composed by different parts or "themes", consisting of ascending and descending sounds, lasting between 20 and 30 min, and can be repeated several times. Recently Suzuki, Buck and Tyack (Suzuki et al. 2006), examining the songs of 16 male humpbacks and thanks to a specifically designed software, have analyzed their basic structures. The algorithm has mathematically confirmed the hypothesis of Payne and McVay which states that humpback whales have their own syntax and that their song, like human speech, is based on a hierarchical language, consisting of lengthy sound blocks with increasing complexity, inserted into each other as in a system of Chinese boxes. A syntactic system, in many respects, analogous to human verbal language with its subdivisions in phonemes, phrases, words, propositions and periods is being found in an ever-increasing number of social mammals and birds. The same conclusions have led to the analysis of the vocal languages of cetaceans such as dolphins and "killer whales", of mice and especially of sparrows and other birds, from which new confirmations come continuously. Among the latest, a study on the song of the *Parus minor* (Suzuki et al. 2016).

References

AA.VV. 2006. *National Geographic. Enciclopedia degli animali. Mammiferi III. Scimpanzé.* Gruppo Editoriale L'Espresso.

Alem, Sylvain, et al. 2016. Associative Mechanisms Allow for Social Learning and Cultural Transmission of String Pulling in an Insect. *PLoS Biology* 14 (12): e1002589.

Allen, Colin, and Marc Bekoff. 1997. *Species of Mind.* Cambridge, MA: MIT Press.

Bailey, E.D., and James Baker. 1982. Recognition Characteristics in Covey Dialects of Bobwhite Quail. *Condor* 84: 317–320.

Best, Steve et al. 2007. Introducing Critical Animal Studies. *Animal Liberation* 5 (1): 4, 5.

Bookchin, Murray. 1980. *Toward an Ecological Society.* Montréal: Black Rose Books.

Bookchin, Murray. 1990. *The Philosophy of Social Ecology.* Montréal: Black Rose Books.

Boyd, Richard, and Peter Richerson. 1985. *Culture and the Evolutionary Process.* Chicago, IL: University of Chicago Press.

Bosi, Roberto. 1994. *Aborigeni australiani.* Fiesole, FI: Nardini.

Brosnan, Sarah. 2010. What Do Capuchin Monkeys Tell Us about Cooperation? In *For the Greater Good of All: Perspectives on Individualism, Society, & Leadership*, ed. Donelson Forsyth and Crystal Hoyt. Palgrave Macmillan Publishers.

Bshary, Redouan, Wolfgang Wickler, and Hans Fricke. 2002. Fish Cognition: A Primate's Eye View. *Animal Cognition* 5 (1): 1–13.

Burt, John, and Michael Beecher. 2008. The Social Interaction Role of Song in Song Sparrows. *Comparative Cognition & Behavior Review* 3: 86–98.

Campan, Raymond, and Felicita Scapini. 2002. *Ethologie.* Bruxelles: De Boeck Université.

Ceballos, Gerardo, and Paul Erlich. 2018. The Misunderstood Sixth Mass Extinction. *Science* 360 (6393): 1080–1081.

Chabout, Jonathan, Abhra Sarkar, David Dunson, and Erich Jarvis. 2015. Male Mice Song Syntax Depends on Social Contexts and Influences Female Preferences. *Frontiers in Behavioral Neuroscience* 9: 76.

Celentano, Marco. 2000. *Etologia della conoscenza.* Napoli: La Città del Sole.

Celentano, Marco. 2016. Funzioni sociali ed espressive del canto. Un caso di evoluzione culturale convergente tra specie diverse. *Animal Studies* V (14): 33–47.

Celentano, Marco. 2017. From Konrad Lorenz's "phylogenetic apriorism" to the birth of Evolutionary Epistemology. In *Readings in Numanities*, ed. Oana Andreica and Alin Olteanu. New York/Berlin: Springer.

Celentano, Marco. 2018. Interspecific Cultural Convergences (ICC) and Interspecific Cultural Studies (ICS). In *Predicatbility and the Unpredictable*, ed. David Ceccarelli and Giulia Frezza. CNR Edizioni: Roma.

Chatwin, Bruce. 1987. *The Songlines.* New York: Penguin Books Ltd.

Dalziell, Anastasia, and Robert Magrath. 2012. Fooling the Experts: Accurate Vocal Mimicry in the Song of the Superb Lyrebird *Menura novaehollandiae. Animal Behavior* 83 (6): 1401–1410.

Despret, Vinciane. 2004. *Our Emotional Makeup: Ethnopsychology and Selfhood*. New York: Other Press.

de Waal, Frans. 2001. *The Ape and the Sushi Master*. New York: Basic Books.

de Waal, Frans. 2016. *Are We Smart Enough to Know How Smart Animals Are?*. London: Granta.

Dugatkin, Lee Alan. 2001. The Imitation Factor: Evolution Beyond the Gene. *The New York Times*, August 15.

Emlen, Stephen. 1971. Geographic Variation in Indigo Bunting Song (*Passerina cyanea*). *Animal Behavior* 19 (2): 407–408.

Englaro, Graziella (ed.). 1998. *Canti degli aborigeni australiani*. Mondadori: Milano.

Feekes, Francisca. 1977. Colony-Specific Song in *Cacicus cela (Icteridae Aves)*: The Pass-Word Hypothesis. *Ardea* 65: 197–202.

Fiorito, Graziano, and Scotto. 1992. Observational Learning in *Octopus Vulgaris*. *Science* 256 (5056): 545–547. https://science.sciencemag.org/content/256/5056/545?maxtoshow=&HITS=10&hits=10&RESULTFORMAT=&fulltext=%2528Fiorito%252CG%20AND%20Scotto%252CP.1992%20AND%20Observational%20AND%20learning%20AND%20in%20AND%20Octopus%20AND%20vulgaris.%20AND%20Science.256%20AND%20545-547.%2529&searchid=1&FIRSTINDEX=0&resourcetype=HWCIT-aff-2Pietro.

Fouts, Roger. 1997. *Next of Kin: What Chimpanzees Have Taught Me about Who We Are*. New York: William Morrow and Company.

Gioia, Ted. 2015. *Love Songs*. Oxford: Oxford University Press.

Gallup, Gordon. 1970. Chimpanzees: Self-Recognition. *Science* 167 (3914): 86–87.

Geissmann, Thomas, and Vincent Nijman. 2006. Calling in Wild Silvery Gibbons (*Hylobates moloch*) in Java. *American Journal of Primatology* 68: 1–19.

Harmon, Luke. 2013. Convergent Evolution. In *Oxford Bibliographies in Evolutionary Biology*, ed. Jonathan Losos. New York: Oxford University Press.

Harris, T. 2000. *Shrikes & Bush-Shrikes*. London: Princeton University Press.

Haslam, Michael, et al. 2016. Pre-colombian monkey tools. *Current Biology* 26 (13): R521–R522.

Helfman, Gene, and Eric Schultz. 1984. Social Transmission of Behavioural Traditions in a Coral Reef Fish. *Animal Behaviour* 32 (2): 379–384.

Heyes, Cecilia. 1993. Imitation, Culture and Cognition. *Animal Behaviour* 46: 999–1010.

Heymer, Armin. 1977. *Ethologisches wörterbuch-Ethological dictionary-Vocabulaire ethologique*. Berlin-Hamburg: Paul Parey.

Hobaiter, Catherine, et al. 2014. Social Network Analysis Shows Direct Evidence for Social Transmission of Tool Use in Wild Chimpanzees. *PLoS Biology* 12 (9): e1001960. https://doi.org/10.1371/journal.pbio.1001960.

Jablonka, Eva, and Marion Lamb. 2005 (2012). *Evolution in Four Dimension*. Cambridge, MA: MIT Press.

Jenkins, Henry. 2006. *Convergence Culture: Where Old and New Media Collide*. New York: New York University Press.

Kawai, Masao. 1965. Newly Acquired Pre-cultural Behavior of the Natural Troop of Japanese Monkeys on Koshima Islet. *Primates* 6: 1–30.

Kühl, Hjalmar, et al. 2019. Human Impact Erodes Chimpanzee Behavioural Diversity. *Science* 363 (6434): 1453–1455.

Laland, Kevin, Jeremy Kendal, and Rachel Kendal. 2009. Animal Culture: Problems and Solutions. In *The Question of Animal Culture*, ed. Kevin Laland and Bennet Galef. Cambridge, MA: Harvard University Press.

Laland, Kevin, N. Atton, and Mike Webster. 2011. From Fish to Fashion: Experimental and Theoretical Insights into the Evolution of Culture. *Philosophical Transactions of the Royal Society B: Biological Sciences* 366 (1567): 958–968.

Langergraber, Kevin, John Mitani, and Linda Vigilant. 2007. The Limited Impact of Kinship on Cooperation in Wild Chimpanzees. *Proceedings of the National Academy of Sciences* 104 (19): 7786–7790.
Lestel, Dominique. 2001. *Les Origines animales de la culture*. Paris: Flammarion.
Lieberman, Philip. 2013. *The Unpredictable Species*. Princeton, NJ: Princeton University Press.
Lockwood, Douglas. 1962. *I, the Aboriginal*. Adelaide: Rigby.
Lorenz, Konrad. 1973. Engl. Transl. 1977. *Behind the Mirror: A Search for a Natural History of Human Knowledge*. London: Meuthen & Co.
Mainardi, Danilo. (diretto da). 1992. *Dizionario di Etologia*. Torino: Einaudi.
Marchesini, Roberto. 1999. *Zooantropologia*. Como: Red.
Marchesini, Roberto. 2016. *Etologia filosofica*. Milano-Udine: Mimesis.
Marler, Peter. 1960. Bird Songs and Mate Selection. In *Animal Sounds and Communication*, ed. Wesley Lanyon and William Tavolga, 348–367. Washington: American Institute of Biological Sciences.
Marler, Peter, and Hans Slabbekoorn. 2004. *Nature's Music: The Science of Birdsongs*. San Diego: Elsevier Academic Press.
Marler, Peter, and Miwako Tamura. 1962. Song "Dialects" in Three Populations of White-Crowned Sparrows. *Condor* 64: 368–377.
Marler, Peter, and Miwako Tamura. 1964. Culturally Transmitted Patterns of Vocal Behavior. *Science* 146 (3650): 1483–1486.
Martinelli, Dario. 2007. *Zoosemiotics: Proposals for a Handbook*. Imatra: Finnish Network University of Semiotics.
Martinelli, Dario. 2009. *Of Birds, Whales and Other Musicians. Introduction to Zoomusicology*. Scranton and London: University of Scranton Press.
Martinelli, Dario. 2010. *A Critical Companion to Zoosemiotic*. New York: Springer.
Martinelli, Dario. 2011. *Quando la musica è bestiale per davvero. Studiare e capire la zoo musicologia*. Roma: Aracne.
Martinelli, Dario. 2016. *Arts and Humanities in Progress: A Manifesto of Numanities*. Berlin, New York: Springer.
Martinelli, Dario. 2017. *Basics of Animal Communication: Interaction, Signalling and Sensemaking in the Animal Kingdom*. Newcastle upon Tyne (UK): Cambridge Scholars Publishing.
Mercader, Julio, et al. 2007. 4300-Year Hold Chimpanzee Sites and the Origins of Percussive Stone Technology. *Proceedings of the National Academy of Sciences* 104: 3043–3048.
Miles, Lyn. 1994. Me Chantek: The Development of Self-Awareness in a Signing Orangutan. In *Self-Awareness in Animals and Humans: Developmental Perspectives*, ed. Sue Taylor Parker, Robert Mitchell, and Maria Boccia. New York: Cambridge University Press.
Naguib, Marc, and Katharina Riebel. 2014. Singing in Space and Time: The Biology of Birdsong. In *Biocommunication of Animal*, ed. Guenther Witzany. Dordrecht-Heidelberg-New York-London: Springer.
Nocella, Anthony, et al. 2014. *Defining Critical Animal Studies*. New York: Lang.
Patterson, Freancine. 1999. *Koko-love! Conversations with a Signing Gorilla*. New York: Dutten.
Payne, Roger. 1995. *Among Whales*. New York: Charles Scribner's Sons.
Payne, Roger, and Scott McVay. 1971. Songs of Humpback Whales. *Science* 173: 585–597.
Pepperberg, Irene. 2002. *The Alex Studies*. Cambridge, MA: Harvard University Press.
Peterson, Dale, and Jane Goodall. 1993. *Visions of Caliban*. Boston and New York: Houghton & Mifflin.
Piaget, Jean. 1976. *Le Comportement moteur de l'évolution*. Paris: Gallimard.
Povinelli, Daniel. 1987. Monkeys, Apes, Mirrors and Minds. *Human Evolution* 2: 493–509.
Premack, David. 1986. *Gavagai! Or the Future History of the Animal Language Controversy*. Cambridge, MA: MIT Press.
Riedl, Rupert. 1980. *Biologie der Erkenntnis*. Berlin: P. Parey.
Rogers, Leseley, and Gisela Kaplan. 1998. *Not only Roars and Rituals*. Sydney: Allen and Unwin.

Rodman, Peter. 1999. Whither Primatology? The Place of Primates in Contemporary Anthropology. *Annual Review of Anthropology* 28: 311–339.

Savage Rumbaugh, Sue. 1977. *Language Learning by a Chimpanzee.* New York: Academic Press.

Stimpert, Alson, Lindsey Peavey, Ari Friedlaender, and Douglas Nowacek. 2012. Humpback Whale Song and Foraging Behavior on an Antarctic Feeding Ground. *Plos One* 7 (I 2): e5 I 2 I 4.

Suchak, Malini, Timothy Eppley, Matthew Campbell, Rebecca Feldman, Luke Quarles, and Frans de Waal. 2016. How Chimpanzees Cooperate in a Competitive World. *Proceedings of the National Academy of Sciences* 113 (36): 10215–10220.

Suzuki, Ryuji, John Buck, and Peter Tyack. 2006. Information Entropy of Humpback Whale Songs. *Journal of the Acoustical Society of America* 119 (3): 1849–1866.

Suzuki, Toshitaka, David Wheatcroft, and Michael Griesser. 2016. Experimental Evidence for Compositional Syntax in Bird Calls. *Nature Communications* 7: 1–7.

Thinh, Van Ngoc, et al. 2011. Concordance between Vocal and Genetic Diversity in Crested Gibbons. *BMC Evolutionary Biology, II* 36: 1–36.

Thompson, William, and John O'Hara Rice. 1970. Calls of the Indigo Bunting, *Passerina cyanea. Zeitschrift für Tierpsychologie* 27 (1): 35–46.

Thorpe, William. 1961. *Bird Song.* Cambridge, UK: Cambridge University Press.

Thorpe, William. 1972. Vocal Communication in Birds. In *Non-verbal Communication*, ed. Hinde Robert. Cambridge, UK: Cambridge University Press.

Tomasello, Michael. 1994. The Question of Chimpanzee Culture. In *Chimpanzee Cultures*, ed. Richard Wrangham, et al. Cambridge, MA: Harvard University Press.

Tomasello, Michael. 2009. The Question of Chimpanzee Culture, Plus Postscript. In *The Question of Animal Culture*, ed. Kevin Laland and Bennet Galef. Cambridge, MA: Harvard University Press.

Tomasello, Michael. 2014. *A Natural History of Human Thinking.* Cambridge, MA: Harvard University Press.

Waddington, Conrad Hal. 1975. *The Evolution of an Evolutionist.* Edinburgh: Edinburgh University Press.

Warren, Denice, Dianne Patterson, and Irene Pepperberg. 1996. Mechanisms of American English vowel production in a Grey Parrot (*Psittacus erithacus*). *The Auk* 113: 41–58.

Weeden, Judith, and Bruce Falls. 1959. Differential Responses of Male Ovenbirds to Recorded Songs of Neighboring and More Distant Individuals. *The Auk* 76: 343–351.

White, Leslie. 1959. *The Evolution of Culture.* New York: McGraw Hill.

Whiten, Andrew, and Carel van Schaik. 2007. The Evolution of Animal 'Cultures' and Social Intelligence. *Philosophical Transactions of the Royal Society B: Biological Sciences* 362 (1480): 603–620.

Wickler, Wolfgang. 1986. Ital. Transl. 1987. *I dialetti degli animali.* Torino: Boringhieri.

Part II
Knowledge, Subjectivity, and Intelligence in Non-Human Animals

Chapter 4
Animal Learning: An Epistemological Problem

Roberto Marchesini

Abstract Learning is one of the most important processes in the development of animal behavioural identity. Since the mid-nineteenth century, and especially during the twentieth century, there have been many explanatory proposals and models aimed at describing this process. In particular, the North American behaviourist school and the Central European school of ethology have produced two strongly structured—but incompatible—traditions in terms of their explicative models. In the second half of the twentieth century, the mentalist interpretation increasingly succeeded in explaining some aspects of expressive intentionality. Today, there are several models that are often forcibly juxtaposed, even when they are incompatible: this is the case with associationist, psychoenergetic and cognitive models. The question I will address is whether this abundance could be replaced by a single model capable of subsuming and resolving the contradictions that still exist in all these explanations.

4.1 Premise

The evaluation of learning processes in animals combines the three following aspects of the ontological analysis of animal beings. (i) The adaptive protagonism of the animal in its interface with the outside world, in its extraction of problems from it, and in its formulation of answers that are helpful in solving the problem. (ii) The meaning of the endowment realized through learning and, in particular, whether this endowment should be considered an automatism that drives the individual and therefore presents a relationship of cogency between structure and function, or if it is rather an instrument used by the animal. (iii) The relationship between phylogenetic (i.e. innate) resources and ontogenetic (i.e. learned) resources, especially if related to different domains and within a ratio of inverse proportionality, or if mutually dimensioned and therefore directly proportional. These are three central topics in the ontological definition of animality, from which one can draw two essentially different ways of considering the status of the animal being: (i) the idea that the animal is a totally passive entity, driven by automatisms that are analytically responsible for its behaviour, with the different elements and factors in question presenting

R. Marchesini and M. Celentano, *Critical Ethology and Post-Anthropocentric Ethics*, Numanities - Arts and Humanities in Progress 16,
https://doi.org/10.1007/978-3-030-74203-4_4

a specific domain of intervention; (ii) the idea that "animal-being" involves ownership of one's endowments, a systematic and emergentist view of the predicates, and protagonism—not only in the intersection with the world, but also in the construction of the endowments themselves.

It is evident that each learning process is placed at the centre of the three points outlined above, tilting the scale toward one of the two possible ontological interpretations. However, it is undeniable that all animals are always called to learn: in fact, phylogeny is unable to achieve definitive adaptation, because the world is constantly changing. The life of every animal is indeed a constant challenge: it is necessary to identify important phenomena as opposed to background noise, to act and be differently involved in various situations depending on their relevance, to know how to interpret events based on their risks and opportunities, to develop strategies and specific tactics for each type of problem, to anticipate events in order to react promptly, to find the appropriate solution to the challenges of each situation, and so forth. It is a matter of creating a kind of interior/exterior mirror, just as the leaf organization in a tree crown depends on the specific lighting conditions, which enable an indispensable adjustment of position even in situations of unchanging environmental conditions foreseen by phylogeny.

Today we know that there is something in nature called the "principle of singularity", as specified by several authors, including Ilya Prigogine: in other words, reality may respect the dictates of similarity and repetition, but never identity. In spite of Laplace's demonic hypothesis, it is therefore necessary to also employ diachronic and historical analysis in physics. Mirrored adaptation is thus complicated by the fact that the environment varies naturally due to various factors. These include: (i) environmental factors because, despite the geological stability of a given environment, there are significant variations in all parameters; (ii) factors related to the niche and the food chain, because here too some mechanisms that influence fitness, like competition, predation, parasitism, are not stable; (iii) factors that cannot be reduced to mutations within the population to which the individual belongs. Phylogenetic selection certainly produces a sort of adaptive textbook: an approximate scheme that puts the organism in the best conditions to face its survival challenges. However, it is also undeniable that organisms must constantly adapt their profile and the various apparatuses that compose it. Ontogenetic adaptation differs from phylogenetic adaptation, even if the two processes are not clearly separated: (i) Today we know that some properly somatic experiences can produce effects that can be passed on epigenetically to subsequent generations; (ii) Furthermore, it is obvious that individuals, through their behaviour and, one might say, through their discoveries and creative results, can influence selective pressure and thus indirectly intervene in phylogenetic screening.

Therefore, individual responses are crucial throughout the adaptation chain, with learning processes, i.e. the subject's ability to introject experiences by modifying their behavioural canon, playing a central role. This brings me back to highlighting the link between the evaluation of the learning process and the ontological considerations concerning the animal condition. Learning is first and foremost about the concept of challenge, that is, a discrepancy between the animal's internal needs, which are oriented towards pleasure, and the characteristics of the external environment, that is, everything that surrounds this individual in terms of risks and opportunities. However, learning has got to do with another discrepancy as the motive for adaptation efforts, namely the gap between what the animal seeks and the means to sustain it. We can then say that every learning process aims to narrow this gap. In summary, the behaviour of the individual is assessed in terms of its parameters of effectiveness/efficiency with regard to the relationship between two aspects. The first is the problematic context with its objective challenges, when it confronts the subject with the urgent need to intervene in order to safeguard their state of well-being, or vice versa, becoming a field of possibilities to be seized, i.e. possible objectives that require solving a subjective problem. The second is the attempt to work on distance, in the sense of moving away from risks or getting closer to goals through responsive or propositional solutions, evaluating the consequence of one's intervention, i.e. how the context has changed following the operation implemented in a given behaviour. Learning means acting on both fronts.

If, instead, we consider the context as objectively problematic—that is, as a stimulating entity in itself—and the response as formulated in an equally objective-causal way, we can hypothesize that learning is the result of a close association between a state S (a stimulus) and an action R (a response) consolidated by a valid consequence in terms of effectiveness/efficiency (reinforcement). This is the explanatory option traditionally preferred in the description and explanation of animal behaviour, which happens in one of two ways. One involves embracing a mainly reflexogenic-associative view, as in the case of the North American Behaviorist school. The other relies on a more species-specific perspective, involving Gestalt and elicitative settings prior to the stimulus and consequent choreographic responsive predefinitions—this is the case with the Central European psychoenergetic view. The two schools, which are divergent and sometimes antinomical in their epistemological model, share the negation of any protagonism or subjectivism in the individual-world interface. Both explanatory mechanisms deny the animal any presence-will in the management of its here-and-now. As if to say: animality is a simple reactive function, like the flow of a river or the fall of a body, regulated by the interaction of internal/external objective variables that lead to rigidly determined flows-waterfalls, sometimes in a linear way, sometimes through exponential progressions or recursive circuits. One of the topics, or perhaps the main topic, of this debate is the animal's ability to learn from experience, or to adapt its behaviour not only to specific circumstances, but also by virtue of previous events, introjecting their past experience. This biographical becoming of animal individuality is what's at stake in the explanatory models of the two approaches mentioned: on the one hand there is behaviourism, which claims to infer a universal law of learning beyond the species boundary; on the other there is classic ethology, which instead regards learning as a further element of adherence to a common ethos, that is, as participation in a social style aimed at intraspecific homologation and trans-specific disjunction. This is undoubtedly where we need to start in order to discuss the desubjectivation of the animal, so as to understand whether these explanatory models can account for the diversity of animal learning processes or, conversely, whether it is time to reconsider this entire explanatory paradigm.

4.2 The Debate on the Animal Mind

Since the 1970s, there has been a substantial debate about the animal mind and the different levels of intentionality in non-humans, mainly thanks to the research in primatology and later also in various animal categories. The debate on animal intelligence had already begun at the beginning of the twentieth century: some of its forefathers include disciples of Charles Darwin like George Romanes (1912). The topic was later developed through observational and experimental sessions and based on very different epistemological assumptions, such as the holistic approach of Gestalt psychologists like Wolfgang Köhler.[1] In the dialogue between the reflexogenic hypothesis and the hypothesis of an intellectual act, the topic of learning has been the focus of discussion from the very beginning, as evidenced by the debate on insight or by the case of Clever Hans (Sebeok and Rosenthal 1981; Despret 2015). However, even following Lloyd Morgan's extension of the parsimony canon, it was preferred to follow the reductionist and reflexogenic path inaugurated by Ivan Pavlov (1927).

Things changed in the second half of the twentieth century, when the elaborative model of the cognitive approach began to take hold in human psychology and the first research was carried out on the resolvent processes put in place by animals in the face of precise challenges. The cognitive approach owes much to the rise of computer science, from Claude Shannon and Weaver's information theory (1963) to the development of computational models and applications by Alan Turing (1950) and John Von Neumann and Kurzweil (2012). The dualism of "hardware and software" brought to the fore mentalist theories that understand the mind as a packet of information or as an organization of the substrate (for example, the synaptic network, but not only) based on a particular connection scheme that, like software, is capable of processing input data. Thus, the point was to move from an associative to a "processing" view of data, which inevitably separates the endowment (the software) from the functional outcome, since several functions can be obtained with the same application. Hence a third interpretative model of animal learning, called to explain all those behavioural modifications that cannot be accounted for by the two traditional approaches. This is the situation we find ourselves in today, and the complications that this entails are beyond question.

The problem is not only descriptive and explanatory, but also epistemological, because there is no doubt that the current tripartite division is a redundancy, in addition to bearing ambivalent and contradictory aspects and obvious contradictions. Therefore, I will try to find a model that brings together all the evidence we have in the various areas of animal learning. As for the choice of model, I will refer to some basic epistemological aspects: (i) the subsiding principle, i.e. a model's ability to explain several phenomena or to keep apparently different events under the same explanatory umbrella; (ii) the guiding principle, i.e. a model's ability to solve

[1]For a presentation of the Gestalt theory, a psychological current developed in Germany at the beginning of the twentieth century, whose privileged field of investigation was perception, see W. Köhler (1970).

inconsistencies found in other explanatory models, bringing exceptions back to the rule; (iii) the parsimony principle, i.e. a model's ability to use fewer explanatory tools or tautological concepts; and (iv) the falsifiability principle, i.e. a model's ability to present its assumptions in a clear way, so they can be subject to rebuttal or counter-evidence.

4.3 The Analytical Approach in the Explanation of Learning

Finding correlation links (every time *x* happens, *y* also happens) or causation links (if I act on *x*, *y* happens) between different stimuli or between stimuli and responses appears to be an adaptive priority for the individual. At first glance, this conjunctive grammar does not seem to follow any other rule than putting together events with spatial or temporal contiguity, as the philosopher David Hume already pointed out (1896). Learning, in this sense, would mean nothing other than establishing connections between contingencies, with the only limitations to total freedom of association being: (i) as regards the stimulating entities, possible sensory accesses; (ii) as regards the responses implemented, the performative potentials made possible by a given somatic structure.

According to a reflexogenic interpretation of behaviour that was en vogue at the beginning of the twentieth century in the light of Ivan Pavlov's research on conditioned reflexes and Jacques Loeb's theories on tropism, learning is interpreted as a mechanical act in which a stimulus is strictly, linearly, and atomically connected to a reaction. The associative-behavioural view undoubtedly has some advantages: (i) it is easy to refute and can be reproduced under controlled, quasi-experimental conditions; (ii) it does not appeal to constructed entities or, better, is based on concepts that strongly adhere to objectively verifiable entities, such as the relationship between reinforcement and food reward; (iii) it is intuitive with respect to the hedonic and survival principles of the individual, since the idea that repeated useful behaviours produce pleasure appears obvious; and (iv) it responds to the canon of explanatory simplicity, in its direct connection between the stimulus input and the response output.

For these reasons, the reaction mechanism became the predominant explanation model in the first half of the twentieth century and still exerts considerable influence today. The stimulus-response link here takes on the title of "behavioural automatism", i.e. an imperative that moves the animal-puppet like a thread, renewing the Cartesian dictate. We find it in the psycho-energetic mechanism, which according to classical ethology produces the innate instinctive expression through the key signal, but above all in the animal learning model that underlies the concept of conditioning. This approach found a coherent and seemingly exhaustive formulation in the epistemological framework of behaviourism, promoted by John Watson in 1913 and then supported in an even more paradigmatic way by Burrhus Skinner (1957).

If we analyze the findings that followed one another from Pavlov to Thorndike (1965) in the late nineteenth century—in the wake of Charles Darwin's continuative proposal made in the essay *The Expression of the Emotions in Man and Animals* (1873) and then resumed in an even more poignant way by his disciple George John Romanes—we will find a need for a reductionist counter-reform. Learning could be traced back to very different events. For example, it could be attributed to the quantitative remodulation of an innate response to given stimuli, that is, not to the construction of new associations, but to intervention in the response index, as is the case with addictive or sensitizing phenomena, hence the term "non-associative learning." Or else, it could be traced back to building new associations, new ways of responding to particular stimuli through links between a stimulus and a response in such a way as to create specific responding and operating associations. The reflexogenic view found a degree of coherence with the reductionist need that, after the case of Clever Hans, gained full legitimacy in Conwy Lloyd Morgan's canon.

In the first decades of the twentieth century, a reductionist view was advocated that ignores the mentalistic dimension of the animal in its different stages of intentionality. This paradigm also embraces an atomistic conception of behaviour, explained on the basis of automatisms that are believed to autonomously produce animal expression through a trigger mechanism. I would like to clarify that this analytical approach not only fails to refer to mentalistic concepts at the various levels of awareness, but starts from the assumption that every behaviour, however complex, is nothing but the result of a sum of switches acting in a disjointed manner.

The real victim of this approach is not consciousness, but the systemic conception of the explanation. Learning is obviously evaluated—and in animals it cannot be otherwise—on the basis of the behavioural changes it produces. However, this model transforms an epistemic need into an ontological principle and into an epistemological canon, hypothesizing that animal behaviour is the result of individual associative elements (atomistic model). These elements are assumed to function autonomously, albeit interacting with one another, ignoring the principles of complexity, i.e. all the predictions that exist in a system—for example, the organizational characteristics of complex systems, the emerging qualities, and the threshold systems. In other words, referring to the mind does not mean ipso facto admitting introspective arguments or appealing to consciousness, but rather considering behaviour in terms of expression—from the Latin exprimo, in the sense of clarifying or bringing out the functional result of a system—using models based on complex systems.

After all, it is obvious that a synaptic network or neuromodulation is more similar in its dynamics to a complex system than the mechanisms of a clock. In a complex system, qualities do not depend only on the ingredients, but also on the information—one could say on the recipe—that organizes or brings together the ingredients, not only in compositional-functional terms but also in an interactive-emergent sense. The elements work together: (i) in a global-partial way, depending on the situation, giving rise to a plurality of functional states; (ii) on different hierarchical levels, and not in a horizontal chain, realizing functional molarities that supersede on molecular pluralities; and (iii) on a multitude of functional conjugations, so that the 1:1 ratio between structure and function is never possible. Among other things, ignoring the fact that information is carried on a network—like the synaptic one—means deliberately not taking into consideration the characteristics of the neurobiological substrate. Therefore, I believe that the real weakness of the analytic paradigm lies in its cybernetic genealogy, which is based on direct feedback and does not take into account what we know about systematics.

The behaviourist model, ignoring all that is not directly observable, considers the mind as a whole and all the variables referable to it as a "black box", i.e. as entities that can only be inferred. Therefore, such elements cannot be part of an explanatory theory aimed at responding to the canons of objective and simple verification—a criterion formulated in the so-called "Ockham's Razor", taken up by Lloyd Morgan's "law of parsimony". Learning, in this sense, would be nothing but the construction of new automatisms that link a stimulus to a response in a direct and imperative way, in a 1:1 relationship between structure and function. Hence the mechanistic cogency and the predictability of inference, in a universal grammar of learning that does not allow for variations of order-degree, but above all makes no reference to a systemics and a functional plurality of structures.

The behaviourist model also excludes: (1) references to the species-specific dimension of behaviour because, as said, it claims to construct a universal grammar and an environmental prevalence; (2) any deductive elements in the performance of actions, also because it denies the resolvent approach to challenges, preferring the reactive one based on random attempts; and (3) the molarity of behaviour, through the denial of any form of teleology in any response that involves complex strategies and dedicated tactics, whereas complexity is seen as a linear sequence of S-R atomic elements. Undoubtedly, this is a very structured epistemological framework, responding to a cybernetic explanatory model: learning is like a course correction made by reinforcing mechanisms that act on blind, purposeless expressions, and through random attempts that are neither directed at anything nor based on pre-existing skills. Behaviourism embraces research on habituation, sensitization and conditional reflexes, but its real workhorse is operant conditioning, first developed by Edward Lee Thorndike.

The American psychologist working at Columbia University formulated the three laws of learning in the first decades of the twentieth century, based on research in animal psychology: (i) the law of exercise, whereby learning is gradual and improves with repetition; (ii) the law of transfer, whereby a response acquired in a given circumstance can be generalized in similar situations; and (iii) the law of effect, whereby learning takes place according to the consequences of a given behaviour (Thorndike 1971). And it is thanks to this last law that behaviourists developed the process known as "operant" or instrumental conditioning, much more coherent with the US cultural climate. The fundamental concept in the explanation of operant conditioning is that of "reinforcement", aimed at assigning a binary arithmetic or quantitative value to the consequence, where: (1) +(positive) reinforcement increases the probability of a given behaviour; (2) −(negative) reinforcement, vice versa, decreases it. This dichotomy was later re-elaborated, transforming the reinforcement into a factor that always increases the probability of a given behaviour, and inserting the word "punishment" to define something that decreases the probability of repetition. Therefore, the

+ and − signs were used both in the reinforcement and in the punishment to indicate the act of administering something (+) or subtracting something (−), thus reaching a fourfold view of the consequential event. The principle of operant conditioning then gave life to a complex set of methodological and explanatory possibilities, such as chaining, shaping, etc.

To understand the foundations of the behaviourist approach it is essential to start from six "explanatory principles", which cannot be traced back to the law of effect, but rather to the conceptual framework of the law itself. The principles are the following. (1) The external world is made of objectively stimulating entities and not of a complex background of entities-events that the individual must investigate, experiencing them in a subjective way. (2) The context is objectively problematic in demanding a response, because problems do not come from a subjective distinction between objectives-expectations and available operational resources, that is, from a perceived gap to be filled. (3) The reinforcement is an unexpected event, the unpredicted final point of the process: envisioning a goal at the alpha/omega of the process would mean falling back into a teleological and mentalistic view. (4) The attempt is always random and not chosen on the basis of the specific characteristics of the problem: claiming otherwise would imply an introspective and resolvent view of the learning process. (5) The grammar of learning is universal and can be applied as such without any species-specific difference: claiming otherwise would entail a non-objective and non-environmental view of the learning process. Finally, (6) the construction of complex behaviours is based on linear chains of stimulus-response (S-R) links, but there are no complex or Gestalt structures with systemic characteristics and no hierarchical levels or behavioural molarities that can be translated into strategies/tactics.

As we can see, so as to avoid falling into mentalistic and elaborative outcomes, the behaviourist model does not only refer to the law of effect, but builds a very rigid explanatory model. The goal is to eliminate any appeal to elements such as: objectives and expectations, evaluation and judgment of one's condition, understanding and attribution of a problem, the choice of resolvent recipes useful to reach an objective, intuitions or solutions that are neither gradual nor approximate, and the species-specific dimension. Consequently, the behaviourist explanation leaves no room for elaborative models, as found in the proposals of the early twentieth century like those made by Jean Piaget (1955), Wolfgang Köhler (1970), Lev Vygotsky (2004), or Frederic Bartlett (1951). For all these authors, learning means evaluating and modifying a framework of interpretation-operativity related to the context, that is, constructing an elaborate endowment and not an atomic association between a stimulus and a response. One can say that the two explanatory models are antinomical to each other, like the Ptolemaic and the Copernican paradigm, because one believes that learning creates automatisms—SR links—whereas the other assumes that learning constructs endowments freely available to the subject.

I wish to underline this aspect because lately we have been trying to somehow combine the two models in an attempt to find a field of convergence or a common domain of validity. However, the hypotheses of non-associative learning, Pavlov's conditioned reflex and the law of effect refer to mechanisms that can fall within a systemic view of learning. On the other hand, this is not the case for the analytical view entailed by the model of operant conditioning, unless we stretch the epistemological framework to such an extent as to make it, in fact, unrecognizable, by failing to fulfill its basic presuppositions. That model, in fact, is not only based on the fact that the world has an arousing effect on the individual, that the latter responds to it through attempts or approximations, or that the consequences of an action encourage or discourage a given behaviour. More fundamentally, it is based on the idea that: (i) learning is carried out in compliance with the six explanatory principles outlined above; (ii) learning gives rise to an endowment that can be compared to an automatism. That is why I say that these models are incompatible—certainly not with respect to all the other aspects and to the evidence found in the experimental studies of the behaviourist school, which on the contrary represent a precious wealth of findings to better understand the law of effect.

At the same time, the North American behaviourist tradition has always been profoundly antithetical to the Central European ethological school, which, while not formulating explicitly elaborative models, underlined the importance of the molarity of behaviour. This predicate was recalled: (i) in the Gestalt structure of the key signal, capable of eliciting specific behaviours; and (ii) in the configured structure (pattern) of the expressed behaviour, which was very different from the atomistic-sequential model of the S-R chain. Moreover, for the ethological school it was fundamental: (1) to refer to the innate endowments structured by phylogeny; (2) to bring learning back to the species dimension, with the trends and constraints determined by it; and (3) to underline the importance of the social context. Another fundamental aspect was the meaning given to the learning process: the behaviourist school emphasized its environment-related individuality while the ethological school focused on its species dimension, i.e. its belonging to a species-specific behavioural style.

In the second half of the twentieth century, both within the behaviourist school[2] and in the ethological school,[3] there was a post-paradigmatic phase which aimed at reconciling the two traditions by revisiting their explanatory models. This attempt, however, seems somewhat incongruent and the reason is very simple: the two models are, in fact, one the opposite of the other in their explanation of both how and why learning happens. In an attempt to integrate the two views, scholars have thus separated the innate and the species dimension, interpreting them according to ethological models, from the grammar of learning to the individual dimension, in which the concepts of respondent and operant conditioning remain valid. Also for this reason, above all given recent neurobiological findings—through the works of Joseph LeDoux (2003), Jean Pierre Changeux (1997) and Gerald Edelman (1987)—I believe we must overcome the old dichotomy between innate and learned by virtue of a recomposition that requires an epistemological rethinking.

---

[2]Cf. Clark Hull's (1943) and Edward Tolman's (1932) views on organic-intervening variables and on purposive behaviour.

[3]Cf. Klaus Immelmann (1980), William Thorpe (1969) and Robert Hinde (1970), who sought to mitigate the psycho-energetic approach.

4.4 The Dichotomy Between Innate and Learned

When reconsidering the explanatory models called to describe and explain animal behaviour, one of the most important aspects concerns the relationship between learning and the phylogenetic dimension. Due to the custom of considering the two spheres—innate and learned—as disjointed, a dichotomy-based determinism has been created, whereby the innate is the outcome of phylogenetic selection based on the principle of fitness, while the learned is the outcome of an ontogenetic selection determined by the environment. This dichotomy is aimed at restoring an analytical view—i.e., a view free from any evolutionary or metamorphic conception of the endowment, within a systemic-overall logic where individual profiling is a conjugation of belonging. Moreover, this conception has also created a further dichotomy between individual singularity and species belonging. The result is therefore twofold: (i) the innate is considered exclusively binding or even hostile to the variability of learning, thus defining a relationship of inverse proportionality between the two terms; (ii) individuality is regarded as a disjunction with respect to belonging and not as a conjugation of it. The question that we must ask ourselves, therefore, is how the species-specific identity is realized in the individuality of each animal, creating both singularity and belonging.

By "singularity" I mean those aspects of identity that point to the uniqueness and unrepeatability of the subject, while by "belonging" I mean the shared qualities by which individuals of a particular species resemble each other. One would be inclined to believe that belonging is due to biological or innate determination—being born a lion and not a gazelle—while singularity is constructed through the individual experiences made by the subject in the contingency of their life. In reality, we know that both these predicates of identity are realized through relational and experiential processes that connect the subject with the outside world and make them undertake a unique evolutionary path. In other words, identity singularity can be achieved precisely because and insofar as one belongs to a species. But what is the role of learning in the construction of this belonging? And how does belonging to a given species influence and define the learning processes?

In behavioural analysis, the traditional approach entails a sort of dichotomy between innate and learned. The first refers to a rigid determination of characters that cannot be changed or refined by interacting with the context. The second is instead left undetermined in the name of a free association between stimulus and response in line with the law of effect—i.e., the consequential value of the stimulus-referenced expression. According to this view, beyond a generic appeal to social learning, the species dimension is detached from the acquisitive grammar and mainly belongs to the context of the innate, whereas individuality is founded in the unspecificity of free associations that accompany the unique and unrepeatable life of the subject. The overlap between innate and species-dimension and, on the other hand, between learned and individual-dimension is not only intuitive, but goes back to the ancient philosophical dispute about the concepts of a priori and a posteriori.

Consequently, even today, while recognizing a degree of collaboration between innate and learned and likewise between species identity and individual identity, we are inclined to consider the two contexts as disjointed dimensions—like water and oil—that at most can work side by side in a mosaic-like view of identity. In fact, rather than following from a lack of investigation or explanatory models, the persistence of a disjunctive understanding of the two areas seems to be due to an epistemological obstacle: that is, the forced cohabitation of two antinomic paradigms—the associationism of the behaviourist tradition and the psychoenergetics of classical ethology. As I said, both schools sought convergence in the second half of the twentieth century, but they are inevitably difficult to integrate, because they are based on divergent and sometimes contradictory assumptions. There is a sense that the aim of this explanatory patchwork is essentially to (1) maintain a mechanistic view of heterospecific ontology as dominated by innate (instincts) and learned (conditioning) automatisms; (2) not to question explanatory models consolidated by a long tradition.

For this reason, even when admitting that innate and learned cooperate in the construction of identity, one inevitably ends up considering them: (a) juxtaposed, that is, resting on each other without mutually changing or affecting each other; (b) complementary, that is, aimed to fill mutual gaps and therefore inversely proportional. Furthermore, learning, as a specific topic within the larger discourse of animal behaviour, has been mainly addressed by the behaviourist school and translated into training techniques. For this reason, there is a certain negligence in assessing the relationship between learning and species dimension. Not to mention that the associative grammar, in its simplicity, seems to be made to toy with Ockham's razor without actually using it. The stimulus-response association mechanism developed by Ivan Pavlov and the selection of useful operants discovered by Edward Thorndike are undoubtedly perspicuous, and yet they risk hiding the most interesting aspects of how the species dimension interweaves with learning, thus falling into tautology. And this negligence is supposedly compensated for by making partial and occasional reference to intervening variables or to releasers, while keeping the associative grammar of conditioning unchanged. In short, the underlying assumption remains that learning, for an animal of any species, means associating a stimulus with a response.

Thus conceived, learning is freed both from phylogenetic information—that is, from the highly detailed specification of the evolutionary matrix of the neurobiological system—and from the species dimension within which the evolutionary subject learns. The greater our knowledge of embryology (that is, regarding the structure of the central nervous system and even the intimate refinement of the synaptic network), the more we discover that it is actually made to collect experiences and be influenced by them through a specific evolutionary organization. Ignoring this species-specific disposition and continuing to view phylogenetic data in a deterministic way—i.e. as antithetical to a universal, nonspecific, associative grammar that only depends on the stimulus-response link—means creating an explanatory discrepancy between neurobiological findings and the models that seek to explain animal behaviour.

Today, this view is contradicted by the new interpretation of genetic translation, according to which the genes' modal expression is strongly linked to the specific context, thus refuting the juxtaposition and complementarity of innate and learned: we are moving towards a dimensional interpretation. What does this mean? First of all, that genes and environment do not act on separate contexts and through disjointed mechanisms, but are part of the same process. For this reason it is not enough to overcome the traditional antinomical view in the name of co-factoriality: it is necessary to reject the idea that the two terms are inversely proportional. The dimensional interpretation considers innate and learned as: (i) directly proportional and (ii) blurred in their intimate structure. It is thanks to the potentiality of the innate that the context can express a high variability of outcomes, so that innate complexity corresponds to complex learning.

Experience thus becomes what makes gene expression possible. So, we find that the experiential quantum of a species is directly proportional to its phylogenetic heritage and its hereditary education. It is certainly important to reconsider inherited semantics as functional modulation and as a process whereby the definition of when/where/how is translated into the related protein and into the role it plays in the environment for the activation of translational cascades, in order to avoid the close and deterministic correlation between gene and character. However, in the same way, it is indispensable to overcome the indeterministic concept that the behaviourist model attributes to the experiential process. As said, experience defines the expressive modulation of the gene, which is not at all like writing on a blank sheet. Indeed, every dispute about attribution—innate or learned—actually hides a misinterpretation of the two terms. Consequently, it is not possible to still view learning in terms of associative grammar and ignoring the species dimension. Rather, it is necessary to approach learning according to evolutionary models, which take into consideration the innate as a matrix and development organization that uses external information as a growth coordinator, but within specific fields.

The complementary view (whereby learning is called upon to fill the gaps of the innate) and the juxtapositional view (whereby learning simply exists next to the innate) reveal a compromise between the classical ethological school and behaviourism: this lies in attributing different scopes to the two explanatory models. In other words: (1) the ethological paradigm is used to describe fairly standardized innate behaviours and learning processes aimed at standardizing the individual style in a given species dimension; (2) the behaviourist paradigm is called to explain individual learning, context-related ontogenetic variability, and the learning of new performances. Observing the combination between the heritage and the information acquired in the morphological process, we can see that not only genes use the context to express themselves, but epigenetically stored information also uses genes to shape a given form. If we then broaden our gaze on the whole ontogenetic process, we will immediately realize how illusory the dichotomous explanatory claim is.

The "innate vs learned" dichotomy can also translate, albeit in a fallacious manner, into the "nature vs culture" dualism, finding further support in the humanist paradigm that separates the human (Promethean, that is, predominantly cultural) dimension from the animal (Epimethean, that is, predominantly natural) dimension. It is no coincidence that humanism, from Giovanni Pico della Mirandola to Arnold Gehlen,[4] had to hypothesize scarce innate equipment in order to celebrate the openness and self-determination of human beings. As can be seen, two hypotheses are taken for granted here: (1) that the heritage determines the expressive characteristics of an individual, acting as a blind, pre-determined, but above all fixed or non-evolutionary wisdom; (2) that heritage is hostile to learning processes and to the singularity that produces individuality.

Thus we have come to define the human condition in terms of freedom, will, and self-determination insofar as it is deficient in its hereditary endowment—unlike the animal, which is totally determined. This logic of explanatory mechanism used to describe animality is fertile soil for the hypothesis of operant conditioning, as a further automatism in behavioural ordinariness. In the second half of the twentieth century, however, the complex behaviour found mainly in the ethographic analysis of mammals and birds[5] induced scholars to hypothesize new forms of animal expression. This was based on cognitive maps, simulations, prefigurations, evaluations, productive thoughts, as well as articulated levels of intentionality—whence the concept of "sentient being". Thus, if the two traditional approaches (associative and psychoenergetic) are not able to explain expression and learning in complex contexts—or else if some behaviours are simply human-like—then cognitivity is magically pulled out of the hat. It's as if, on all other occasions, the mind abdicated from its functions. The tripartite model thus encloses the animal being within disjointed expressive fields pertaining to specific domains: the innate as instinct, the learned as conditioning and the processing capacity as strictly bound to the levels of intentionality.

Conversely, I wish to note a few points. (i) The innate does not necessarily indicate an already predefined expressive model but a range of possibilities that take one configuration or another depending on the specific experiential context. (ii) Many of the species' behavioural standards require an apprenticeship phase, i.e. the innate has to go to the school of experience to realize the species dimension. (iii) Any form of experience and learning is realized within the range of possibilities established by innate information and development processes, which are coordinated by the epigenetic dialectics within the species dimension. Finally, (iv) learning is more like the re-adaptation, re-finishing, assembly, and configuration of present elements rather than the ex-nihilo construction of resources. This means that explaining learning

[4]Just think of Pico della Mirandola's manifesto of humanism. In his *De hominis dignitate*, the author compares man to a chameleon characterized by freedom and changeability. The human being, therefore, would be endowed with free will, which would allow him to break away from the transience of earthly life so as to rise up, without constraints and conditions, to God. Cf. Pico della Mirandola (2012). This conception of the animal as complete from a predicative point of view, and therefore totally immersed and confined within a precise existential dimension, underlies both Arnold Gehlen's theory of incompleteness and Martin Heidegger's idea of the animal as poor-in-world. Cf. Gehlen (1988) and Heidegger (1995).

[5]But not only: think of the studies on cephalopods, as well as research on social hymenoptera by Randolf Menzel (2012) and Giorgio Celli (1962), who engages in several studies of the kind over the years.

while maintaining the dichotomy between the two paradigms means ignoring the points outlined above, namely that learning depends on the phylogenetic legacy that establishes the dimension of learning itself. I therefore believe we need a paradigmatic rethinking that can bring together these different contexts, keeping in mind that the innate is a processing matrix that can evolve, and not a simple drive. Indeed, the innate gives rise to new elaboration endowments that are realized through different levels of awareness, up to the unconscious or preconscious, where consciousness is never the conditio sine qua non through which the elaborative event takes place. But to do this it is first necessary to refute the behaviourist epistemological model of operant conditioning.

4.5 Critique of the Behaviourist Model

So, let's examine once again the model and the presuppositions of the behaviourist school. According to the traditional theory of instrumental conditioning, learning is based on the two following principles. (i) If a behaviour achieves a pleasant outcome (reinforcement), the animal is more likely to re-propose it. (ii) If the consequence is, vice versa, unpleasant (punishment), the expressive probabilities of such behaviour will diminish. The law of effect is linked to the intuitive principle that whatever yields pleasure is re-proposed and, on the contrary, unpleasant situations are avoided, based on the need for hedonic orientation. What dictates the instructive coordinates is therefore the environment—as already underlined by Watson—while pleasure becomes the mark of conformity. Needless to say, it is not possible to question the "phenomenon of effect" as such—i.e., learning by consequence—as it is confirmed by innumerable findings in observations and training, which also apply for the human being. The objection that I wish to raise here is about the explanatory model called to explain this phenomenon, translating it into a paradigmatic construction that sounds somewhat tautological.

I'll explain. The hedonic principle is undoubtedly the basis for the orientation of every species and is founded in the satisfaction of needs. Life itself, in its most elementary expressions and in its constitution as an open and dissipative system, requires a continuous fulfillment of energy needs and the safeguarding of vital endowments. It follows that all the systems of monitoring of and acting on the world have been regulated by phylogeny according to the utilitarian dictate, which can be summarized in the maxim "increase the good". Therefore, it is quite obvious that there is a hedonic orientation as well as a mnestic system capable of remembering what behaviours have favoured the achievement of said good (obtaining pleasure or escaping danger). What I want to challenge here is the epistemological framework of behaviourism.

Let's analyse its underlying assumptions. They are based on three hypotheses. The first is that (i) the law of effect is exhaustive i.e. able to explain all the processes of operational learning and never falsified by the findings: amplifying effects result from reinforcement and mitigation results from punishment. The second is that (ii) making an attempt is characterized by responsive passivity and randomness, as it is not oriented by the species dimension; also, the individual is a neutral entity under the cognitive profile of perception, only influenced by its sense organs and therefore exposed to the stimulating context. Finally, the third assumption is that (iii) the outcome of the learning process is an automatism that is strongly correlated to the context and is stimulus-referred: it is not an endowment that the subject can use in other analogous situations. If these assumptions fall, the framework as a whole needs rethinking. In this section I will analyze the first assumption, i.e. the alleged exhaustiveness of the law of effect—the cardinal principle of the behaviourist paradigm. I will then discuss the other two points in the following paragraphs.

The assumption "i", in its strong form, implies two consequences. (1) Whenever an expression increases its probability one can infer that it was reinforced: there must have been a dialectic between a problematic condition, a number of random attempts and a pleasant consequence to the magnified expression. (2) If a behaviour is reinforced it should increase, while if it is punished it should decrease. These two aspects form the basis of operant conditioning, and represent different aspects in terms of the validation of the framework and its domain of validity. In relation to point (1) I will ask whether there are learning conditions that do *not* follow the law of the effect, i.e. if there can be operational (non-reactive) learning that does not fall within instrumental conditioning. In relation to point (2) I will ask if there are situations in which the concepts of reinforcement and punishment are falsified, so that reinforcement is followed by a decrease while punishment by an increase in a particular behavioural expression. To be clear, should there be significant evidence contradicting points "1" and "2", I will not speak of a falsification of the law of effect per se but rather of its domain of validity, which necessarily implies that the law is not exhaustive and, consequently, the behaviourist model is not immune.

(1) The law of effect, as formulated by Edward Thorndike and then summarized in the canon of instrumental conditioning, does not deal with a mere response to a stimulative situation, as in the case of classical or Pavlovian conditioning. In the latter case an unconditioned reflex (salivation in front of food as an unconditioned stimulus) is associated with a new type of reflex (salivation at the sound of a bell) with the stimulus to be conditioned (the bell) preceding the unconditioned stimulus (the food). Instrumental conditioning is also defined as operant because the subject puts in place an operation to overcome a challenge or reach a particular target. Of course, in its paradigmatic formulation—that is, in accordance with John Watson and Burrhus Skinner's refusal of any mentalistic assumption—there is no mention of orientation or of some objective and, consequently, of tactics or strategies. Any useful behaviour—which, producing pleasure, is reinforced—is achieved through gradual approximations expressed at random. In the paradigmatic or orthodox behaviourist tradition, reinforcement is neither an objective nor an expectation, and attempts are not based on orientations, but are absolutely random. Furthermore, the subject has no understanding of the problem nor any awareness of getting closer to the solution, since these aspects would imply a mentalist-cognitive element.

As early as the 1940s, this interpretation was revised, producing the idea of a post-paradigmatic behaviourism in authors such as Clark Hull and Edward Tolman, who emphasized the importance of intervening variables and expertise in formulating responses. Obviously this has paved the way for a review of the theory in general. In particular, in addition to emphasizing the role of intervening variables, such as to question the direct link between the stimulative condition and the formulated response, Tolman noted the molarity of the behavioural response. This explanatory shift had important consequences. It introduced (1) the Gestalt structure of behavioural expression, a concept that not only could resonate with the ethological idea of pattern but also with the systemic emergent effects included by the response, in line with Kölher's idea of insight. It also posited (2) the hierarchical difference among behavioural expressions, subdividing the responses into molecular acts and molar acts, and paving the way for the "tactics and strategies" dialectic which would then be addressed by proponents of the cognitive approach.[6] Furthermore, Tolman emphasized prefigurative aspects in the idea of purposeful behaviour, which is therefore intrinsically oriented: for all these reasons, Tolman can be considered a precursor of cognitivism.

In his adaptive view of the organism, in addition to underlining the importance of independent variables, Clark Hull also stressed the need to consider the objective-dependent meaning of the learning processes. One should also address the structure of behavioural expertise, which plays a role in the formulation of responses—this idea would resurface in the cognitive concept of "heuristics" introduced by Herbert Simon (1961). With Tolman and Hull the law of effect lost its crystalline mechanicity, because intervening variables, purposes or objectives, the hierarchical structure of responses and behavioural expertise had narrowed down—through specifications and corrective measures—its domain of validity. But the most interesting aspect concerns Tolman's observation that mice, if put in a maze, learn its configuration without receiving any reinforcement, something that Tolman summarized in the concept of "latent learning" (1951). Here it is not only the concept of reinforcement that is being questioned, but also that of synchronicity in the affirmation of the operant.

At the same time, ethological research shifted the attention on several other aspects of learning that inevitably further delimited the behaviourist claim of an acquisitive universalism based on the stimulus-response association. First of all, starting from the concept of "imprinting",[7] that is, an innate preadaptation of newborns to follow a model, it was highlighted that the learning process develops along precise innate evolutionary coordinates. These therefore not only orient ontogenesis towards some learning paths, but also go on to select useful stimuli, disproving the idea of a blank sheet. Furthermore, ethologists highlighted other forms of learning, not linked to the reinforcement mechanism, such as: (i) learning by mimesis, i.e. adopting a style by observing the other members of the group; (ii) learning by vicariance, or the tendency

[6]As for example in the cognitive model TOTE, standing for "test – operate – test – exit". This model was described in its functional aspects by psychologists George A. Miller, Eugene Galanter, and Karl Pribram in 1960 as signifying the testing between subject status and target status, with the two resolvent functions being defined with the word operate (on) and exit (off). According to the TOTE model, the system does not close through reinforcement but through the achievement of the objective, as Noam Chomsky already emphasized in 1957 by questioning the behaviourist concept of reinforcement. Cf. Chomsky (1957).

[7]Defined by Oskar Heinroth (1911) and Konrad Lorenz (1952).

to choose winning behaviours with respect to objectives set by other individuals. In other words, in addition to demonstrating pre-adaptation to forms of learning, ethology underlined the importance of social learning (Bandura 1986).

(2) Research has therefore acquired important evidence that some learning processes cannot be properly classified within the behaviourist frame, in its claim of a pure association between a stimulus and a response. This has progressively narrowed down the latter's domain of validity. At this point one may ask a second question: is there evidence capable of falsifying the amplifying assumptions of reinforcement and inhibitory ones of punishment? The idea that complex behaviour is nothing but a domino series of S-R atoms in a linear sequence—such that R1 in turn becomes S2 for the next response—was questioned already in the 1950s by physiologist Karl Lashley (1955). He demonstrated that this model could not explain some phenomena, such as the reactivation of a response and anticipatory errors. Equally important objections were made by Noam Chomsky from the methodological point of view, as regards the terms *stimulus*, *response* and *reinforcement*. In particular, he noted: (i) the lack of clarity as to how the term *stimulus* differs from environmental complexity and other entities defined as neutral; (ii) some generalism in the definition of response and in the generalization of the operants; and (iii) the risk that the concept of reinforcement may be an elegant way to hide a petition for intentionality.

In other words, according to Chomsky, Skinner's theory was vague and circular, in that it was based on what it purported to prove. The principle of space-time contiguity of the stimulus-response-reinforcement triad, similarly to the one formulated by David Hume, encountered criticism in many views. These included the concept of latency of the evolutionary process, Piaget's dialectic between assimilation and accommodation, and the emergence of insight as a sudden composition of elements introduced earlier in the Gestalt approach. In all these views, the emphasis was not so much on the existence of other types of learning: rather, what was being questioned was the very validity of the explanation based on an alleged self-sufficiency and objectivity of the stimulus-response-reinforcement triad.

We can therefore summarize these criticisms in the following points. (i) The definition of stimulus is tautological: everything that causes a stimulation is considered a stimulus, as opposed to everything that is accessible to the individual's sensory organs; therefore the stimulus is not an objective reality but a construct made by the subject. (ii) The response does not emerge from random attempts but follows well-defined patterns that are presented in a molar form in discreetly organized configurations (not through atomic and continuous graduality); these patterns are used as such in new contexts or are redefined through adjustments. (iii) There is no temporal contiguity in the stimulus-response relationship, especially in complex behaviours, because a response often emerges by merging previously introjected elements (insight) and because, in sequential behaviour, an act can be anticipated, demonstrating the existence of a mental level (preactivation). (iv) In a challenge, the reinforcement is nothing but the overcoming of the challenge or the achievement of the target and therefore reinforcement is a different name—tautological and redundant—for the target.

Now, a central aspect of the behaviourist framework is the fact that reinforcement increases an operant whereas punishment diminishes its expressive frequency: if this presupposition falls, the whole law loses solidity. Let's start with the objective property by which reinforcement increases behaviour. All trainers know that if a behaviour is followed with certainty-punctuality by a rewarding event, said behaviour will follow a bell-shaped curve: at the beginning it will show a rapid increase, then it will stabilize itself and finally decrease. This is the reason why, after the first phases of training, we follow the path of the so-called variable reinforcement instead of the fixed one. This is a phenomenon that the law of effect, in the behaviourist framework, cannot explain and that, vice versa, is understandable if instead of the word "reinforcement"—as suggested by Noam Chomsky—we use terms like "objective" (possible target to be reached) and "expectation" (presumable target to be achieved).

For the subject, both the objective and the expectation are goals that respond to the mechanisms of value attribution. The first mechanism involves (i) correspondence to the specific needs of the subject's here-and-now and therefore to the "organic dependent variables", to use Hull's definition (1943), but also to the mental state variables. The second concerns the objective's, (ii) importance in the subject's here-and-now, in terms of emotions and connected representations, its coherence with other prefigurations, and its standing out with respect to other distracting factors. The third one is about the (iii) value attributed to the objective-expectation based on its availability, its rarity, the languor it arouses, and lack of saturation events. It should therefore be obvious why fixed reinforcement not only loses its value as a behaviour amplifier, but also paradoxically takes on a negative value.

The point is that if in place of reinforcement we use terms like *objective* or *expectation*, we inevitably refer to a prefiguration of the subject, i.e. we are forced to formulate a mentalistic explanation. However, it is not possible to stretch the law of effect by means of explanatory epicycles—and the notion of variable reinforcement, in fact, is exactly that—when the factors involved demonstrate a clear insufficiency in the causal determination of a phenomenon. We find a similar problem with punishment, which, strictly speaking, should decrease a behaviour if it systematically follows its expression. However, it is easy to see that many punished behaviours not only do not diminish, but paradoxically even go so far as to increase. The research on attachment carried out by Robert Hinde and John Bowlby (1988), as well as the harsh experiments conducted by the Harlow on small macaques (1958), have shown that the behaviour of search for a "secure base" and shelter did not diminish but grew if followed by punishment.

In the same way, research in animal behavioural medicine has shown that if we punish particular expressions, such as phobic, anxious, hyperactive, or conflicting ones, we are going to increase them and not decrease them. Even the discouragement of behaviours on a motivational basis, which should be seen as a form of punishment, actually amplifies behaviour rather than inhibiting it. From this we can infer that the law of effect, as it appears, is not wrong in terms of associative possibilities, but is certainly mistaken in its epistemological framework. In other words, it is wrong in conceptual terms and in the causal presuppositions called to explain the associative event. And this also applies to the post-paradigmatic formulations provided by neo-behaviourists, as it is the explanatory core of this view that does not bear scrutiny. I therefore believe that the law of effect can be subsumed within a broader theory of learning, capable of better defining its domain of validity and therefore of explaining both the situations where the law of effect is falsified and the paradoxical effects.

4.6 Learning and the Species Dimension

Moreover, it is not possible to understand the learning process without considering the species coordinates—that is, those development events regulated by the translation of the information contained in the heritage. Indeed, the species coordinates establish the evolutionary stages and how the subject deals with the experiential context and responds ontogenetically by forming new neural networks. A fortiori, consider the following points. (i) The orientation and susceptibility to external stimuli is phylogenetically determined in the different evolutionary periods (sensibility) and in the consequent structure of species-specific belonging, as shown by the research carried out in ethology and psychology on prototypical stimuli and key signals. (ii) The responses implemented by each species concern collections of operants and heuristics strictly defined by the phylogenetic heritage, such that some behavioural expressions not only emerge without reinforcements but also take place regardless of the contingent presence of stimuli.

Of course, identifying a theory of learning able to combine dimensional coordinates (belonging to a given species) with adaptive needs (correlating to the characteristics of the context)—without separating the innate and the learned—makes it essential to start from the species-specific identity, that is, from the elective parameters that it defines. By electivity, I mean four aspects. The first is (1) susceptibility to particular elements present in the external context (perceptual electivity). The second is (2) a tendency to implement specific responsive and proactive behaviours with respect to the problematic context (expressive electivity). Then there is (3) specificity in joining particular contextual findings to very precise and defined behavioural expressions (conjunctive electivity). And finally there is (4) a mode of identification of challenges in the context (epistemic electivity). But before going into the examination of these four points, it is necessary to better specify what is meant by species dimension and how this dimension is realized by keeping the individuals that belong to it within a precise range of expressive variability.

The species dimension, also defined by the term "ethogram", should not be seen as an ideal or standard species model, but rather as a field of ontogenetic possibilities, defining the index of variability that the subjects of a given species present. It can first be noted that the more complex the phylogenetic heritage of a species, the greater the field of ontogenetic possibilities, or the value of the variability index—a further argument to emphasize the direct proportionality between innate and learned. On the other hand, as we shall see, it is precisely the species with a wide range of variability (for example mammals and birds) that require the greatest external information to strengthen their belonging, decreasing the identity drift for example through parental care. This is especially true in some areas, such as social behaviour.

In this sense, one can say that the species dimension in these animals is reached through complex heritage information, which in turn requires a complex external contribution not only in terms of occasionality but also of apprenticeship. This evolutionary articulation can be ascribed, in addition to the previously mentioned cascade effects of the development environment, also to more properly intersubjective factors. These include: (i) the prenatal care of gestation and hatching; (ii) parental care that also transfers species-specific behavioural patterns; (iii) the self-active and dialogic-interactive dimension of social play; (iv) the sedimentations of cultural models within a particular population.

If we took away even just one of these elements, which could be defined as "intraspecific dialogics", from the ontogenetic process, the species dimension would be incongruent. Reaching the species dimension is therefore an evolutionary goal, much like adapting to the context: for this reason we can conclude that learning does not always aim to achieve individual variability by singularly specifying the expression of the heritage. In mammals and birds, learning also aims to reduce individual variability, limiting the virtuality or range that would be determined if the individual were exposed exclusively to environmental contingency. In this case one can say that a complex innate not only allows for complex learning, but "requires" standardized learning, capable of favoring the development of codes of belonging and limiting the uncertainty of experiential processes.

So let's go back to the parameters of electivity. To belong to a given species means: (i) to have a body conformed in a peculiar way, which predisposes it to favour some behaviours over others; (ii) to be endowed with sensory organs that enable a particular access to the world and therefore a specific immersion in reality; (iii) to be interested in some targets and inclined to implement particular behavioural styles; (iv) to have a given socio-relational dimension as part of a dialectical-referential domain; (v) to be equipped with a neurobiological system capable of processing the findings that come from the external world and from the body in a specific way, giving rise to an emerging representation. Electivity therefore means reconstructing one's world starting from the elements available, where concepts such as stimulus, response, association and reinforcement cannot be objective terms.

(1) *Perceptual electivity*—As far as this point is concerned, one can say that a subject belonging to a particular species is already predisposed by the evolutionary dialectic between phylogenetic and ontogenetic contributions to pay attention to very specific stimulatory patterns, i.e. to data accessible to the sensory organs that present specific configurations in space-time. In line with Gestalt psychology, studies in the neurobiology of perception show that the stimulating body is reconstructed inside the brain on the basis of particular compositional grammars. And these grammars have got more to do with species-specific adaptive objectives—i.e. with the history of the species—than with an objective translation of the external world. The perceptual translation system is therefore wired to provide a functional reconstruction of the external world. Since each species has been calibrated according to specific selective pressures, alongside common functional overlaps—for example it is useful for

many animals to have a processor able to realize an amodal completion—there are also specific cognitive processes at play in the functional reconstruction of the data. What recalls and evokes a given behaviour is not an objective reality, but an elective reconstruction starting from non-stimulating elements.

It is not only the sensory organs that define filters that make only some stimuli accessible, while obscuring others. It is the very act of processing and, more generally, emotional and motivational susceptibility that does so. This specificity does not only affect the reconstruction of the data—for example the attribution of illusory margins, completion, the definition of continuity and correlation, etc.—but also other aspects such as salience or persistence as well as the evaluation of the order of magnitude or perspective. In psychology we define "prototypical" a stimulus that can emerge from the context and is perceived more easily than others: this stimulus corresponds to a sort of ideal unit of measurement, and is retained more easily in memory. There are also stimuli capable of provoking an emotional response both in terms of arousal, that is, activation, and of appraisal, that is, specific evaluation or attribution: for example, the silhouette of a hawk causes fear in hens, partridges and pheasants. Moreover, perceptual electivity can be related to motivational systems which can be referred to the lifestyle of the species. These systems define different aspects, including: (i) the subject's preferential orientation towards the key signal or hypersignal, which can be achieved with a puppet by magnifying some aspects of the key signal; (ii) the close correlation between perception and expression of particular behavioural patterns, most often strongly standardized in the innate dimension; (iii) the subject's varying susceptibility to the key signal according to the overall motivational state experienced at the given time.

Also, we must not forget the relational dimension of perception, especially in animals that engage in parental care and carry out their evolutionary process within attachment. In these cases, the secure base is not only a perceptual counterpart, but a filter capable of orienting-configuring the perception of the young. The world is perceived through the mother, who defines the different entities on the basis of three aspects. The first is (i) orientation, i.e. which phenomena must be paid attention. The second is (ii) expression of arousal, i.e. with respect to which phenomena we must activate in a particular way. Finally, there is (iii) attribution of meaning, i.e. the type of decoding that is assigned to a given phenomenon. The relationship with the mother therefore allows the young to lean on the secure base to build a first experiential interface capable of operating a stimulatory matrix. Next to this there is the direct experiential process which, by equipping the subject with knowledge, in turn modifies their access to the world (knowledge being a tool for perception)—whence the recursive nature of experience. All this shows that the behaviourist claim that the individual is exposed to the stimulating context—understood as an objective element to be simply acquired—is mistaken, because in fact it is the subject that builds their own world.

(2) *Expressive electivity*—It is equally evident that each species has an elective behavioural expression, both when it is called to respond to external solicitation and when actively proposing operations on/towards the world. This expressive electivity can be traced back to three main coordinates. The first is (i) the way in which

the subject responds to stimuli that affect the emotional or motivational environment. Then there is (ii) the way in which the subject relates to their own kind. Finally, there are (iii) the behavioural manifestations that the subject puts in place even in the absence of specific evocative stimuli. Looking at individuals of different species, it is unquestionable—as shown by all ethological research—that there are behavioural configurations (displays, patterns, choreographs) typical of each species, whose origin can be traced back to the serial and process-like dialectics of ontogenetic configurational flows. To think that the behaviour expressed in a particular situation—especially if it represents a challenge—is random, like a mutation in the transcription of the genome, means not taking into account that each species has its own responsive, proactive and resolvent a priori in terms of operant and heuristic responses, can be attributed to the phylogenetic heritage.

The behaviours that a subject puts in place are always species-specific and never start from scratch, not even when the experimenter places the individual in front of an unusual problem-context for their species dimension. In these cases the subject, through a sort of *exaptation*, uses or tests the operants or heuristics that seem congenial to the problem. Obviously this implies a series of design, evaluation and reconfiguration processes of the operants and heuristics, processes that go beyond simple random attempts. Even though it ought to be redefined altogether, a much more coherent explanation is provided by Piaget's model of "assimilation", that is, the use of a pattern for new situations or targets, and of "accommodation", i.e. the redefinition of the internal scheme of the pattern.[8] If an animal were to solve a problem by using expressive randomness it would never reach a target or overcome a challenge.

When we confront an individual of a given species with a problem, that is, with a situation capable of evoking goals or expectations, we find the implementation of peculiar behaviours that are then adapted through successive approximations, up to reaching the solution. These attempts are never purposeless and random, but always purposeful and made on the basis of internal coherence as "useful resolvent recipes". Furthermore, the chosen behaviour does not appear as a motor atom, but follows a precise Gestalt pattern, which is random but depends on a pattern borrowed from the style of the species. As we will see, the final configuration becomes an "operating model" capable of working next to the heritage operants and heuristics, ready to be used whenever a similar problem-situation arises.

If a person teaches their dog to sit down—approximating a behaviour that is already present in the dog, though not as a social but as a parental heuristic—when they present the dog with a new problem, as a first option, he will not make random attempts but will try to sit down and will then continue with the other heuristics learned in the relational context. The idea that expressive or resolvent behaviour is random is only justified within an epistemological framework which, in accordance with the absolute principle of the stimulus-response association, denies several

[8]In assimilation, any information present in external reality can be assimilated to the cognitive system where, in a complementary manner, the existing cognitive structures are accommodated, though a sort of internal recognition that allows the system to be reset by virtue of the new integrated elements. See Piaget (1955).

aspects. These include: (i) any form of reflection on the part of the animal, even a simple review of the operational options; (ii) a fortiori, the concepts of objective, evaluation, and reconfiguration. It is clear that in such a paradigm there is no room for appeals to processing activities, whether they are intended to give a meaning or a judgment, to formulate internal simulations, or indeed to account for sudden solutions (insight).

For this reason, even if it is true that the achievement of a goal produces pleasure, thereby stabilizing the effective behaviour in relation to the given purpose—hence the confirmation of the reinforcement effect—, it is the epistemological framework around it that cannot be accepted. The behaviour that is reinforced by success was not random, but due to a cognitive selection between several useful operational-resolvent options to be reconfigured ad hoc: alternatively, it even emerges from a creative process aimed to produce a new solution altogether. Furthermore, the expression of a given behaviour involves a process of prefiguration (definition of a goal-state) and evaluation (definition of the gap between the present state and the goal-state). Also in this case it is useless to say that both the prefiguration and the evaluation depend on the species dimension that defines the range of possible objectives and the coordinates for calculating the gap. In this sense one can say, borrowing the concepts of Piaget's theory of learning, that every behavioural expression is always an epistemic evolution: the emergence of the new starting from the old, though not ex-nihilo.

(3) *Conjunctive electivity*—With respect to this point, the behaviourist hypothesis imagines a free associative process, linked only to contingency and therefore to sensorial relevance or temporal proximity. This involves two things: (a) the stimulus to be conditioned immediately before the unconditional stimulus in the corresponding conditioning; (b) the reinforcement/punishment comes immediately after the behaviour to be crystallized in operant conditioning. Instead, ethological observation has shown that there are elective conjunctions at play in the process. The associative behaviourist principle is based on the free associative grammar between the world of stimuli and the world of responses, but while it is very easy to teach a rat to push a lever to get food, it is almost impossible to teach it to press the lever to avoid an electric shock.

We can imagine conjunctive electivity as an a priori and sometimes exclusive way of bringing together the stimulative and responsive elements. This occurs following four main coordinates: (i) combining two stimuli to build forecasting prospects (y follows x) to act promptly; (ii) defining useful predicates as referential operators (if x then y) in order to extract an entity through a reference; (iii) creating implicative correlations (only if x then y) able to determine distinction processes with respect to similar but not significant events; (iv) identifying effective interventions to achieve given objectives (if x, then act y). The conjunctive relationship emerging from these coordinates can refer to four behavioural functions that are indispensable to the survival of the subject. These are: (i) anticipating the event (y) so as to reach an opportunity or escape danger faster; (ii) finding some categorical qualities (x), i.e. possible prey or predator, to make generalizations about other animals (y); (iii) identifying a quality (x) to distinguish between similar entities (y); (iv) understanding what the best strategy (y) is every time a situation (x) arises, assuming a cause-effect relationship between the two variables.

Also in this case, the conjunctive factor does not depend on contingency. However, on the one hand (a) the time factor facilitates the composition, since there are biosemiotic universals—such as the synchronous movement of body parts—which, beyond the species-specific perspective, create associations based on the temporal frame, able to establish relations of antecedence (1), synchronism (2) and (3), and consequence (4). On the other hand, (b) the stimulus is striking, for example in its novelty or in its respondence to sensory characteristics: these elements are capable of activating both the arousal and appraisal systems, and inevitably facilitate the acquisition of the elements to be joined as well as the very act of composition. On a neurobiological level, the synchronous ignition of a neural set makes it more likely to evoke the same set in the future. This increases: (i) the possibility that stimulating an element of the set leads to the reactivation of the entire set; (ii) the possibility of synchronous evocation of the same set even in the absence of evocative elements. In the same way, we know that the compositional intervention can have different levels of stability—from the operative memory to the hippocampal one and from the latter to consolidation. Also, memory retention is deeply linked to the arousal/appraisal stimulation caused by the event. In other words, I do not deny the importance of the temporal factor and the degree of strikingness of the stimulus in the conjunctive process. What I deny is something else: i.e., the idea that *any* conjunction is possible or that the elements at play do not have precise conjugative joints.

The alleged associative equipotentiality professed by the paradigmatic approach has been questioned mainly by ethologists. The latter have shown that the conjunction of a stimulus to a consequence attributable to a given behaviour or the link between a stimulus and a response are very different in various species: in other words, it is not possible to associate any stimulus with any response. In fact, we find that (i) some associations are carried out even in the absence of present reinforcement, as if the link were set up so that the mere presence of a stimulus acts as a releaser for the conjunctive process. On the contrary, (ii) some associations are never realized even though the subject is repeatedly exposed to the same experience and all the rules in terms of time and strikingness are respected. However, (iii) associations respond to the dictates of an overall compositional scheme which is always species-specific: the final result of a learning process that can be superimposed on teaching varies in different species. Finally, (iv) associations are realized in different ways depending on the state of the subject—for example age, motivational status, emotional set-up—giving rise to different outcomes.

Points "i" and "ii" represent the polarities of a phenomenon that is always present, i.e., the varying probability that a particular association be realized—a parameter that is always species-specific and which therefore cannot be referred to simple environmental variables. Points "iii" and "iv" refer to the fact that the behavioural outcome is never a simple association between two atomic elements, normally defined by the terms "stimulus" and "response", but a composition of elements that can only be defined within a complex choreography. This composition implies: (i) elements of positional status, such as the level of arousal, the emotional attitude, and the active motivations of the subject; (ii) elements of operational status, such as praxis and heuristics; (iii) elements of bodily motion in the given context and situation. This final scheme responds to two basic coordinates. The first is (1) the species-specific dimension of the behavioural choreography that results from the conjunction of the elements, such that even when the didactic situation overlaps there is no overlapping of results in individuals of different species. The second is (2) the individual-specific dimension of the outcome that is always correlated to the evolutionary age, to the positional state of the subject, and to the cognitive and experiential status of the subject. In other words, the behaviourist claim of an association that is primarily dependent on the external environment is denied by the evidence we have. Also, it requires explicative epicycles that are often forced to define "noise" that which instead marks the specificity of the outcome.

(4) *Epistemic electivity*—Let us now consider the alleged objectivity of the setback to which individuals would be exposed regardless of their species. The law of effect envisages the presence of a stimulative condition, capable in itself of activating a need for a response. This, in my opinion, is a source of many errors of assessment since, most of the time, those who do not respond to a state that is considered objectively problematic are immediately stigmatized as stupid. When we speak of a setback or a challenge we refer to a problem *for the subject*, with respect to the canons of desirability or to the objectives of the subject itself. The setback can therefore differ according to different parameters: (a) its valence, in terms of risks (which the individual must avoid) versus opportunities (which the individual must be able to grasp); (b) the type of problematicity, for example if it is an obstacle to be removed or a gap to be filled; (c) the mode of intersection for the individual, ranging from simple distracting search up to connotations of urgency or even cogency; (d) the type of operative and resolvent actions that it involves, in terms of praxis, attitudes, cognitive functions; (e) the consistency of the possible resolvent practice with respect to the normal expressive canons of the species.

Behaviourism starts from an assumption that cannot be ignored, i.e. that the context objectively puts the subject in the condition to operate and therefore to formulate a response. In reality, the context is simply a space of agency for the subject and not an objective setback condition, if not to a very partial extent. The setback arises from the relationship between the internal condition of the individual—in terms of motivations/emotions on the one hand and operational resources on the other—and the constraints-opportunities that the context presents *potentially*, not objectively and blatantly. For this reason it is necessary to measure the setback from two perspectives that are strongly focused on the subject. The first (i) is the subject's positional dyscrasia, i.e. the distance between the motivational involvement (which translates into desire, projection, languor, participation in the here-and-now), and the resources in terms of knowledge-competence that can be translated into useful actions to achieve the objective. The second (ii) is exploratory availability, i.e. the tendency to go beyond the appearance of the context to bring out its opportunities and understand the constraints that the environment presents.

We can essentially admit that there is an epistemic principle that brings out the challenge: this principle is the protagonism of the animal, not only in the ability to solve a problem, but also in constructing and bringing out the problem in the first place. Each species has its own epistemic dimension because each species has different motivational and emotional characteristics capable of investigating reality, as well as equally different resolvent tendencies. It is obvious that a dog approaches problems in a very different way compared to a cat, and that animals that are highly prone to manipulation, such as primates, face situations of challenge in a different way from, say, ungulates. Often there are differences in approaching the problem even among apparently similar animals, such as chimpanzees and bonobos.

One can say that in order to learn, even before developing a congruent-effective response, it is necessary for the subject to be in a challenge condition which solicits an investigative action. However, neither the emergence of this problematic condition nor the way in which the problem is faced can be traced back to a universal grammar, regardless of the species dimension, the biographical-individual profile and the here-and-now of the subject. As said, the knowledge acquired modifies the subject's proximal plane of experience, that is, the way in which the individual epistemically intersects the external world. Therefore, to think that the individual responds passively and randomly to objective stimulations of the environment means to ignore that there is no problem without a goal. However, objectives are not universal: claiming otherwise would mean throwing away the notion of behavioural diversity among different species along with the very principles of adaptive ecology.

4.7 The Cognitive Approach to Learning

At the beginning of the 1960s, a new epistemological framework was established thanks to authors such as Ulric Neisser (2014), George Miller (1956), and Kenneth Craik: the cognitive approach. This was in line with the Information Processing paradigm, which saw learning as a modification of the subject's processing schemes when intersecting the external reality. Two main ideas emerged: that of "subjective experience" in the interaction with the world and that of a "functional representation" of reality, taking up the Kantian dictate of internal or non-objective processing of external reality. As I said, this approach is rooted in the psychological field (besides the already mentioned Gestalt psychologists, one has to mention the legacy of Piaget and Vygotsky) and in the philosophical field (besides Kant, one cannot fail to refer to Nietzsche). On the other hand, the elaborative view proposed by the cognitive approach can claim a degree of continuity with Konrad Lorenz's essay *Behind the Mirror* (1978), as well as with Karl Popper's interpretation of the episteme in the hypothetical-critical approach. According to the cognitive conception, the ontological view of the animal being is very different: instead of being exposed to the world and passively subjected to an associative grammar, it has its own perspective and is epistemologically oriented towards the world.

What changes is not only the idea of perception as a reconstruction of one's own world—which in the convergence between von Uexküll (2010) and Nietzsche becomes no longer the acquisition of data but the emergence of agency—but also the idea of orientation, i.e. of construction of the problem. This is done thanks to a deep immersion in the meshes of reality (*intus-legere*) in order to bring out: (i) present or possible opportunities; (ii) the constraints or risks that separate the projection from the here-and-now of the subject; (iii) the nature of the problem, both of attributive order and in its structural requirements; (iv) potentially useful solutions to be used in a heuristic way; (v) the evaluation resources with respect to the consequences of the operation. One of the first models to be developed, as we have seen, was the one defined by the acronym TOTE (which stands for: test-operate and test-exit) elaborated by Miller, Galanter and Pribram in 1960 to define the binary cognitive action of execution and monitoring. The model adopted a representative logic that replaced the behaviourist principle of random attempts with an interpretation of behaviour based on orientation, expectation and monitoring by which to measure the result of the action performed by reducing the gap.

We can therefore say that the animal strives for an expected goal, which is prefigured and not just manifested in the target to be reached. Indeed, the animal has an expectation, a sort of "map of the territory" that allows it to choose the most appropriate action and to assess the consequences of its actions. In this sense, rather than requiring a response from the subject, the eveniential elements are subjected to internal processing—one could say, a cartographic reconfiguration—on the basis of the subject's endowments, which can be thought of as utilities or applications that organize the external reality according to a need and by virtue of map contents. These representational structures, in turn, do not respond either to the dictates of objectivity or to those of knowledge as an end in itself. Rather, they are instrumental to the orientation, so that they fall into the flow of subjective experience and respond to it. This does not translate into some sort of epistemic relativism, but rather into the subjective tendency to probe the depths of reality so as to bring out possible configurations and viable ways to satisfy the phylogenetic motivational orientations.

The elaborative-representational paradigm inevitably proposes a view of learning based on a systemic conception, where the individual, the bearer of subjective experience, cannot be fragmented into independent atoms of reactivity. The set of elaborative endowments actualizes a singular overall perspective, whose genealogy must be identified throughout the diachronic path of the individuation process—to take up Carl G. Jung—which leads to the convergence between the "genius of the species" and personal experience. These endowments bring out an inner state, of which the given behaviour is nothing but an expression. The cognitive model therefore is not based on an imperative and direct connection between a stimulus and a response, as implied by associationism, but assumes the presence of internal—elaborative and evolutionary—entities that, like maps, predispose an external framework "fit for the subject", i.e. reformulated on the basis of meanings or representations that are attributed to some situations.

The experienced world is therefore no longer a collection of stimuli that objectively require a reaction, but becomes a field of viability, investigable in terms of opportunities and constraints with respect to objectives and purposes inherent in the individual, understood as an intrinsically desiring entity. In the cognitive approach the animal plays an active and protagonist (though not necessarily conscious) role in its orientation towards the world and in its active research activity in the world: it is therefore part not only of the solution to the problem, but also of the construction of the problem itself. The protagonism of the animal depends on the heritage, i.e. on the elaborated endowments, and on the positional conditions referred to the here-and-now. Both are responsible, on the one hand, for the specific orientation of the desires and objectives that give the world a representation of practicability; on the other hand, for the resources required to address a challenge or face a problem.

Learning, according to the cognitive approach, means immersing oneself in a problematic condition. (i) The subject is a desiring being, since the problem lies not in the world but in the relationship between the objectives that one sets oneself and the resources available to overcome the obstacles. (ii) The subject resorts to internal endowments capable of providing evaluation, understanding and solution frameworks. In this sense, the internal endowments cannot be automatisms, but are tools or structures that can be used for multiple functions, therefore not in a 1:1 ratio between structure and function. Learning, in other words, does not produce an automatism that drives the subject—as posited by conditioning—but rather a useful *tool* for the individual that has functional ownership of it, using it in a range of possible situations. All this points to a shift in the very concept of knowledge.

It is true that the cognitive approach of the second half of the twentieth century suffered from the influence of the earlier psychological and philosophical perspectivism, which had been hastily banned by behaviourist extremism. However, it is equally true that the authors who, starting from the 1960s, have offered explanatory models in this sense, relied on very concrete simulation systems provided by the computational revolution. The computer model can, in fact, use machines to repeat processing events, where the input data takes on a different meaning depending on the internal processing that it receives. That is, the data loses its objective relevance, because the centre is no longer the environment with its response imperatives, but the subject with its objectives and maps. And these elements inevitably transform the context into a field of viability and the intersection with the world into a genealogical perspective. According to these authors' approach, learning is a process aimed at reaching a proposed solution with respect to a setback experienced subjectively by the individual. The endowment therefore does not only have a map value, instrumental to the realization of a function but not superimposable to the function itself—like a word processing software with respect to a text. It is also an evolutionary entity, a bit like the concept of "genetic algorithm". Learning occurs when the internal endowment requires a structural review—very similarly to what Piaget suggested in his model of genetic epistemology (1997)—because it would be otherwise unable to find a solution and lead the operational work to the state of "exit".

We can then see how the cognitive approach is able to subsume the two traditional models—behaviourist and psycho-energetic—under some common ideas. In fact, it believes (i) that the innate represents the basic endowment of the individual, i.e. the set of evaluational (appraisal) and operational (coping) resources thanks to which each subject intersects the external reality, transforming it into an experiential field of agency. It also believes (ii) that learning takes place not through an ex nihilo construction of resources, but following processes of epistemic enlargement of the endowments, which are adapted by virtue of the individual's specific experiences. And the cognitive approach did not only redefine the model of what learning produces—a tool versus an automatic response—but also the very way in which learning takes place. According to it, in order to learn it is necessary to do a number of things. The first is to (i) construct a representational image of external reality, based on a transformation of the context into a map of useful and individual-related information. The second is to (ii) set objectives with respect to individual goals that emerge from the various dimensions of subjectivity, such as the species dimension, memories, the present motivational status, etc. Then it is necessary to (iii) evaluate the structural requirements of the problem, i.e. the distance from the target or the obstacles that must be overcome. After that, one has to (iv) propose solutions coherent with the characters of the challenge using resolvent recipes previously experimented as effective (heuristics) and producing new resolvent models (insight). Finally, one has to (v) carry out a continuous test with respect to the results obtained from the execution of resolvent operations.

This revolution undermined the interpretative models of human learning, but had a low impact on the explanation of animal learning, which remained strongly tied to the models of behaviourism and psychoenergetics. Due to a mistaken obedience to the parsimony principle, it was considered more convenient to continue to explain animal learning processes through the associative model, appealing to cognitive interpretation only for complex behaviours or high levels of intentionality. This led to a confusing cognitive model of intentionality, so that cognitive ethology was believed to assume different states of awareness. In reality, the term "cognitive" does not mean "conscious" or "aware", but refers to a different epistemological model concerning: (i) how the learning process is carried out, i.e. through a phase of problematization and solution; (ii) what produces the learning process, i.e. an elaborative tool available for several functions, and not an automatism.

The intense discussion about animal consciousness that developed in ethology and primatology starting from the 1980s, especially thanks to the work of Donald Griffin (1994), thus determined a misunderstanding that equates the cognitive model with a reference to consciousness. This, in my opinion, has considerably delayed the epistemological revision of the descriptive-explanatory paradigm of animal learning. By "knowledge"—whence the etymology of the term "cognitive"—the cognitive approach does not mean the intentionality of the process but the following explanatory principles. (1) The individual is an entity that dialogues with the outside world, based on inherent orientations, and therefore presents a sort of intrinsic purpose. (2) Innate and learned endowments are instruments, i.e. maps or applications, where the ratio between structure and function is no longer 1:1 but 1:range, like a map that can produce several routes. (3) Every learned endowment is an evolution of a previous endowment, through a process of adjustment, so that the innate-learned relationship is one of direct proportionality. (4) Learning means broadening one's experiential horizon through greater agency on the context, which not only produces a more effective-efficient resolvent capacity, but also broadens the very interface between the individual and the world. Finally, (5) experience has a strong subjective connotation, since the endowments give the external world a functional representation.

Unfortunately, focusing on levels of animal awareness did not help to question the prevalent paradigm. Thus a third explanatory model was inserted in ethology manuals, next to the two previous ones—with the exception of habituation and sensitization processes. As a result, today animal learning is explained by referring to three different paradigms, depending on the type of acquisition. The first is (a) associative learning, with reference to the two types of conditioning (respondent or Pavlovian and instrumental or operative): this model is therefore still informed by behaviourism. The second is (b) social learning, linked to the relationship with conspecifics, which contemplates imprinting, mimesis, vicariance, and is explained by referring to traditional models of ethology. Finally, there is (c) intuitive or resolvent learning, which takes up Köhler's insight, Tolman's construction of cognitive maps, formation of representations, memory research, etc. Frankly, this solution seems redundant and misleading, in addition to being nominalist, considering that learning processes take place within the neurobiological substrate itself. Furthermore, these three models cannot be integrated, creating a situation of explanatory confusion with respect to the topic that was supposed to be described and explained.

4.8 Reviewing the Explanatory Framework

So, it is not possible to demolish an explanatory model without providing a coherent alternative that meets certain requirements. As mentioned, the first is (1) "explanatory completeness", i.e. the ability to explain the phenomena of the learning process that are inexplicable according to the associative model. Then there is (2) "explanatory parsimony", i.e. the ability to reduce inconsistencies through a framework that, as a whole, is more parsimonious than the epicycles added to trace any incongruity back to the model. Finally, there is (3) "explanatory subsumption", i.e. the ability to keep within the same explanatory framework learning events that are currently explained by different explanatory paradigms. It is therefore necessary to assess whether these critical issues are present in the behaviourist paradigm, which today presents itself as the reference model for animal learning—even though it comes alongside other secondary forms of learning and is sometimes partially amended through intervening variables. Secondly, it is necessary to present the cognitive model in its basic structure and show how it is able to overcome the critical issues present in the associative approach.

Explanatory completeness—Let's start from the indisputable fact that the associative model is totally incapable of explaining some forms of learning that do not lend themselves to the stimulus-response link. This incapacity can be deduced from the simple fact that important aspects, such as social learning and intuitive learning, are always treated separately by the behaviourist paradigm. Moreover, while associative learning is explained to the smallest details, constructing a model that aspires to transparency and quasi-mathematical completeness, when it comes to these two forms of cognitive acquisition it is merely stated that the phenomenon is variable. This yields a very modest explanatory result: an animal learns by observing its conspecifics or by inventing new solutions. If one analyzes the texts dedicated to social learning or intuitive learning one will see that descriptive aspects abound, for example highlighting the different forms in which a transfer of knowledge between a demonstrator and an observer takes place. On the other hand, no explanatory model is given; at best an attempt is made to include the phenomenon within an improper explanatory model—the attention paid to the stimulus, the reinforcing effect of the demonstrator, etc.

In fact, it is evident that even in the simplest forms of social learning, such as motivational stimulation or orientation on a particular object, it is not possible to magically include the phenomenon in a paradigm that posits the objectivity of the stimulus and the randomness of the response. Not even by referring to intervening variables—such as the demonstrator—can one explain an event that involves the intervention of emotional and motivational components, the attachment process and a social relationship, or the predetermination due to innate components. For example, we have seen that the learned mobbing behaviour was significantly lower if a plastic bottle was used as a target (Poli and Prato Previde 1994). In the same way, other studies have shown that conspecifics can induce the fear of snakes in other macaques, but the same does not hold for other targets (Cook and Mineka 1987).

Then of course there are more complex social learning phenomena. There is (i) mimesis, i.e. the acquisition of an observed pattern; or (ii) vicariance, i.e. the proposition of winning behaviours. There is (iii) emulation, i.e. learning the objective to be pursued; or (iv) tradition, i.e. stabilizing a species-specific style. Finally, there is (v) culture, i.e. acquiring the peculiar style of a population within the species. It is really difficult all these aspects work within the behaviourist paradigm. First of all, such behaviours are learned in a molar way, i.e. as a whole, and not through sequential assemblages of molecular elements; secondly, the acquisition is always immediate and does not happen through gradual approximations, therefore it cannot be attributed to the trial and error model. Also, it is hard to find reinforcing events in mimesis, and it cannot be denied that there is an evaluative and decision-making element in vicariance. Likewise there is an inventive performance in emulation, which is never the simple repetition of the observed behaviour. In many apprenticeship events, for example in chimpanzees' attempts to break nutshells with a rock, the first approaches are often followed by consequences that should be assigned to the category of punishment, since most of the time they are hurt by crushing their fingers. Yet, the behaviour is encouraged.

In the same way, intuitive learning cannot be traced back to the behaviourist scheme for a set of reasons. First of all, (1) there is the molarity of the process, which does not lend itself to being translated into a linear sequence but presents a Gestalt structure and involves hierarchical plans. Then, there is (2) the timing of the resolvent processes, which do not manifest themselves by approximation but according to "canons of expertise" referring to specific situations such as heuristics or insight. Furthermore, one should consider (3) the non-direct link between the present stimulus event and the response (provided one can even use such terms) as well as the occasional nature of reinforcement mechanisms. Finally, there is (4) a need to appeal to complex memory processes that relate operational memory (also referred to as short-term memory) to noetic or conceptual memory (also referred to as long-term memory). In no case does an animal approach a problem through random attempts because, if it did so, it would never get past the challenge.

It is therefore clear that a behaviourist explanation cannot account for a solution reached thanks to heuristics. An even more complex situation arises in relation to insight because, in this case, there isn't even a gradual approximation by trial, but rather a predefined and global solution. We find the same situation when we consider basic logical functions in the life of an animal, such as categorizing or distinguishing. In the case of categorization by family, that is, when more targets are correlated not because they present a common character, but because they manifest a collection of characters that are variously shared—such that the evocative response input cannot be referred to by a common denominator—it is clear that the association is of no help. To understand how a categorization by family occurs, it is indispensable to resort to a "scheme" model, that is, a structure capable of bringing out the response whenever a given number of elements are present. It therefore emerges from an internal pattern like the hidden entities of connectionist models.

As we know, categorizing means giving the same response to different stimuli while its opposite, distinguishing, means giving different responses to similar stimuli. Also in the case of distinction it is essential to construct a discriminative scheme, meaning the ability to isolate those configurations of elements that allow one to attribute the target to a specific element requiring a particular response. If then we consider conceptualization, i.e., the response to abstract elements such as same or different, big or small, old and new, it is clear that we need to refer to concepts of hidden or internal entities, that is, representative structures and not associations. In the construction of a problem-context to which to give a resolvent response, where learning means building a new resource to remove the setback condition, it is clear that the solution to a problem implies certain things. First of all, there is (i) the ability to set goals even when these are not directly perceptible, and therefore the ability to adopt a specific behaviour, which entails the negation of the behaviourist model. Secondly, there is (ii) the property of assessing the challenge in terms of space and structural requirements of the problem. Then, we find (iii) the ability to test the evolution of the challenge based on the resolvent behaviours implemented. Finally, there is (iv) the need to considerably reduce the field of useful resolvent options: a random attempt would lead to a combinatorial boom of possibilities, therefore the individual needs to recall useful resolvent recipes, which even in this case cannot be explained in a non-mentalistic theory. From all this we can deduce that the behaviourist model inevitably leaves out two very important chapters of the learning process, thus being forced to resort to new explanatory frameworks.

Explanatory parsimony—The second objective of an epistemological paradigm is to reduce inconsistencies through parsimonious models. The behaviourist approach has been modified over time to account for the many incongruities that piled up in experimental tests. Latent learning, pre-activation, targeted behaviour, intervening variables, the presence of molar expressions, etc. are only some examples of the inconsistencies that have been resolved by correcting the paradigm, without however questioning its basic model, which remained the one formulated by Burrhus Skinner.

Even the phenomenon of overshadowing or blocking demonstrates the absence of the alleged evocative objectivity of the stimulus in the learning processes as explained according to the behavioural canon. Indeed, the decrease in a learned behaviour that is continuously reinforced is itself an incongruity, because such a behaviour should instead become more likely.

The proposals that have been made to bring these inconsistencies back to the behaviourist paradigm recall the "epicycles" introduced by ancient astronomers to reconcile the inconsistencies of the Ptolemaic paradigm with the actual motions of the planets. Just like the Ptolemaic epicycles, these corrections weigh down the behaviourist explanatory model, making it far from parsimonious. The misunderstanding about the parsimony level of the cognitive model is based on the idea that "cognitive" means conscious and, likewise, that functions of awareness imply a high level of explanatory resources. I am not going to discuss consciousness, the resources used in the explication processes and the different levels of intentionality. However, without denying the importance of these aspects in the learning process, I want to clarify that the cognitive approach does not ipso facto refer to different levels of awareness. It is rather an explanatory model that can be used also and above all with regards to elementary functions, based on the concept of elaborative scheme.

A model is always an approximation and there is no doubt that feedback, within a given epistemological paradigm, is extremely useful to go beyond the paradigm itself. The associative view has brought to light some evidence that must certainly be dealt with and that must be accounted for by any alternative explanatory model, including so-called exceptions. Indeed, it is precisely the numerous inconsistencies between the associative explanation on the one hand and the observational and experimental evidence on the other that point to the need for a new epistemological framework. The latter must be capable of combining: (i) neurobiological findings, which thanks to neuroimaging techniques and clinical research are increasingly consistent with the phenomenal investigation of behaviour as an objectively verifiable manifestation; (ii) individual and species-specific identity. This model should therefore avoid any juxtaposition between innate and learned: in the light of the genetic-epigenetic process, doing so appears anachronistic and misleading but, unfortunately, cannot be avoided in the traditional explanatory tripartition.

For now I will refrain from analysing complex functions such as awareness and self-awareness, which—as reiterated by the psychology studies on the cognitive unconscious (Pelli and Tillman 2008; Berlin 2011)—are not necessary to construct elaborative schemes. I believe that the explanatory model of learning processes is always the same: i.e. the construction of a processing scheme as a readaptation of a previous scheme available to the subject. In this sense, the phylogenetic (innate) legacy, as a set of elaborative endowments predefined by natural selection on the basis of species-specific adaptive needs, represents a sort of evolutionary matrix originating the knowledge that the subject acquires through the various experiential events. At the same time, each learning process changes the condition of the matrix itself and therefore the range of further evolutionary possibilities of the subject. Ultimately, I believe the time has come to completely abandon the epistemological and psychoenergetic associationist frames, so as to construct a new explicative synthesis that considers learning as a modification of the subject's processing interface.

In my opinion, what invalidates the full affirmation of the cognitive framework in the explication of animal behaviour and learning is a misinterpretation of the principle of parsimony itself. The appeal to cognitive models only to explain learning processes concerning complex areas—such as identifying solutions through inventions, learning symbolic or arbitrary meanings, using or constructing tools, building a representational map of a particular environment, etc.—starts from a specific assumption. This assumption can be referred to Morgan's canon, i.e. the idea that it is possible to appeal to higher faculties only if the explanation that refers to lower faculties is not sufficient to account for the event in question. One can then note an epistemological distance between Ockham's razor and Morgan's canon, since the former refers to the explanatory model while the second transforms or superimposes the explanatory structures with the evaluation of the faculties in question. The parsimony principle, as it is usually applied, would make sense if we indissolubly bound together a cognitive explanation and the appeal to awareness. In reality, the cognitive paradigm, first developed in the 1960s, has nothing to do with the introspective and projective mentalistic perspective of the early twentieth century.

In this regard, let's ask ourselves, which is more parsimonious: (i) a paradigm that uses three models to explain the same phenomenon or one that is able to subsume them? (ii) a paradigm that needs to resort to explanatory epicycles, such as the concept of variable reinforcement, or one that can do without such things, by simply resolving its inconsistencies? (iii) a paradigm that creates a sharp disjunction between innate and learned or one that is able to bring the two back to a single domain? So, which is more parsimonious: (iv) the analytical and horizontal model, which proposes an endowment for each function (which is the case of automatism), or one that assumes that every endowment can perform more functions, as in the case of the tool? In other words, which is more parsimonious: a map or a collection of itineraries? (v) a model that does not take into account the characteristics of the substrate, limiting itself to phenomenal findings, or a model that is coherent with the network structure of the substrate, based on a systemic-functional conception?

Also in terms of falsifiability, it is evident that today it no longer makes sense to refer to the mind by using introspection, since information technology, while not being superimposable on neurobiological functions, helps us propose models that can be subjected to confutation. The explanatory model based on the "elaborative scheme", in fact, can find in computation (that is, in data processing schemes) a model that does not necessarily require recourse to intentional states. The "elaborative scheme" model therefore can replace the "associative scheme", not only because it is effective in accounting for complex behaviours, but also because it is more efficient (more parsimonious) in explaining two main things The first is (1) *how learning is determined* because, if we consider the act of learning as a way to deal with a setback, a resolvent structure that works through heuristics rather than through random attempts is much simpler, as it involves a smaller number of operational sessions. The second is (2) *what learning produces* because, if we consider the functional availability, a map from which to derive an itinerary is much simpler than a set of linear routes. Moreover, the elaborative approach allows us to explain (i) the passage from an innate to a second-order processing structure through readjustments of the first based on new processing information. It also explains (ii) the compositional or multiple-possibilities character of the configurations of representational processes as is the case in connectionist models of neural networks. Finally, this approach accounts for (iii) the relationship between the neurobiological system and the other physiological systems, because every organic process, due to its systemic character, can be transformed in an informative pattern rather than in a simple stimulus.

Last but not least, the traditional approach creates a sharp disjunction between the behaviour of the human being and that of other animals—a dichotomy that can only be considered laughable by scholars of comparative anatomy and that no longer makes sense if we consider Darwinian thought, and in particular the concepts of homology, analogy, and universals. Instead, the cognitive model is consistent with the principles of similarity and difference that characterize the phylogenetic specialization. While the behaviourist approach presupposes the idea of a universal grammar of learning, the cognitive approach considers learning as one of the many expressions of the genetic perspective capable of introjecting phylogeny and ontogenesis. The endowments, in fact, are the result of a programming or development of processing resources that inevitably assimilate the experiential and adaptive differences over time. In other words, the cognitive approach replaces the SR atomic model with a scheme structure that some define as a representation, although not necessarily iconic, i.e. strictly related to an entity. This structure is capable of: (1) working with other schemes to give rise to systems of a higher hierarchical level, such as mental states or mental flows; (2) assuming multiple configurations, such as the hidden entities of the connectionist systems, varying the internal conjunction between the parts; and (3) changing over time and creating new schemes.

The scheme is blatantly more parsimonious than a linear S-R sequence, and at the same time it is able to account for the inconsistencies previously mentioned without having to appeal to further elements external to the model. If one admits implicit learning both in humans and in other species, without having to refer to the "awareness" factor, the scheme model is undoubtedly more parsimonious, simply because it resolves inconsistencies in a more direct way than the associative model. This means that it does not make sense to relegate the cognitive explanation only to those phenomena that cannot be explained through the associative model: this approach is based on the false assumption that the cognitive model is more complex. Explaining animal learning processes through the scheme model does not imply any reference to more costly elements than those of the associative model. At the same time, indeed, it dramatically reduces the number of factors involved in explaining the learning process.

Let us now consider the most important inconsistencies that can be solved by the scheme model. (i) The molarity of behavioural expression is explained by the systemic character of the scheme, whose value is not given by the sum of the elements but by the systemic predicates which the endowment takes as a whole. (ii) The hierarchical value of behavioural acts—for example between strategies and tactics—can be explained through the compositional effect of the schemes, for example the TOTE subunits. (iii) Some effects such as overshadowing and the blocking effect can be explained by the irrelevance of the additional stimulus with respect to the construction of meaning. (iv) The drop in the reinforcement value if proposed in a continuous way is explained if, as Chomsky suggests, instead of the term reinforcement we use the word goal or expectation, whose value obviously decreases with repeated achievements. (v) The preactivation effect is explained if the behaviour put in place does not follow a linear sequence, continually dictated by the consequences, but a scheme or plan, where some steps can be put forward in advance. (vi) Latent learning can also find a convincing explanation if one admits that the subject learns by constructing patterns of reality, such as a cognitive map. Finally, (vii) the factor of the intervening variables is explained if one admits that the expressed behaviour is the result of a specific configuration that translates input factors of different nature, and not S-R automatisms that, given S, produce R.

Explanatory subsumption—A good explanatory model must be able not only to be parsimonious and to explain otherwise unexplained events, but also to subsume the other explanatory models or to offer a summary framework capable of reducing the gap between observed and described phenomena. In this sense, behaviourist experiences and the law of effect may very well fall within a cognitive scheme; indeed, in many application proposals there is more and more talk of a cognitive-behavioural model. Also, many cognitive theories are derived from previous behaviourist approaches, and some cognitive models, such as the connectionist one, are sometimes confused with the associative model. The cognitive approach does not conflict with most behaviourist experiences: it simply interprets them by using a different model than the S-R link and by redefining the concept of reinforcement. On the other hand, as we have seen, the cognitive model does not deny learning by consequences but sees it as an explanatory review. (i) The stimulus is not an objective entity that has value in itself, but is built based on specific schemes of the interface endowments and of the here-and-now of the subject. (ii) The behaviour put in place is not a random attempt, but something that the subject does by addressing the given challenge with the possible resolvent recipes available. (iii) The achievement of the objective is what reinforces a given behaviour, and the reinforcing effect has value to the same extent as the value of the objective. (iv) The process gives rise to a new cognitive or applicative endowment (assimilation), i.e. a scheme finds new areas of applicability or a new structural order (accommodation), being modified in accordance with a new utility of the system.

Even with respect to the model of classical ethology, the cognitive approach, unlike the behaviourist one, enables considerable overlaps and subsumptions. Behaviourist models are undoubtedly foreign to the ethological analysis, and in their claim of non-specificity they hinder the development of new explicative frames based on species-specific cognitivity. Phenomena like imprinting, expository learning, the releaser effect, learning by mimesis and vicariance, insight and other forms of resolvent learning require a new explanatory framework that puts autopoietic coordinates at the centre of species-specific learning. Here the logic is not that of building links but that of transforming innate patterns. Every individual learns by flexing reality within their own species dimension and by adapting their innate styles to the particular situations of ontogenetic development in which they find themselves. Therefore, more than a simple associative logic—where an objective "x" entity comes with another equally objective "y"—this is a transformation of the interpretative and operative schemes prepared by phylogeny.

The cognition of every species has been calibrated by natural selection, exactly like all other functions, in order to define specifically adaptive performances. One is to (1) monitor the context, or acquire findings and interpret the events occurring around the individual with respect to a model of existence that intersects said events in a peculiar way. Another is to (2) refine the operational practices, that is, the modalities of intervention in reality and of solution to challenging situations, starting from the innate styles and adapting them to the situation of the given context. Every species, therefore, is born with its own interpretative and operational schemes, which should not be considered as rigid, unchangeable structures but as "evolutionary schemes" that are capable of adapting to the conditions of growth and, indeed, were designed to be adapted. In recent years, the cognitive model has developed tools, such as genetic algorithms or neural networks, which show how new software can evolve from a starting configuration and adapt to the specific conditions of a particular context. The cognitive model therefore allows one not only to account for innate functions, such as elaborative schemes configured by the natural selection process, but also to explain how these innate configurations can evolve or be readapted and generate new processing functions.

If the concept of pattern, inherited from Gestalt psychology, lends itself to a cognitive interpretation—that is, to a scheme model—it is not so for the behaviourist paradigm. The problem of the SR model lies in two general aspects. The first is that (1) it does not take into account the starting matrix, considering learning as the construction of an association based on nothing and not as a modification of a scheme, taking the species and individual dimensions as simple variables intervening in the process. The second is that (2) it does not give rise to a systemic-molar endowment, and therefore it is incapable of interpreting learning as the realization of an organized scheme that allows the individual to monitor the world and act in a comprehensive manner. To deal with reality it is essential to transform it into a set of useful coordinates intersecting the specific ontic condition of the subject, in its species-specific and individual dimensions. Association, unlike a scheme, is not able to transform the data into information—something that any software can do—but merely creates a mechanical link between two entities.

One of the strongest inconsistencies between the ethological and the behaviourist interpretation concerns the concept of motivation. In psychoenergetics we consider motivation—I speak of course of intrinsic motivation—as an internal drive or energy (hence the psychohydraulic model) that demands to be used by virtue of a phasic or latent accumulation pouring onto a chosen target. The psychoenergetic view was questioned already in the 1970s by research in neurobiology, since the expression of a behaviour is the result of the activation of a synaptic network that leads not only to the behavioural output, but also to the strengthening of the network itself, making that behaviour increasingly probable over time. And it is not just a question of evaluable feedback on the substrate. If it were true that a behaviour is mitigated through expression by consumption, we should expect an inverse effect with respect to the exponential growth of a trend through repetition. If, on the other hand, we consider motivation as a compositional entity that is increasingly rooted through experience, i.e. if we apply a connectionist model, the facts are easily explained.

On the other hand, being intrinsically oriented—for example, a predator is oriented towards any moving target—motivation can explain the intrinsic teleology of the animal being, i.e. its desiring condition that underlies its continuous formulation of objectives, which in turn are the flywheel of setbacks and challenges to address through learning processes. The cognitive approach allows us to consider motivations as "compositional copulas" or binders which, just like verbs-actions, give rise to propositional expressions, resulting from several predicative elements of a representational and emotional order. A motivation, in fact, in order to be translated into action, must connect expressive specifications—for example in the four coordinates of what, how, where, and when—and specifications of an emotional and arousal state. A simply associative view, unlike the connectionist model, is not able to translate behaviour into a propositional attitude.

Finally, we must not forget neurobiological research, which increasingly shows that the learning process involves several brain areas and not only the organs responsible for gathering sensory information. It also shows that the learning process gives rise to neural configurations that are more reminiscent of a composition—a painting or a map—than a direct link between the collection areas of sensory stimuli and the areas of motor processing. If one thinks, for example, of memory mechanisms—also involved in the acquisition and not only in the retention and the subsequent evocation of data—one will realize that several areas of the brain intervene in operative memory, in the transformation that involves the hippocampus and in the subsequent cortical consolidation. Memory is perhaps what best suits the interpretation of the scheme proposed by Frederic Bartlett and Endel Tulving (2014). Learning processes also involve emotional and motivational areas, not as variables but as essential components of the memory configuration. Also in this case, the cognitive model appears to work much better than other explanations with the characteristics of the neurobiological structure and its functions, which today can be viewed through the techniques of neuroimaging.

4.9 In Conclusion

To conclude, I believe that the behaviourist paradigm must be overcome not only in the so-called "not merely associative" learning areas, but also in the explanation of the learning processes that are ordinarily placed within the category of "associative learning" and defined as respondent conditioning and instrumental conditioning (Marchesini 2013). Learning implies knowledge, not conditioning: the difference between these two terms lies in the automatism of the second and in the elaborative endowment of the first. An elaborative endowment is not a thread that cogently defines a function but a map structuring coordinates that lend themselves to different types of functional outcomes. Furthermore, I believe that the "innate vs learned" dichotomy must be abandoned in the name of a dimensional and non-complementary understanding of the two terms. Learning is called to re-adapt the innate input by processing schemes, and at the same time its very configuration involves the intervention of the learning process.

A processing-compositional theory is better suited to explaining these processes, which are both functional and evolutionary, since they structure a specific configuration when they are activated, so that each output becomes in turn an input. The schemes are called upon to process the two main types of information: (1) that about monitoring, which concerns both the acquisition model, since not everything is useful and important, and the interpretative model, i.e. what a particular phenomenon means; (2) that of an operational nature, which concerns both the assessment of the challenge, i.e. the structural requirements of the problem, and the identification of the operators useful for achieving the subject's objectives. Rather than making use of atomic links between a stimulus and a response, the individual therefore uses type 1 processing schemes coupled with type 2 schemes, thus achieving greater flexibility and molarity in the overall response. However, this model implies a noticeable shift with respect to associative epistemology. First (i) the dual action of the two schemes breaks the automatism between acquisition and response, that is, it sees them as two processes with their own autonomy, based on the expertise reached by every scheme in data processing, so we can talk about knowledge as opposed to conditioning. Secondly, (ii) the processing scheme model considers species-specificity, as well as other dimensions of subjectivity (individuality, particular physiological states, mental images in the here-and-now) not as variables that intervene in the SR link but as specific configurations of the processing schemes (of type 1 and 2) and as specific couplings of the same, explaining some highlighted predicates such as peculiarity, variability, flexibility.

Ultimately, the scheme model represents an explicit overturning—exactly like the Copernican revolution—of the traditional view of animal learning. In the scheme model, the centre of learning is the subject, with its functional-configurational state of mind and its ability to build an interface with external reality. Just as the sensory organs represent a physical filter of access to reality, in the same way the cognitive endowments define specific methods of processing the findings. An aspect that is not always taken into consideration is that if the animal were nothing more than a set of innate and learned automatisms, certainly its consciousness—in its intentional character of "being aware of"—could not bring out any kind of subjectivity: a torch cannot shed light on missing furniture in a room. In order for subjectivity to be possible, it must precede consciousness. I therefore believe that it is meaningless to maintain the analytical approach, based on automatisms, and then add a phantom consciousness to an entity otherwise explained as a puppet moved by threads. And it is equally clear that only a model that does not overlap the structure of an endowment with its function can explain the ownership of endowments that is the basis of subjectivity.

References

Bandura, Albert. 1986. *Social Foundations of Thought and Action*. Prentice Hall: Englewood Cliffs, NJ.

Bartlett, Frederic Charles. 1951. *The Mind at Work and Play*. Crows Nest: George Allen & Unwin.

Berlin, Heather. 2011. The Neural Basis of the Dynamic Unconscious. *Neuropsychoanalysis* 13: 5–71.

Bowlby, John. 1988. *A Secure Base: Parent–Child Attachment and Healthy Human Development*. Minneapolis: University of Minnesota Press.

Celli, Giorgio. 1962. Contributi allo studio degli Imenotteri parassiti di Insetti minatori. II. Nota preliminare sui parassiti di Paraleucoptera sinuella Reutti. *Bollettino dell'Istituto di Entomologia dell'Università degli Studi di Bologna* 26: 207–216.

Changeux, Jean-Pierre. 1997. *Neuronal Man: The Biology of Mind*. Princeton: Princeton University Press.

Chomsky, Noam. 1957. *Syntactic Structures*. Indianapolis: Bobbs-Merrill.

Cook, Michael, and Susan Mineka. 1987. Observational Conditioning of Fear to Fear-Relevant Versus Fear-Irrelevant Stimuli in Rhesus Monkeys. *Journal of Abnormal Psychology* 98 (4): 448–459.

Darwin, Charles. 1873. *The Expression of the Emotions in Man and Animals*. New York: Appleton & Company.

Despret, Vinciane. 2015. Who Made Clever Hans Stupid? *Angelaki* 20 (2): 77–85.

Edelman, Gerald Maurice. 1987. *Neural Darwinism: The Theory of Neuronal Group Selection*. New York: Basic Books.

Gehlen, Arnold. 1988. *Man, His Nature and Place in the World*. New York: Columbia University Press.

Griffin, Donald. 1994. *Animal Minds*. Chicago: University of Chicago Press.

Harlow, Harry. 1958. The Nature of Love. *American Psychologist* 13: 673–685.

Heidegger, Martin. 1995. *The Fundamental Concepts of Metaphysics: World, Finitude, Solitude*. Bloomington: Indiana University Press.

Heinroth, Oskar. 1911. *Verh. 5th Int. Ornithol*, 5, 589. Berlin: Kongr.

Hinde, Robert. 1970. *Animal Behaviour: A Synthesis of Ethology and Comparative Psychology*. New York: Mcgraw Hill.

Hull, Clark Leonard. 1943. *Principles of Behavior: An Introduction to Behavior Theory*. New York: Appleton-Century/D. Appleton & Company.

Hume, David. 1896. *Treatise on Human Nature*. Oxford: Clarendon Press.

Immelmann, Klaus. 1980. *Introduction to Ethology*. Dordrecht: Kluwer Academic Pub.

Köhler, Wolfgang. 1970. *Gestalt Psychology: The Definitive Statement of the Gestalt Theory*. New York: Liveright.

Lashley, Karl. 1955. *The Behavior of the Rat in Latch Box Situations*. Baltimore: Johns Hopkins Press.

LeDoux, Joseph. 2003. *Synaptic Self: How Our Brains Become Who We Are*. London: Penguin.

Lorenz, Konrad. 1952. *King Solomon's Ring*. London: Methuen & Co.

Lorenz, Konrad. 1978. *Behind the Mirror, a Search for a Natural History of Human Knowledge*. New York: Harcourt Brace Jovanovich.

Marchesini, Roberto. 2013. A Re-examination of Epistemological Paradigms Describing Animal Behavior in 8 Points: 'Animal Consciousness and Science Matter': A Reply. *Relations: Beyond Anthropocentrism* 1 (2): 69–77.

Menzel, Randolf. 2012. *Animal Thinking: Contemporary Issues in Comparative Cognition*. Cambridge, MA: MIT Press.

Miller, George Armitage. 1956. The magical number seven, plus or minus two. *Some limits on our capacity for processing information, in Psychological Review by the American Psychological Association* 101 (2): 343–352.

Neisser, Ulric. 2014. *Cognitive Psychology*. New York: Psychology Press.

Pavlov, Ivan. 1927. *Conditioned Reflexes: An Investigation of the Physiological Activity of the Cerebral Cortex*. London: Humphrey Milford, Oxford University Press.
Pelli, Denis, and Katharine Tillman. 2008. The Uncrowded Window of Object Recognition. *Nature Neuroscience* 11 (1): 1129–1135.
Piaget, Jean. 1955. *The Construction of Reality in the Child*. London: Routledge.
Piaget, Jean. 1997. *The Principles of Genetic Epistemology*. London: Routledge.
Pico della Mirandola, Giovanni. 2012. *De hominis dignitate*. Pisa: Scuola Normale Superiore di Pisa.
Poli, Marco, and Emanuela Prato Previde. 1994. *Apprendere è sopravvivere. L'apprendimento animale tra psicologia ed etologia*. Milano: Raffaello Cortina.
Romanes, George. 1912. *Animal Intelligence*. New York: D. Appleton & Co.
Sebeok, Thomas Albert, and Robert Rosenthal. 1981. The Clever Hans Phenomenon: Communication with Horses. In *Annals of the New York Academy of Science*, 364. Apes & People, Whales. New York: The New York Academy of Sciences.
Shannon, Claude Elwood, and Warren Weaver. 1963. *A Mathematical Theory of Communication*. Champaign, IL, USA: University of Illinois Press.
Simon, Herber. 1961. *Administrative Behavior: A Study of Decision-Making Processes in Administrative Organization*. New York: The Macmillan.
Skinner, Burrhus Frederic. 1957. *Verbal Behaviour*. New York: Appleton-Century-Crofts.
Thorndike, Edward. 1965. *Animal Intelligence: Experimental Studies*. New York and London: The Macmillan.
Thorndike, Edward. 1971. *The Fundamentals of Learning*. New York: AMS Press.
Thorpe, William Homan. 1969. *Learning and Instinct in Animals*. London: Methuen.
Tolman, Edward. 1932. *Purposive Behavior in Animals and Men*. New York: Appleton-Century.
Tolman, Edward. 1951. *Behavior and Psychological Man: Essays in Motivation and Learning*. Berkeley: University of California Press.
Tulving, Endel. 2014. *Memory, Consciousness and the Brain*. New York: Psychology Press.
Turing, Alan. 1950. Computing Machinery and Intelligence, in *Mind. A Quarterly Review of Psychology and Philosophy* LIX (236): 433–460.
Von Neumann, John, and Ray Kurzweil. 2012. *The Computer and the Brain*. New Haven, CT and London: Yale University Press.
von Uexküll, Jakob. 2010. *A Foray into the Worlds of Animals and Humans: A Picture Book of Invisible Worlds*. Minneapolis: University of Minnesota Press.
Vygotsky, Lev Semënovič. 2004. Imagination and Creativity in Childhood. *Journal of Russian and East European Psychology* 42 (1): 7–97.

Chapter 5
The Obscure Object of Animal Subjectivity

Roberto Marchesini

Abstract The issue of animal subjectivity has been addressed in many ways over time. On one hand there is the anthropomorphic assimilative interpretation, supporting a projective view according to which animality is just a conjugation of the human being. On the other, we find the break operated by René Descartes, who assimilated the animal condition to a mere *res extensa*, thereby annihilating any hint of subjectivity and turning the non-human animal into a bundle of deterministic mechanisms. In a way, the Cartesian postulate has remained on the background for centuries, despite being criticized by many philosophers (especially Michel de Montaigne); it then regained some epistemological and applicative strength as a counter-reformist reaction to the continuity theory supported by Darwin and his followers. In the twentieth century, the reductionist project—promoted, among others, by Martin Heidegger, misinterpreting or stretching Jakob von Uexküll's analysis—gave birth to a mechanistic paradigm according to which behaviour is due to innate and learned automatisms, de facto reviving Cartesian assumptions. Thus, animals were subjected to a de-subjectivizing process that annihilated any trace of self-presence. On the one hand, (i) Darwinian evolutionism, with its inherent morphopoietic plurality and the activity of the organisms, was amended in the sense of the applicative dichotomy between "chance and necessity", as put by Jacque Monod. On the other hand, (ii) Konrad Lorenz's heritage was purged of any aspect of self-presence in the construction of one's epistemic perspective; the only thing that was kept was the Psychohydraulic Metaphor. Nevertheless, in the second half of the twentieth century, something seems to have changed, particularly in the ethological debate: intellective activity is now recognized as the centre of animal ontology, admitting various levels of intentionality when talking about sentience. In this essay, I would like to deal with the issue of animal subjectivity while avoiding any recourse to mere projectivity—i.e. banal anthropomorphism—or to any explanatory cycles or forms of *petitio principii*, by which consciousness is the decisive element that makes subjectivity appear out of nowhere, like a rabbit in a top hat. I think that the issue of subjectivity today is the main topic of ontological analysis, not only for non-humans, but also in relation to the challenge of human ontology itself: traditional dualistic explanations keep losing ground, whereas the performative concept is erasing any distance between living beings and machines. This is why it is necessary to focus on the concept of subjectivity: all too often it is made to overlap with other predicates.

R. Marchesini and M. Celentano, *Critical Ethology and Post-Anthropocentric Ethics*, Numanities - Arts and Humanities in Progress 16,
https://doi.org/10.1007/978-3-030-74203-4_5

5.1 Premise

The post-Cartesian view of animals as automata undoubtedly represents a great break in the Western tradition: for the first time, human beings were seen as opposed to animals in a sharply disjunctive sense, because of their exclusive access to the dimension of *res cogitans*. The Cartesian presupposition, in its alleged mathematization of animality, provides a rational and applicative tool to the attempts at assigning non-humans to the predetermined and deterministic domain envisioned by humanist anthropocentrism. However, the first humanists still posited a *fil rouge* of underlying continuism between man and animal. The Cartesian paradigm, instead, undoubtedly marks a "qualitative leap", albeit in the wake of a process of separation already initiated by its conceptual precursors starting from the Platonic, Aristotelian and above all Stoic schools—think of the genealogical division between the Promethean and the Epimethean, or the definition of man as a "zóon logòn échon".[1] This trajectory was rooted in a rigid and predefined rank attribution, depriving the non-human of any life-protagonism and eventually reaching the Cartesian de-subjectivization, which questions even the animals' ability to suffer and desire. Only by retracing this path is it possible to understand the explanatory frames that, in spite of Darwinism, took hold in the twentieth century.

The explanatory assumption does not dare to challenge the alleged passivity of the animal condition—Morgan's canon goes beyond the intentions of its inventor—so the evolutionary process itself is brought back to the dialectic of chance and necessity. This entails the annihilation of any animal protagonism, whereby the individual loses any value in the dynamics of selection, stuck between an *a-priori* heritage and an environment aimed at dictating possible ontopoietic changes. There is no doubt, however, that even Konrad Lorenz's thought has been extensively amended by the prevailing reductionism, in order to trivialize its epistemological contents, in line with the mechanistic idea of instinctive automatism and psychohydraulics. In reality, the Austrian ethologist had left us some unforgettable pages on the epistemic protagonism of animal subjectivity, responsible, in his opinion, not only for the construction of a single perspective on the world, but also for actively intervening in the evolutionary dynamics of fitness and selection.

The mechanistic view therefore denies the animal's presence: in this perspective, its behaviour would be nothing but a domino sequence of automatisms, a blind flow with no will or purpose. Indeed, Descartes warns us that any attribution of protagonism to the non-human is an undue form of anthropomorphism. The reductionist project, in the humanist parable of Western culture, has at least three powerful allies on its side. The first is (i) anthropocentrism, which transforms the human being into the ultimate ontological fulcrum, making other entities instrumental to man's full realization and therefore lacking any inherence of their own. Then there is (ii) the scientific paradigm which, ever since the seventeenth century, has favored a

[1] It was Aristotle who defined man as *zóon logòn échon*, meaning animal endowed with language. Indeed, language has historically been one of the intrinsic qualities of the human rational animal, cutting off any continuity between humans and non-human animals.

mechanistic, determinist, and reductionist approach, epistemologically oriented by Ockham's razor to the simplification and rejection of any form of teleology and anthropomorphism. Finally, there is (iii) the stigma of pantheism that, since the Middle Ages, has been placed on anyone who attributes a creative force to nature, as demonstrated not only by the misfortunes of some philosophers (think of the emblematic cases of Baruch Spinoza and Giordano Bruno), but also of naturalists like Jean-Baptiste de Lamarck. In this sense the denial of animal subjectivity was achieved not so much by finding a predicative (ontic) difference between the human being and the other species—aimed at establishing the specificity of the human condition—but rather by establishing a metapredicative (ontological) distance between the human dimension and animality. The goal was to affirm the special nature and primacy of the human, translating it into the idea that the human being is not an animal.

The metapredicative analysis conducted by humanism has always privileged oppositional meaning, useful to bring out the human through a background oscillation rather than trying to understand what the animal condition means. Animal being, in its various understandings, has always been interpreted by focusing on some distinctive ontological foundation of the human, leading to several views. One sees (i) animals as machines, deprived of that presence and decision-making power which represents the human's *proprium*. Another regards (ii) animals as a reproduction of characters predetermined by nature, lacking self-determination: this is viewed in opposition to human Prometheanism, the character of becoming which inaugurates the historical dimension. Then there is (iii) the animal as poor-in-world, stunned in fruition, incapable of contemplative distancing or of prospective evaluation and epistemic multiplicity. Another paradigm is that of (iv) the animal as an expression of ancestrality, a mirror of the past, a resonance indicating the old path of the human being, not as a *specialization of* but as the *emancipation from* the animal condition, and consequently synonymous with regressiveness and repression. Of course there is also (v) the animal as a brute, that is, as devoid of yearning, of verticalization with respect to physiological instincts and needs, destined to impulsiveness, irrationality, aggression, selfishness. Finally, there is the view of (vi) animals as devoid of language, representation, explication and declaration of its own state, stuck in the instantaneousness of its action, whence their absolute lack of retrospection on past events, of any real reference to the present and of any projection on the future.

Therefore, it is far from easy to make a case for animal subjectivity: too many conceptual biases have been accumulating in our culture over the centuries, fueled by different sources or intentions. These biases force those who venture along this path to operate a long and complex deconstruction. Indeed, it is necessary to remove the canvas from such rooted, infiltrative and convoluted conceptual frames—and this is so hard that it is easy to lose one's thread of argumentation and even one's central speculative theme. If, for some, the reference to an animal subjectivity appears to be almost an oxymoron, for others it is taken for granted. And yet it would seem so simple to observe the parental care of a mammal, the aerial acrobatics of a bird, the impressive coordination of a school of fish, the meticulous work of a spider and to deduce, even just intuitively, that an overall design is undeniable in such behaviours. Surely, one feels, there is an evaluative and decision-making subject in place, pursuing an objective within a vast field of viability. But the simplest way is not always the correct one. Indeed, one could say that, if the world followed the coordinates that are dictated by immediate intuition, we would not need science. Are we really sure that performative qualities are a useful indicator to discuss subjectivity? Machines should make us think twice about that.

So let's try to take up this analysis from where it has run aground with Descartes, i.e. in the relationship between the entity and the predicates that define its manifestation. In Aristotelian philosophy, the subject is what lies below the predicates (*subiectus*) and in so doing supports accessory qualities, interpreted as accidental and contingent; in this sense the subject is constitutive (*substantia*), i.e. that which lies 'underneath' its connotations and makes up their ontological foundation. I consider this view important to define an animal condition that describes animal-being as a general condition, beyond the characterizations of species and individual. To go deeper into the theme of animal subjectivity we will therefore have to move beyond the traits that diversify the various species, so as to focus on the common condition that defines animality. In other words, considering biodiversity as a collection of entities endowed with predictive differences, one could say that the *subiectus* of an animal being defines a metapredicative condition, and it is precisely this dimensionality that the discourse on subjectivity is about.

In modern thought, the term *subject* is contralateral to *object*, referring to the activity of the first (which acts, owns, obtains, thinks) as opposed to the passivity of the second (which is enacted, owned, obtained, thought of). This dichotomy is clarified in the Cartesian dualism of *res cogitans* and *res extensa*, charging subjectivity with further metapredicative meanings such as the concepts of freedom, interiority, certainty, transcendentality, as opposed to the counterparts of necessity, exteriority, uncertainty, unknowability. Furthermore, Descartes opened up the path of ontological anthropocentrism by claiming that participation in the *res cogitans* is reserved for humans only. Hence the mechanical character of all that is not human, combining animality with other natural phenomena such as the gravitational fall of a body or the flow of a river. But can we honestly believe that there is nothing but mechanistic articulations behind the behaviour of an animal? Consider a deliberative act expressed by an animal, which for the most part includes processes of planning, evaluation, simulation, feedback-based correction, and knowledge acquisition: can we really say that it amounts to a simple cascade of switches turning on and off?

In fact, overlapping the performative expression—whatever the level of excellence or intentionality that characterizes it—with subjectivity as the actual presence of an animal in its present leads to a series of interpretative drifts and to often deceiving epistemic results, such as tautology and infinite regress. If, in fact, we follow the dualistic assumptions, we inevitably fall into extrinsic teleology, having to imagine an external entity that presides over the—programming, evaluative or decision-making—projection. Indeed, it is obvious that, when we speak of a being who intersects the external reality in a perspective and copulative way, the *res extensa* must necessarily rely on a second entity, however it is defined. Every layer of the onion we remove, defining it as a variable environmental stimulus—be it the world, the body, the brain or whatever—supposes a nucleus that does not exist or, better, a nucleus that is not identified and that, nevertheless, supports the whole line of reasoning.

Overlapping subjectivity with predicates does not help either, because if it were so, then a hypothetical machine capable of implementing and even surpassing our own performances should lead us to the conclusion that said machine is more human than we are—whatever the meaning of this statement may be. And yet we are certain that an elevator does not worry about our health when it keeps the doors open to let us through, that a thermostat ignores the temperature that it maintains within an environment, that a computer does not know the laws of mathematics despite being able to carry out complicated calculations better than us or, more prosaically, that a coffee maker does not have the goal of greeting us in the morning. We experience this everyday: there are functions related to contexts or to results, and therefore mechanically intelligent, which do not necessarily imply a mind in action. There are machines that work by feedback mechanisms, according to the coordinates of cybernetics—the thermostat is an example of this; others perform the dictation of an algorithm on the basis of hardware and software (for example, all computer systems work this way). But when it comes to analyzing animal behaviour we face a sort of explanatory crisis, polarized between a mechanical simplification and trivial intuitionism.

Undoubtedly, regardless of all the operations of distancing and repression, we feel that this is something that strongly concerns us. In this sense we are inclined to look for a mind in action wherever there is some impressive performance: from the radial perfection of a spider web to the precise orientation of migrating birds, from the architectural wisdom implicit in a beaver dam to the mimetic oscillatory movements of a stick insect. But what if it wasn't like that? What if the mind is revealed more in uncertainty than in fluent realization, in hesitation more than immediate execution, in approximation and imprecision more than in full correspondence? What if it is found in the flexibility of functional co-optation—i.e. in an expedient that roughly adapts to a singular need more than to a coherent and standardized response? Certainly, it is not in the function itself that the subject emerges—i.e. that the individual shows a glimpse of their subjective presence in reality—but in the ability to *master* these functions, adapting them to specific situations. So, if it is true that performative perfection arouses marvel, it is also undeniable that even an embryogenic event marks executive phantasmagories, seemingly pursuing a final objective and not simply being the result of a cascade of events prepared by information packages accumulated in phylogeny. And yet we would never feel authorized to see a mind in action behind the formation of an organism, explaining it in teleological terms.

At this point, a first doubt concerns the very nature of the explanation: when we speak of subjectivity, do we mean perhaps to bring back extrinsic teleology? Or is there a third way, able to avoid this risk, while rejecting the paradigm of the animal as a machine? This, in my opinion, is the challenge that opens up before us and that should be food for thought to anyone who aims to question the Cartesian concept of animal-automaton. For a serious critique of explanatory mechanism, it does not help to appeal to feelings or to complex elaborative functions: Morgan's canon is not the real obstacle to the admission of subjectivity. The real enemy is the explanatory model itself. Perhaps it is a question of following in psychology the same path indicated by Gödel in mathematics, demonstrating the incompleteness of the mechanistic formalization. On the other hand, I want to immediately clear the field of possible misunderstandings. In speaking of subjectivity and rejecting the mechanistic paradigm of Cartesian origin, in order to explain the expressiveness and above all the condition of animal-being, I do not mean to refer to any form of dualism by assuming that there is some transcendent principle at play in machines.

My intent is, on the contrary, to understand the animal condition in two ways: (i) in a systemic view, i.e. in a mentalistic reading that takes into account the evo-devo coordinates (Minelli 2009); and (ii) in an active conception of the individual as a protagonist, instead of considering the animal dimension as the condition of being totally stunned in one's function. The mechanistic assumption, in fact, does not deny the complexity—let alone the intersection between the individual and the world—of the animal, but proposes to attribute the latter's expression to an endowment-mechanism that acts like a switch, determining the function itself. Therefore, the focus of the discussion does not lie in the complexity of the function: we are not endowed with subjectivity if we are able to achieve impressive performance results, but if and only if we are the protagonists of our intersection with the world—if we are able, that is, not only to be in the moment but to own it. To overcome the machinic reduction, i.e. the reduction of animals to objects, it is not necessary to appeal to some other or transcendent entity, but it to admit a system that denies the claim of analytical exhaustiveness, and involves various levels of supervenience with respect to the functional canon or to the heritage recorded in the individual. As can be seen, the mechanistic view denies both point "i" and point "ii".

In order to speak of subjectivity, on the contrary, it is indispensable (i) to admit the globality of the individual, in that it cannot be divided into single autonomous elements; this means rejecting the analytical conception of expression and condition. It is also necessary (ii) to assume the factive presence of the individual, a being-there in the intersection with external reality, so that the entity is capable of not simply undergoing reality, but of creating, redefining and conceiving a new perspective on the world, albeit on the basis of their heritage. This view obviously requires a different level of analysis than the ordinary one which has implicitly taken on the Cartesian dualist dialectic and simply extended it to other animals. In other words, rather than an appeal to a transcendent *res cogitans*, albeit revisited perhaps in the concept of *res informatica* or in a dichotomy between automatisms and consciousness, my intention is to bring back cognitivity—i.e., knowledge as a "creative and inventive" rather than passive and proactive act—to global embodiment. My goal is to trace the intersection back to the realm of protagonism, but not by hypothesizing some control system watching over the whole thing—a sort of *homunculus*—but rather by placing cognition within a systemic and dynamic conception of animal presence. The latter coincides with the condition of "being-a-body", being capable of emergent predications which go beyond the endowment repertoire and the already-given. There is subjectivity wherever there is a surplus in the comparison between the individual's expression-condition and the synchronic correspondence to their inherited endowment. A subjective being is expressed not in coherent or performative excellence, but in any singular and creative expression, even if ineffective.

We are therefore talking about a non-overlap between the heritage, i.e. what that particular individual has received from phylogeny or matured with experience, and their way of dealing with their here-and-now. Emergence, in fact, is what allows for the supervenience of the specific characteristics of the single endowments—that is, the internal functional and structural contents or the contents prior to the act of intersection itself. An animal being, in its intersection with the world, goes beyond the already-given, going beyond itself every time it dialogically interacts with external reality and consequently redefines its perspective. Cognitive emergence can explain the different levels of intentionality that characterize the non-stunned presence of the animal, from the processing unconscious to self-consciousness. Now, any paradigm that wants to bring the principle of subjectivity back into the animal expression-condition must be aware of two things: on the one hand, it is misleading to bypass the issue of embodiment through the deus-ex-machina of dualism; on the other, it is insufficient to trace consciousness back to a set of endowments formalised according to the coordinates of the *res extensa.*

To understand the shift required by a subjective view of animality, in my opinion, it is necessary to compare the animal condition with the machinic one, in order to bring out the differences between them. We should move from the humanistic opposition of "human vs animal" to the post-humanistic one of "animal vs machine", trying to understand the differences between the two terms. However, there are some "metapredicative qualities of animality"—that is, qualities that are not about belonging to a species but to the condition of "animal-being"—which are not found in the machinic conception. I will examine three of them in this essay: (1) *ownership*, that is, self-belonging and owning one's endowments; (2) *desiring*, that is, being open towards the external world, not being ontologically contained; (3) *sentience*, i.e. being a body that, on a multiplicity of levels, is connected to and depends on dialectic factors. If a machine can ever be said to have these characteristics, then maybe it will be an animal. But make no mistake: if that moment comes we will have no power over that machine. This is why the animal-being has an intrinsic finality, i.e. self-belonging and being-there as a projection towards one's own realization. It is a self-ownership that forces us to re-read subjectivity in terms of will, inherence, partiality, infidelity, dialogue—in a word, life. Let's start with these metapredicative characteristics to address the topic of subjectivity—without, of course, claiming to be exhaustive.

5.2 The Ownership Principle

We need to overcome the anthropocentric prejudice, which attributes a surplus to the human condition with respect to animality—that is, we must avoid tautologies or epicycles to attribute our *Dasein* to something external/extraneous to the animal condition, be it the *res cogitans*, the soul or language. On the other hand, we still see the human as a specific conjugation of animality. Therefore, it follows that we need a new model with respect to the *res extensa*, i.e. the mathematical and mechanical translation of being. Animality, in fact, becomes something that concerns us globally, no longer representing a backdrop or a partiality, a vestigial presence or an origin to be placed in a remote past. Animality thus takes on the role of the matrix of our human-being, the principle we use to explain the predicative characteristics of the human. The humanistic opposition is basically a convenient trick, because in the dialectical game of opposites it allows one to stay on the surface of the predicative argument, i.e. the question where the human comes from. If, in fact, we talk about freedom or autopoiesis as opposed to a rigidly determined condition, we avoid looking either into what it actually means to be an animal, or into the inspirational or even basic coordinates of anthropopoiesis.

Instead, we should embrace an "animalist" ontological perspective—paraphrasing humanism one could exemplify it as "nothing animal is alien to me". In other words, we must understand that the human is not a counter-term of the animal but one of the many expressions of this dimension, and that our human predication must be sought within the potentialities present in animal-being. It is therefore evident that we can no longer limit ourselves to the game of opposites. It is now essential to have a model capable of describing and explaining our ontological dimension within the framework of animality. What must be explained within the dimension of animality is our free will with respect to the possibilities offered to us, placing a distance between us and the conditions of the world, i.e. the will that projects us into the future and makes us dig into the folds of reality. We could summarize these concepts in Heidegger's existential statements, but we would have to attribute them to the animal condition in general and not to the human being alone. In other words, it is necessary to go from a "contralateral view", which describes the animal as what we are not, to a "subsumptive view" that sees the human as an expression of animality. On the other hand, it is obvious that, if we follow this path, the generative dichotomy that allowed humanists to derive the human through background oscillations—using the animal as a counter-term—is no longer acceptable. We need new epistemological operators and new discriminatory backgrounds.

Let's start from the descriptive dimension: the simplest—though far from obvious—aspect, which refers to the predicates. It is clear that a wolf is different from a bear, or a cat from a dog: they have different predicates. However, it is also clear that there are basic needs—to reproduce, to learn, to find food, to escape potential dangers, to monitor one's environment—which do not concern belonging to a particular species, but the animal condition as such. One can say, then, that being a wolf is nothing more than a "predicative conjugation" —a specific one—of the metapredicative condition of animal-being. Is this condition similar to the mechanism that runs a clock or any other machine? In other words, the question that arises concerns the relationship between an animal, as a systemic entity that acts performatively on the world, and its single functional endowments. If, in fact, there were a total overlap between the functional endowment and the performative result, it is obvious that the difference between an animal—or a human being—and a clock would only exist in terms of complexity. The species or individual conjugation does not solve the problem of subjectivity, since the question concerns precisely an aspect that could be seen as prior to the conjugation itself (hence its metapredicative nature). In other words, there is subjectivity only if there is an overlap (and therefore the expression can be treated analytically and without presupposing a systemic emergent condition), or if we are faced with a surplus that entails subjectivity.

Let us consider, then, the ordinary behaviour of any animal that conjugates its presence in the interface with the world through its knowledge-competence, but does not limit itself to functionally translating the contents of its endowment, as a machine would. An animal must not only correct the course of its action, but must also have the freedom to use its endowments, because reality creates ever-changing conditions, and success depends on the animal's ability to go deep into the world. An animal cannot stop at appearances, but must organize a level of reality—that is, transform the context into a field of agibility—through a dialogical-reconfiguring action. The intersection with reality therefore does not lie in grasping what is present or discovering what is hidden, but it is literally a way to bring out what is possible, just like a sculptor does not simply extract a hidden statue from the marble but realizes one of its possible shapes. The intersection and the cognitive act are therefore nothing but one of the many expressions of the interface: it is not a disclosure or an approximation, but an actualization process referred to the range of virtuality present in a given situation.

The expressive singularity of an animal does not arise *ex-nihilo*, nor is it a mere application of pre-existing contents: the singularity is a creative act just like a sculptor's statue, an act that is realized through the subjective use of one's endowments. If this is the case—i.e. the intersection is the actualization of a form taken from the virtuality of reality—an animal can only do so because "it owns its endowments". In other words, it obviously uses them as "creative tools", and is not "driven" by them. When we stick to a descriptive analysis of the predicates (i.e. this is a bear or a bee), and to an explanatory analysis of predication (i.e. whether it functions by association of environmental triggers or by internal energy), we do not actually address what in fact is an animal, as we assume the validity of the Cartesian paradigm. The appeal to consciousness or to the different levels of awareness is no valid solution. Let me be clear: I do not deny these states in other animals. I only say, taking up Brentano, that these are "levels of intentionality", that is, levels of "being aware that": in a word, levels of "shedding light on something that is in progress". It is not the torch illuminating the room that creates the furniture, so it cannot be consciousness (the torch) that creates subjectivity (the furniture), only because it highlights its characteristics. Subjectivity must therefore *precede* consciousness. Attributing awareness to a machine can only allow it to shed light on its mechanisms, not give rise to the free will and creativity that are the basis of subjectivity—let alone reveal its ownership of its automatisms.

Hence the problem of subjectivity: what alternative do we offer to Descartes's paradigm? As a first step it is certainly indispensable to operate an emancipation exercise and free the animal from the role of counter-term to the human. It must be recognized that our animal-being does not lie in ancestral times or in some partiality (the animal that is in us) but in our full adherence to the animal condition (the animal that we are). This emancipatory process is the only premise to any serious reflection on animal being, starting from the recognition of our total belonging to it: everything we express is our animal-being. We may be as specific as we want, but we are not special. As long as we take refuge in ontological difference, attributable to a thousand conceptual pretexts, we will not feel constrained or solicited to question the algorithmic reduction of animality. And doing so means not only admitting a mentalistic dimension with levels of intentionality—as Donald Griffin's cognitive ethology (Griffin 1992) has done in an exemplary way—but re-examining the ontological principle of animality. It is not enough to admit multi-level awareness or self-awareness, if we then continue to explain behaviour with mechanistic explanatory models that deprive the individual of any form of ownership of their own endowments.

This is the great challenge that confronts us, one that cannot be solved through the appeal to consciousness or intellectual abilities, or in any case to any merely performative quality. In fact, an animal's being-there does not express itself in the adequacy or power of the function, but: (i) in the ability to conceive the function itself; (ii) in the continuous emergence of new functional shifts. But to face this challenge it is necessary to acknowledge the animal's ownership of its functions, its metapredicative character—because it does not depend on the structural and functional predicates—and the fact that we cannot find these in machines (nor, for explanatory completeness, are they part of mechanistic models). In this sense it is a question of overcoming the machinic or algorithmic conception in all the expressions of animality—not to simply recognize some functions of excellence. The background I will use is therefore the mechanism which characterizes the concept of *res extensa*, albeit through a conceptual shift that includes the human being and is based on the non-overlap between functional endowments and the resulting expressions. I will do this through two operations that are profoundly interconnected: (1) moving from a contralateral view of animality to one that is inclusive of the human dimension; (2) using the concept of *res extensa* or machine/algorithm, traditionally considered inclusive of the animal condition, as a contralateral to define animality.

Indeed, it is a matter of "emancipating animality" (Marchesini 2017)[2] from all those prejudices, discriminative views, and biased interpretations that our culture has built to misconstrue what animal-being really means and reaffirm the diversity and superiority of the human being. It is therefore necessary to emancipate it from the captivity of the prejudice that throws animality into shapelessness so as to exalt the human. However, first I would like to make some methodological considerations. I regard the background oscillation—which is used to focus on something so as to make it stand out—an epistemic artifice, which therefore does not have the final word in the discussion, but simply serves to better understand some significant emergent differences that can be seen today. As I said, nothing prevents us from imagining a machine with emergent characteristics such as to assign it some degree of ownership; in fact, the guiding principle of mechanical philosophy, like the mathematization of

[2]Emancipating animality does not mean liberating the animal that is in us but rather the animal we are, having understood the overall character of animality. Humanism has set itself the goal of bringing out the human being and to do this it had to define animality as its opposite: therefore, any liberation must be above all emancipation from prejudice. The emancipation of animality therefore does not only involve the redefinition of a term or the re-recognition of non-humans, but above all the liberation of the animality *of the human being*. I have discussed this concept in Marchesini (2017).

findings, is based on the control of the entity, i.e. on the exhauthoration of the ownership characters. Thus, we are faced with a mechanical dimension that, while always being in technopoietic evolution, never betrays the principle of the exhauthoration of ownership and of certainty about the performative standard. If these presuppositions change, the way in which animality stands out from the background will also change. Furthermore, to focus on the metapredicates, it is essential not to be confused by the performative conjugation analysis nor to follow the projective and assimilative procedure: the point is not to humanize the heterospecific, but to animalize the human.

A reflection on subjectivity must therefore reconnect the various conjugations of animality, not in the name of banal anthropomorphism, but rather overcoming the discontinuity prejudices which profess the special ontological relevance of human beings. The first question to be asked cannot concern the description of the predicates that characterize a given animal, nor can it refer to the explanatory principle of predication—dwelling, in the first case, on ethographic findings and, in the second case, on behavioural explanation. The metapredicative issue refers to a prior question, which one could simplify in these terms: what does it mean to be an animal?[3] Generally, we do not ask ourselves this question, which implies a philosophical reflectionì. We implicitly assume the Cartesian dictate, i.e. that the animal is a "machine", a set of mechanisms that operate by trigger. On the basis of this assumption we usually define our questions: (i) of a descriptive order, which refer to manifest or ethographic predicates; (ii) of an explanatory order, concerning the model called to explain predication, such as behaviourism or the psychohydraulic paradigm. But my question precedes these two, and it does so because it does not simply presume that Descartes' view is valid.

Besides, it is obvious that, if the animal were a machine—that is, a sum of single automatisms that follow one another like domino pieces—subjectivity would be pure appearance. No one today dares to express this view with Descartes' clarity and harshness. But then again, when one speaks of an innate automatism (instinct) expressed with energy consumption—pressure cooker model—on a target predefined by a Gestalt deterministic pattern, what does that amount to? And when one speaks of a learned automatism (conditioning) expressed as a reflection of a stimulus—light switch model—on an equally predefined association between input and output, how is this different from the words of the French philosopher? A machine has functions, but inevitably lacks subjectivity or "ownership of its functions". Therefore, the first prerequisite of subjectivity is not *awareness* of one's functions, but *ownership* of them. We often try to solve the problem of de-subjectivization (implicit in the

[3]The term "metapredicate" indicates the non-specific character of animal-being, i.e. what it means to perceive, communicate, project and react, to resume the above examples. The analysis of the metapredicate therefore indicates how we address the issue of animal being. The question "what does it mean to be an animal?" is the starting point of any philosophy of animality that intends to criticize the anthropomorphization of the animal and the Cartesian reduction of the animal to a "machine". Sentience, desire, subjectivity are metapredicates as they are transversally found in all the conjugations of animal-being. See Marchesini (2016b).

res extensa model as a mathematizable entity) by looking for the famous deus-ex-machina to solve everything. Descartes appealed to the *res cogitans* to safeguard subjectivity in the human being and could not have done otherwise; the *res cogitans* only makes sense as the other side of the *res extensa*: without one, the other also falls.

Personally, I am not too bothered by the idea of being, as an animal, nothing but a machine. I just wonder: is this really the case? Many things - indeed too many things—do not add up. I cannot ignore the arbitrariness of animal behaviour and above all its continuous creativity. Animal behaviour is not comparable to a mechanism, although it has expressive prevalences, orientations towards the world, and a range of ontogenetic possibilities. Reality is a source of never-ending unique challenges: perhaps related by similarity, but never identical. Being creative is therefore not a plus, it is not a luxury, but a basic condition. Ownership is the only condition that allows one to implement the epistemic approach to reality that Konrad Lorenz called a "working hypothesis", i.e. not a passive expression of random attempts, but the active implementation of problem-oriented heuristics which, precisely for this reason, presuppose ownership of one's endowments. In other words, it is not possible to conjugate the paradigm of the automaton-animal with the conception of an animal that approaches a problem in a hypothetical form and that, on the basis of the obtained results, builds a singular perspective on the world: it is either the one or the other. But let's take a step back and examine the very concept of mechanized function, which does not concern the level of complexity of the function itself, i.e. the number and gradient of articulation of causal cascades, but the principle that informs it.

A machine summarizes all its possible functions, and not necessarily the ones it has at a given moment: in fact, it is like a computer or smartphone, where it is always possible to download new applications. This is because: (i) it is informed by an external operator in terms of functional design; (ii) it has a close relationship between structure and function precisely by virtue of the operational reliability that it must allow, ie. its predictability of use. A machine therefore lacks ownership because: (i) it does not design its own functions, especially not in terms of setback emergence or construction of the problem on which it has to operate; (iii) it does not separate the functions from the structure, that is, it does not allow for functional emergence, even if the latter is conceived in an evolutionary way, as in the case of the genetic algorithm. In this sense it must be correlated to a context that does not have any singularity—any unforeseen and non-objective setbacks—any predefined input and output variables, and any performative variability, being based on the principle of possible requests. We could say that a machine is a structure or a model that is based on the functional assumption of potential operational completeness, but precisely for this reason it is incapable of managing the singularity of reality. In other words, a machine, even when implementing context-related or result-related functions, is always self-referential: it does not need explanatory additions.

When we observe the way in which an animal intersects the world, we realize that the surrounding environment does not simply represent a set of objective problems that challenge the individual, called to give adequate resolvent answers to them. For an animal, the world represents a field of action where, even before finding a solution, individuality manifests itself in the singular emergence of challenges that are neither objectively given nor already present before the intersection occurs. A machine is an entity called to address a problem through a function and to perform a function through an endowment. An animal, on the contrary, is: (i) a problematic entity, which conceives problems because it seeks opportunities that are not already given; (ii) an entity that makes use of its endowments in an *exaptive* way, so as to bring out new functional capacities that are not already given. Animal subjectivity, unlike machine performativity, indicates a sort of executive suspension; it develops a singular presence that brings out its functions—through the ideation of challenges and functional exaptation[4]—and does not simply perform them.

The expression "to bring out a function" is loaded with meanings that require explanation and that unfortunately often constitute a reason for explanatory circularity. The risk, in fact, is that of falling into a sort of dualism or tautological appeal to some coordinator, which would lead to infinite regress: this is the case with the famous *homunculus*. But there is no doubt that, for the individual to manage the singularity of the here-and-now, there needs to be a continuous operation of emergentiality with respect to the already-given. Indeed, the concept of functional emergence that characterizes the animal condition implies further connotations, next to the two shown above, which can only be partially found in a machine: (iii) every expression always requires a preliminary "performative adaptation" of the system to be related to the context; (iv) each expression determines a "configurative redefinition" in both structural and functional terms, so that no expression of the system can ever be considered a mere function or output but rather an input, i.e. a process of reorganization of the endowment.

The four principles outlined above—(i) ideation of the challenge, (ii) functional exaptation, (iii) performative adaptation, (iv) configurative reconfiguration—entail the system's ownership of its functions. This assumption raises two forms of explanatory criticism: (1) the revisitation of the model called to explain the functions; (2) the problem of the concept of ownership, avoiding dualistic drifts. As for point 1, it is not the performativity-adequacy of a function that constitutes the litmus test of a mind in action; however, to admit the subject's ownership of its functions it is essential to rethink the explanatory model of the behavioural endowments themselves. It is necessary to switch from an "automated" view (given a structure I directly obtain the corresponding function) to an instrumental one (a given structure defines a range of possible functions, but it is the subject that prescribes the type of use). Obviously, in order for the function to be adaptable to a given circumstance, the structure of the endowment which is responsible for it must implicitly enable a functional plurality: in a word, it cannot be a form of automatism.

[4]The degeneration of biological structures not only amortizes any dysfunctions of the system—for example, our genetic system has three different triplets to encode the same amino acid—but also allows for exaptation processes in the different organs. It enables their cooptation to new adaptive territories, as is the case of the branchial peduncle of crustaceans, which translated into the multiform varieties of organelles in insects, or of the swim bladder which, from a balancing organ, became a structure in charge of gas exchange. Cf. Gould and Vrba (1982).

The ownership argument requires a series of reflections on the expressive model of animal being (point 1). Unlike the Cartesian principle, based on automatisms, these inevitably lead us to the non-overlapping between endowment and expressed function, because the former does not imply the function, but only enables it. In my opinion, an endowment is a "virtual entity" from a performative point of view, just like a map that virtually presents (and not simply subsumes) several itineraries. An endowment, so conceived, is potentially able to carry out a particular range of functions and to evolve on the basis of the results obtained. But above all, just like a map, it has the ability to bring out from reality performative opportunities that were not already given. If endowments, for example the a-prioris of a given species, were automatic, they could not be "working hypotheses", as they would lack the contribution of a systemic entity able to evaluate their respondence, approximation or creative potential. The automatism model not only fails to envision any ownership, but also struggles to evolve with respect to the task that Lorenz's evolutionary epistemology entrusted to it.

On the other hand, can we think of an endowment as an entity that rigidly superimposes the structure on the relative function and is trigger activated? I believe that this is the fundamental question and, at this point, given my belief that the human being is part of the animal dimension, I do not think it is inappropriate to reflect on the way in which we ordinarily use our knowledge. Is it an automatism that guides us or a tool that we use, even unconsciously, to bring out new opportunities? It would be enough to observe a dog's behaviour after learning the "sit" command: from that moment on, it will be the dog who will use the new knowledge to obtain whatever he wants from us. This, moreover, requires us to admit a surplus with respect to the contents, without transcending them—a surplus which, however, must be attributed neither to an external entity nor to consciousness, but to the systemic condition of the mind itself. To speak of animal subjectivity it is therefore necessary to have the courage to question the analytical model, otherwise the animal will continue to be a puppet moved by strings, a device that may be very complex but is still essentially a machine. I therefore propose to consider: (i) both innate and learned endowments not as automatisms but as instruments, thus modifying the basic model of behavioural explication; (ii) subjectivity as an emerging condition with respect to the factors that determine it, so that subjectivity can only be a systemic result.

Let's start with the first point. What is an automatism? It is a mechanism that maintains a 1:1 ratio between "how it looks" (S = structure) and "what it does" (F = function). In the face of this, it follows that: (i) S is exhaustive in the explanation of F; (ii) S is imperative in the production of F; (iii) S and F are mutually predictive, i.e. "if S then F" and "if F then S". If this is the basic model that explains animal behaviour, then the animal is de-subjectivized, even if a linear or recursive S-F chain is used. And, as we have seen, consciousness (torch) does not help to create a subjectivity (furniture) that does not exist. When we talk about "automatism" we inevitably consider the endowment to be imperative for the function. In other words, it is the endowment that drives the animal: therefore, the individual lacks ownership of the endowment. On the contrary, considering an endowment as an "instrument" means that S envisions a "range of Fs", that is, S does not strictly imply F. And it is not just a question of considering the relationship between S and F in terms of the functional multiplicity of S. The point is the virtuality of F with respect to S—just as a statue is the result of atransformation of the marble, not an expression of something already given.

For example, the map of a city (S) provides a range of routes (F), but does not imply or reveal already given paths. Rather, it allows one to find some routes within a given range of possibilities. The S:F ratio is therefore 1:range. A map is a tool that the subject uses to create an opportunity within the virtual range of possible routes. Of course, a map is not free of constraints, that is, it is neither absolute nor a sort of *apeiron* containing all imaginable itineraries. This is because every range has limits, some more probable itineraries, and possible vicariances; but still, every range potentially contains an infinite number of possibilities. This is why, functionally, a map is much more economical than an itinerary. As mentioned, a dog that learns the "sit" command will also use it to try new applications, that is, "new relational routes with his human partner". A map is a tool and as such it does not "activate functions" but "makes functions possible". A map is a "functional representation" of a reality, that is, a utility, like a word processing software that allows for, but does not imply, a given text. To claim that (innate and learned) endowments are tools—and not automatisms—obviously requires a subject that uses them: the workshop needs a craftsman, it does not run on its own.

Seeing behaviour as the result of single triggered automatisms is based on an analytical conception—but also, one might say, on the presumed exhaustiveness of the analytical explanation—which, moreover, negates the systemic structure of the neurobiological apparatus. The latter, in fact, involves many systemic levels: for example, synaptic connection or wireless neuromodulation. The systemic view, based on instruments that can be brought together just like a network, not only grants the different endowments a connective character at more than one compositional level, but is also able to foresee, due to an emergent effect, events that supervene on the very repertoire of endowments. On closer inspection, behaviour never resembles a light switch: the link between stimulus and response, as Konrad Lorenz already pointed out, denies everything that can be found around what is prejudicially taken into account. A behaviour is more like a proposition, or the expression of a composition: more like a painting, to be clear. Its components give rise to ever-changing compositional frameworks, interconnected in various ways. These compositions—which involve the combination of emotions, motivations, representations, logical functions and cognitive meta-components—give rise to "propositional attitudes" that then are manifested (though not necessarily) in behavioural expressions. Subjectivity, therefore, is not given by a *homunculus* nested somewhere in the brain, but is the systemic result of propositionality itself.

The mechanistic paradigm, on the other hand, in order to avoid the so-called ghost in the machine, had to model said endowments (innate or learned) in such a way as to make them autonomous in their functional causality, i.e. exhaustive in the explanation of animal expressiveness. If we analyze the constructs of instinct and conditioning, we realize that they are designed for a sufficient explanation. The lordship of the subject over its functions therefore implies a review of the explanatory model of endowments.

Nevertheless, the theme of the subject's ownership—point 2—also implies a complex and articulated reasoning. It is, in fact, a dark and elusive lordship, which leads us to a crossroads that we might consider aporetic, if it does not force us to choose a path that punctually denies subjectivity itself. This is, in fact, what has happened historically in the analysis of the predicates of ownership, such as, for example, free will. Whatever the choice, in both cases, the requirement of inherent subjectivity is not met: (i) the dualist view cannot but appeal to a transcendent principle like the Cartesian *res cogitans*, if it does not want to fall into the infinite regress of the *homunculus*; (ii) the full identification of subjectivity with its functions, instead, inevitably leads to superimpose the individual on the machine and to transform its expression into an articulated sum of mechanisms. As we can see, it is precisely with regards to subjective embodiment[5] that both explanations fail.

What I have defined as hesitation, uncertainty, approximation, bricolage, flexibility, and functional co-optation in reality involves a plurality of expressions of subjective adaptation, which we cannot exclude from the expression. Itdenies both: (i) dualism, i.e. total detachment from one's functions, and (ii) reductionism, i.e. full identification of the subject with the functions expressed. In this sense, to speak of subjectivity it is necessary to envision a third path that grants the animal being itself, as a prerequisite condition, the ability to manage its here-and-now through a free, flexible, exaptive and, last but not least, creative use of its endowments. On the other hand, it is clear that it is necessary to take this path through a clear explanatory model, avoiding any vagueness and tautology—these flaws become mortal when referring to heterospecifics. Indeed, we can still feel the weight of Morgan's canon, which, like a trap, leads to rejecting any appeal to animal subjectivity, even if in so doing one is forced to adopt explanatory epicycles. As a result, animal expression is deprived of all that does not fall within the reductionist canon.

The problem lies in asking whether it is possible to formalize the animal condition, i.e. whether the complete mechanization of the animal is really able to eliminate the ghost from the machine. In my opinion, the answer is no. The concept of *res extensa* is not able to stand on its own feet and needs the counterpart of the *res cogitans*: so as long as we maintain the Cartesian paradigm we will be forever haunted by some kind of ghost. The ownership principle tells us that an animal owns itself and is not owned, it is the bearer of internal coherence and self-ownership, and we must inevitably face this fact, even if we insist on denying it. A machine belongs to its function, it is a passive entity, while when dealing with an animal it is always necessary to obtain some consensus or involvement to induce it to do something. It is therefore essential to ask the animal for permission, and this is not an ethical concern, but something that regards the factuality of the animal itself. Indeed, an animal is the owner of its own being, becomes the master of the knowledge acquired, and actively uses its inherited a-prioris, making them available to its own free and creative way of being in the world.

[5]Subjective embodiment is stretched between ideative somatization and the emergence of singularity, the latter being understood as a unique identity that is unrepeatable in belonging.

5.3 The Desiring Principle

Moreover, animal subjectivity is not limited to this function of supervision over its endowments, i.e. what I have defined as the "ownership principle". Otherwise, one would not understand the reason or need for such supervenience. In fact, subjectivity manifests itself as an act of volition on the world, as the subject's change of the coordinates of reality on the basis of an imposition that starts from the individual. Subjectivity is primarily the expression of one's own projection, a sign of an intrinsic teleology—not reducible to teleonomy[6]—that has its internal engine in "contentless desire".[7] We could define this object-less desire as a sort of will to power, but we would thus assign it a semantic value that barely grasps its scope. Any attempt to focus on the oscillation of the desiring animal—whether as a primordial lack or, on the contrary, as an exuberant-energy that asks to be consumed—has, in my opinion, captured only a part of this a-finalistic desire, of this purposeless languor. But, before addressing this concept, which is apparently contradictory or, if you like, empties desire of its very meaning—as an act of projection towards something or rather as a sense of lack inevitably referred to something—I would like to describe this involvement which makes animals languid, passionate, involved, restless, and nostalgic.

To be desiring means not to contain oneself, it means to have copulae or binders that bring back and conjugate the individual's here-and-now to external reality, giving rise to propositional attitudes. This makes animality a "peripatetic dimension". In this sense, animality is: (i) a state of constant restlessness, an inner movement characterized by the oscillation between languor and gratification, and a constant attraction towards the world, whence wonder as the first emotional and cognitive condition of animal being. At the same time, animality is (ii) an exploration in dialogue with external reality, able to bring out a personal perspective through a process of actualisation of reality, and also to transform the a-finalistic desiring condition into real

[6]For an overview of the teleology of nature after Darwin and in particular on the concept of "teleonomy", I refer to the important work of the philosopher Francesca Michelini. For years, Michelini has explored the territory of the philosophy of life proposing a bio-philosophical reflection that deals with the crucial question of the nature of the living but also of "our nature" as human beings. Cf. Michelini (2011) and Michelini and Davies (2013).

[7]Animality is an open dimension as it is characterized by the metapredicate of desire, by going beyond itself, by inventing new predications, by placing itself in a state of non-equilibrium, and by transcending causal proximity. This condition underlies and comes from the evolutionary process, where there is no place for completeness—let alone perfection, a legacy of creationism. The animal multiverse is made of pleasure that comes from languor, of anxieties that arise from a projective redundancy: the animal knows no boundaries, it always pierces the bubble of its heritage. If, on the other hand, we think that the human being is lacking compared to the completeness of other species, we are simply talking nonsense. Indeed, being in a state of lack is the cornerstone of animal predication, and the animal moves and breathes precisely because it is constitutionally desiring. To be an animal means to realize one's life in desire and one's predication in referring in a given way to something external. It follows that the full realization of the specific animal is also its peculiar expression.

desires. We are born with this remote inexplicable hunger, which is capable of satisfying the most concrete physiological needs, such as eating and reproducing, but at the same time is unrelated to them. This is the case with a kitten that, despite a full bowl, still needs to chase after any moving object: it is taken by amazement and involvement in front of a rolling ball or a thread, as if eating were nothing more than a consequence, a side effect, of something deeper, of a passion that makes its whole being vibrate.

Being-desiring precedes the encounter with the world: it is what makes external reality full of charm, attractive and mysterious in its multiple layers, which are there for us to explore with amazement. Desire is what makes the outside world a field of action: without it, the individual would have no reason to act. Desire is also the cornerstone supporting the protagonism of animal-being. Being-desiring is conjugated through a fractal of phylogenetic and ontogenetic legacies. This does not happen in the form of teleonomies predisposed to a sequence of triggers towards pre-established objectives, but as internal redundancy that asks to be unfolded, like a leaf that opens up to the world and only then acquires a given positioning with respect to the light. An animal's being-desiring cannot be explained by the mere reference to the stimulatory here-and-now, and its principle can only find explanation in the physiological conditions of the body. Being-desiring is not a deprivation of protagonism, a stunned condition that hinders one's contemplation of the world: on the contrary, it is what feeds the *thaumàzein*, what provokes that amazement and thrill that underlie the cognitive process itself. Immersion in desire strengthens the presence of the individual, its *Dasein*—it does not diminish it. Being-desiring is therefore passion, involvement, languor, sense of insufficiency, which nevertheless feeds our will, our exposure, our action, our perspective partiality. When we fall in love, for example, what we feel is not a deterministic force that deprives us of protagonism, but a condition that increases this protagonism, making our presence stronger in the here-and-now.

Desire is the most immediate expression of subjectivity, a redundancy that creates languor, a projection that goes beyond need and in a sense gives it meaning: I am alive because I desire, I am alive when I feel involved, I am alive because I am projected into the outside world and not self-contained/referred, I am alive because I am taken by amazement, because I am sensibly attracted to and awestruck by the world, because passion invades me and gives shape to my presence. All the endowments of an animal being are only tools for the unknown and obscure expression of a desiring being. It is desire that gives colour to the world, that fills a child's eyes with wonder, that underlies the chaotic games of a puppy, that gives meaning to the events of the world. If this condition disappears, life goes to waste in a colourless vegetative timelessness. This is why we need a new philosophical ethology, one that reconsiders the hasty formula of teleonomy with which the animal has been deprived of a leading role in the intersection with the world, precisely by denying its desiring condition in favour of an algorithmic cascade of drives.

For example, we are used to thinking of motivations as innate automatisms, But what if they were more like the feeling of falling in love? What if they were passions? Far from being already predisposed towards a given result, they recall copulative verbs, transitional coordinates—like chasing or gathering—that go into the world to connect with specification contents. And how important is reasoning in such conditions? Even when we fall in love, as we well know, consciousness takes note and reasoning seeks justification, but it is our animality that palpitates in us: we generally call it the voice of the heart, but it is the animal being that vibrates throughout our bodies. A playful puppy, a female who cares for her young, a predatory carnivore, two males that rival each other: there are many examples of this passion that does not annihilate subjectivity, but makes it emerge. We think because we desire, not the other way around: and it is in the desiring principle that we find in every animal. Therefore, research on the metapredicate is also a reflection on this condition, based on the dialectic of three principles: projection, search and pleasure. Indeed, animality is: (i) *projection*, i.e. continuous oscillation between languor, which arises from projecting oneself beyond one's state, and satiety-satisfaction in reaching the goal; (ii) *search*, i.e. a tendency to look for opportunities, for which the animal being embodies the peripatetic condition, because it does not passively find opportunities in the environment, but uses an internal movement to actively build them; (iii) *pleasure*, i.e. exposure to the dialectic of enjoyment and suffering, a continuous state of non-equilibrium, based on the condition of being-a-body.

The three principles of desire presented above view animality not as a mechanical expression of automatisms but as "involvement", as a being-a-body that attends its here-and-now. The animal "is present", it has its own here-and-now. While a machine is not positioned in time—its function is isochronous—an animal is the protagonist of its present: it rises above the flow of time, presenting its expression. Translating this into two concepts, we will see that: (1) an animal does not occupy a space but takes a positionality; (2) an animal is not in time but builds its own time. Borrowing a Heideggerian concept, one could say that the animal has a *Dasein*. And, perhaps, the metapredicate that Heidegger sought in vain—the Being—is precisely animality. It is in this protagonism, in this desiring condition about the world, that an animal expresses its subjectivity. *A fortiori*, therefore, it is necessary to revise the explanatory paradigm based on projection. When I say that a cat has a predatory motivation, I draw attention to a trend-character that can be explained, taking up Niko Tinbergen,[8] based on four explanatory lines: (i) what advantages it has produced in terms of fitness; (ii) how it has evolved during phylogeny; (iii) how it has developed during ontogeny; (iv) what physiological or environmental causes have provoked it. These are four important explanations, without any doubt. However, we must also analyse how that trend-character integrates with the principle of involvement and pleasure, i.e., with the "feeling" of that particular individual.

[8]It is important to underline that first Ernst Mayr and then Niko Tinbergen had already made a significant suggestion in explanatory or causal attribution terms: a motivation is present in a subject of a particular species because it was able to increase its fitness or to fulfill a need (remote or phylogenetic cause); a motivation is manifested by the individual because it is driven by an elicitative need and because it is capable of producing pleasure through expression alone (proximate or ontogenetic cause). We must never confuse these two causal matrices, since while the former is regulated by the principle of natural selection, the latter is commanded by the pleasure principle. See Mayr (1982).

If one stuck to these four explanations when speaking of human falling in love, one would not be strictly wrong, but rather not exhaustive: something would be missing, because in the explanatory chain the act of falling in love would have lost all its feeling, its involvement, its languor, its pleasure, assuming a mechanical connotation. The emotional and motivational states of an animal determine a feeling, i.e. an involvement, which is more like falling in love than the ticking of a clock. It is on this point that we should focus our attention. A character is present because it is the result of phylogeny and because it has passed the fitness test, that is, because it has received the go-ahead from remote causes that have established its compatibility. In the same way, a character takes a given conformation during development and is affected by intervening physiological and environmental activations. This is also true. But the feeling it produces arises on a different explanatory level, albeit based on the underlying ones: it is a fifth supervening principle, which does not deny or take away importance from the coordinates of the previous explanations, but should be considered nonetheless.

As Konrad Lorenz said: life is made of physical and chemical laws, but it is not just physics and chemistry! When analyzing complex systems we find that there are various predicative levels: water is made of oxygen and hydrogen, but presents different predicates compared to its ingredients, and the laws that govern the oceans cannot be explained simply based on the qualities of water. Well, the same goes for involvement! The variables, acting on the predicates present, create a "systemic involvement level" that emerges from the background of its ingredients. A cat, in its predatory dimension, expresses its subjectivity: its being-desiring. This condition places it in a dialectic of languor-orientation and expression-contentment that has got to do with pleasure and ownership. A cat does not hunt to eat, but to express its being-in-the-world, the pleasure it gets from being there. Eating is just the consequence of all this, which in the long run feeds back on its fitness.

It is not always easy to understand this undefined condition of a desire, which paradoxically does not seek a given content, but creates it at the very moment in which it languidly brings itself into the world. Traditionally, we have an ambiguous relationship with the desiring principle, most likely because of the contralateral translation of animality that has transformed the body, and its copulative capacity, into something to be rejected or from which to take leave in order to achieve true humanity. On the one hand, the languor of the body is considered something unseemly and lascivious, often seen as the fundamental principle of our captivity in the telluric which, on the contrary, we should overcome in order to follow the righteous process of ascending emancipation. On the other hand, the desiring coordinate is focused on possession, giving rise to a sort of existential bulimia centred on the consumer individual, so that desire increasingly overlaps with its content, flattening onto the desired entity. Apparently, these two directions seem to oppose and contradict each other, but in reality they complement each other well in the attempt to annihilate the nature of the human being, i.e., to deny the human copulative inherence, which is not based on the desired entity but on the implicit restlessness of the animal condition. By gravitating towards the thing—desiring an object or the object of desire—we lose track of the very condition of desiring: that is, the projection and the movement, the verb-action that sustains the subject's languor, which can only be satisfied by expression and which is implicitly somatized in the act.

De-somatizing desire therefore means denying the body its projection, thus transforming desire into need, i.e. into a void to be filled. It also means rationalizing desire or turning it into ecstatic aspiration—something that does not provide for a somatic sharing in its totality, since the body becomes the instrument that allows the subject to reach the object of desire. There is no longer the pleasure of collecting, i.e. the motivational involvement that is the purpose of the very act of "behavioural expression", as one could say in ethological terms. Instead, there is the projection of a disembodied mind onto the object of collection, which becomes the only explanatory factor, i.e. the only reason justifying the action. In the same way, the behaviours of help, assistance, nurture and care are no longer the expression of a body that expresses its ephemeral languor, but the disembodied moral act of one who rationally acts for the good of the other. Anthropocentrism, through the double action of compression of the somatic expression of desire and of de-somatization of world-oriented proactivity, traces a difference between the human being who desires because of rational contemplation and the animal who is totally absorbed in its needs.

This is an important disjunction, not only for its marginalization of animal otherness, which cannot sit at the high table of the convivial dialogue with the human being, but also for the very way in which human subjects perceive their own somatic dimension and reflect on their own animality. On the one hand there are human beings, rationally concentrated on the object and therefore separated from it, transforming it into an extrinsic explanatory principle that denies the inherent need for a copulative order. On the other hand there are animals, totally stunned in the object and therefore devoid of any desiring protagonism. Transforming desire into a non-copulative act, that is, into contemplation or fruition, means de-somatizing the desiring condition in the human being and denying its cognitive aspect in the animal. If the body is simply the place of need, it is evidently a prison, so that the disembodied mind devotes itself to projection and wandering as well as to the pleasures of dreaming, planning, realizing, participating. All verbs—i.e. acts—are de-somatized; the body and animality are considered, at most, to be hostile to, and not the driving force of, desire. Embracing this view also means misinterpreting the somato-expressive meaning of desire, i.e. the body's pleasure to unfold along the supporting axis of an action that moves its internal flows.

The object of desire is just an excuse, like a ball for a cat: the ball is neither the real target of desire nor the real source of pleasure, as demonstrated by the fact that, if the ball stops moving, the cat hits it so as to set it in motion. The cat's desire lies in projecting itself along the axis of chasing-grabbing, and its pleasure comes primarily from the act of stretching its body along that verbal and copulative coordinate. Desire is therefore a somatic state of projection-expression, which produces languor in projection and pleasure in expression. The body has a very precise menu of copulative modalities: these are assessed by the mechanism of phylogenetic compatibility, but are not bound to the fitness aprioris in the expression, so that the subject finds these axes of projection-expression to be the motors of their pleasure. For this reason, a verb—such as collecting or chasing—that is selected for a specific adaptive purpose can derive from play or from any other expressive occasion. Pleasure lies in expression, in copulative somatization, and only secondarily does it concern the object itself. A mushroom finder is more satisfied with the activity of searching and picking than with a mushroom-filled basket, just as a boy enjoys picking and collecting figurines more than owning a completed album. And that's why a full bowl doesn't extinguish a cat's predatory desire.

Tying the languor to the object of desire, i.e., reifying, rationalising and de-somatizing the desiring condition, also determines a stunned state, because it leads us to choose the shortest way to reach the given target, whereas we find greater satisfaction if the target is hard to obtain. This entails a series of misunderstandings and two very serious consequences (i) It increases the need for objects, which become substitutes for actions in that they try to appease, though in vain, a languor that can only be satisfied by expression. (ii) It prevents us from understanding the vastness and multiplicity of the motivational triggers of the human being, in the false belief that satisfaction can come from an external entity or be achieved through appropriation-consumption. The reification of desire misleads us because it projects our pleasure onto the object and not on the expression of the body, de-somatizing it and thus denying it, maintaining us in permanent dissatisfaction and distorting our very expressiveness, which ends up caged within a sort of straightjacket. This does not mean that results achieved by action have no value and give no gratification, but that this pleasure does not satisfy (appease) the languor, because the latter lies within the expressive process or somatic expansion in the world, like the acts of chasing, collecting, competing, imitating, caring, feeding, protecting, collecting, exploring, patrolling, collaborating, etc..

If the achievement of the objective is placed within an expressive exploration, then we will have three kinds of pleasure: (1) satisfaction, which can be traced back to expressive satisfaction, i.e. satiety in the somatization of the action; (2) gratification, which can be traced back to the enjoyment of the object-target; (3) self-efficacy, which can be traced back to the assertion of the expansive will-capacity. On the other hand, I would like to note that type 3 pleasure is much more connected to type 1 pleasure than to the mere fruition or attainment of the object, i.e. type 2. This is demonstrated by the fact that the more arduous and tiring the process of reaching the target, the greater the response in terms of self-efficacy. We can therefore say that reducing desire to the desired object gives rise to a bulimia of discontent, with an excess of objects trying to compensate for a lack of action. And the problem also lies in understanding the somatic meaning of action as the "result of action". In other words, we believe that the propulsive thrust of our unfolding in an action is to be found in an a-priori need of the body that activates the movement, or in an external stimulus that drives it, according to a homeostatic conception of inherent quietness of the body, where every movement would indicate a condition of disturbance.

In reality, the body is an expansive entity: it is not placed in space–time but takes possession of space–time through its implicit proactivity. The body is action for its own sake, i.e., it's the need to express one's own motivational potentialities, regardless of the needs or targets that the action presides over. A cat wants to engage in predatory behaviour—chasing and grasping—but not necessarily because it is hungry (need) or in view of an object to be achieved (target). A cat's predatory behaviour is therefore a somatic expansion or a conjugation of the body's expressiveness on the basis of its protagonism. No doubt, desire can be influenced by need and may lead to achieving a target, and it has undoubtedly evolved because it brings fitness, that is, adaptive results. However, being inherent in the somatic dimension of the subject, it acts for itself, as a distension of the body in the world, because the body is yearning. Anthropocentrism, on the contrary, is based on the denial of this somatic expansion in the name of a de-somatization of the existential condition of the human being, based on the absolute value of the results achieved and of the objects owned. Conversely, the pleasure one assigns to results and things lies precisely in the expansive effort, i.e. in the kinetic distention of the body. It is not the pedomorphic conformation of a puppy that gives us satisfaction in expressing attitudes of care and protection towards it: it is the epimeletic potentiality, the somatic distension towards care, that arouses tenderness in us and gratifies us in caring for it.

We could consider taking care of a puppy as a value in itself, but that would be misleading: it has value because our entire endocrine, neurobiological, metabolic, immune body is inclined to spread somatically in the act of care. This means that, if we do not express its epimeletic yearning, the body will look for ways to express it in the world: not because the world needs our care, but because our body is overflowing with it—it expands in this sense. Also in this case, assigning absolute value to care means trying to outsource it from the body, de-somatizing the care-giving behaviour. The reason for this attitude is easily explained. Ontological anthropocentrism is based on the rejection of contiguity and somaticity: it does not accept the condition of animality *qua* being a body and therefore refuses to hear the reasons of the body: it refuses to remain within a full existential somatization. It therefore operates in the two directions set out above. First, (i) it punishes the body in the name of something spiritual, de-somatized, abstinent, so that desire is considered not as an ontological expression but, on the contrary, as a prison for the self, a misleading magnet, a receptacle of irrational and blinding passions. Secondly, (ii) it translates desire into a result and an object-target, de-somatizing it and subtracting it from the expression itself, from its verbality, from its expressive specificity, in order to project it onto something external that can be subject to rational control. Once again, this process shows an evident wish to make a clear distinction between animals, totally immersed in their somatic dimension and in need of expressive consumption, and human beings, driven, vice versa, by the rational choice of their objectives.

When I say that being-desiring is not based on the contents preceding the projective act, but *builds* the contents precisely in the epistemic-copulative process, I do not mean that there are no orientative coordinates—referred to the what, how, where, and when of the expression. Rather, I mean that the orientation defines the broadest possible predicates, that is, a range of possible specific contents. In other words, a cat's running refers to "a moving entity", and the human act of gathering is about "an entity that emerges from the background for its iconic structure". In the same meadow, a cat is interested in butterflies, while a human being is interested in daisies. Often, the orientative model also presents general Gestalts, such as a string that recalls the tail of a mouse for a cat or a ladybug that resembles a berry for a human being, and we can also find references to places or to expressive choreograms. However, as said, the orientative model does not define a specific content, but only a range of possibilities, so that an animal, when expressing its copulative tendencies, is in fact giving specific contents to the desire itself. This is a perspectival subjectivism at the intersection with reality, that is, an a-priori protagonism, already foreseen in the initial conditions of the animal condition-expression.

The events of desire therefore play the score of heritage as the basis for a representation whose task is to go beyond the sheet music, because the relational subject by definition cannot remain faithful to itself. So it's not a question of denying the heritage, but not even of remaining faithful to it, as if the predicates expressed were nothing more than the emanations of an essence. The heritage is a virtual condition of being, which however does not mean (i) an absence of form or total willingness to assume the form imposed by the external world, as if it were water and the world were a container. However, the heritage doesn't entail either (ii) a power or complete autonomy in the construction of the shape to be taken on each occasion. The virtual condition contained in the heritage arranges the forms of the copulative process and the field of possible outcomes in the definition of contents. Within it, however, there are infinite possible conjugations. The virtual condition therefore indicates a willingness to dialogue that is supported by the desiring condition of the subject, so that each outcome always results from the integration of multiple copulative Is that are unpredictable because they are the product of the creative and singular protagonism of the individual.

Desire, therefore, is the glue, the tension towards dialogue with the world, the push that makes such an encounter possible. To use the metaphor of the foliage of an oak tree, one can say that the copulative process is what allows the oak to translate a given part of the world into itself. The virtual condition, on the other hand, is an expression of excess, of a projective process, and not of lack. Thus, desire is first and foremost a tension that leads to the somatization of external reality; in this sense, it is to be viewed as metapredicative yearning, a sort of generic libido that projects itself in a centrifugal way from the subject to external reality, and as a predicative conjugation that specifies different animals, including humans, through the peculiarities of verbs-copulae. Thus, while cats are prevalently connected to the world through the copula of chasing, human beings desire through the copula of collecting. The object chased or collected is of relative importance and offers a very partial gratification, because the true yearning lies in the expression of chasing or collecting as a somatic projection of the subject in the world. I speak of somatic projection because it is in the body that the copulative matrix is realized, even before the action is weighed, thought or decided. Desire, therefore, is the expression of our animality and also what transforms animality, in the all-inclusive sense, into protagonism. At the centre of desire, therefore, lies bodiliness as a whole, and desire is a process of somatization—not of emancipation from the body.

Animality, therefore, is an open, flexible, participatory, creative dimension whose leitmotif is singularity, because it is characterized by the metapredicate of desire, by going beyond, by inventing new predications, by placing oneself in a state of unbalance, by transcending one's heritage.[9] This condition is the basis and the outcome of the evolutionary process: in it, there is no place for completeness—let alone perfection, a residue of creationism. The animal plurivers is made of pleasure born of languor, of anxieties that arise from a projective redundancy: the animal knows no boundaries, it always pierces its *Umwelt* bubble. Animality is a metapredicative dimension that does not differentiate us from other animals, but brings us together with them, despite the different predicates that characterize us and connote every species in a peculiar way. Animality, therefore, does not oppose our "becoming human" through ontopoiesis, but rather makes it possible through predications, i.e. verbs-copulae, that set the coordinates of our orientation. It is therefore precisely in the phylogenetic heritage that we must look for those coordinates of orientation that transform the desiring condition into acts of desire specified in their what, how, where, and when. Once again, it is important to underline the direct proportionality between phylogenetic heritage and ontogeny.

Placing oneself in a condition of languor—and therefore embracing a sensation of a lack, which is indefinite (devoid of specific content) and actually results from redundancy rather than deficiency—is the backbone of all animals, which live and breathe precisely because they are constitutionally desiring. Desire as projection-expansion of the body is one of the most important principles of the metapredicate of animality. To be an animal means to realize one's experience in desire and to realize one's predication by referring in a given way to something external. It follows that the full realization of the specificity of each animal is also the peculiar—or "somewhat" copulative—expression of desire through that particular animal. We could say that the heritage is nothing more than the manifestation, in that particular animal, of the "vicissitudes of desire" that preceded it; on the other hand, the heritage is only the starting point of desire, which necessarily leads the animal to take unknown paths of predication. In belonging, that is, in adhering to the heritage, an animal realizes its own singularity: it uses it to open up new existential paths. If we analyze the singular positioning we discover that the here-and-now of the individual is nothing but the result of a flow of somatizations that took place in different times and places through copulative processes. What we consider a static adherence to a well-established form once represented a singularity, a leap into the dark, an opportunity seized. However, that singularity was reflected in the world, bringing out an existential possibility through the tension of its own life, by making room for itself in reality. We are still blinded by fixism and unable to understand that life flows in space and time precisely thanks to its openness to the singularity that uses heritage to transcend it, and that builds existential forms-conditions through desire.

The body desires because it is always projected onto another condition. By virtue of desire, it does not only occupy space, but it makes room for itself in the world, and

[9]For a deeper understanding of the link between subjectivity and desire, see Marchesini (2016, p. 63). Through bodiliness—the body being understood as an experiential field—subjectivity is freed from the chains of the narcissistic self and stretches out towards the other. The driving force behind this process is desire, conceived as the "urgency of participating" and the "hospitality towards the other".

in doing so it inevitably gives rise to the creative process of singularity. Each heritage, therefore, is the result of a daring desire, which in interacting with the world—in predicating through relationship, since desire is a copulative act—has given rise to a mirroring; we call it adaptation, for its performative value, but in reality it represents the elastic mat for an animal's leap into the dark. Rank, understood as the occupation of a presence dimension, encompasses the habitus and the environment of life, defining a correlation that is also a somatic mirroring: gills speak of oxygen dissolved in water, while lungs speak of something else. Mirroring is the way in which desire has transiently informed the animal form. One can say that heritage reminds animality of its flow, its undulatory nature, its emergence from other forms, its preceding the corpuscular birth of the individual. Animality, therefore, is the continuous expression of a copulative process of somatization of one's desire, of feeling projected into the world, of perceiving an implicit lack.

The sense of deficiency, however, does not arise from a condition of poverty but from the animal's projective impulse to expand into the world, interpreting it freely just as an actor does with a script. Animal metapredication is therefore a creation of worlds that is expressed in predicative singularities. We feel lacking because we want, not the other way around; we feel deficient because we are captivated, as animals, by the unstoppable impulse to flow into the world. Heritage, as a prelude to desire, is a boundary in that it defines a given copulative mode; however, there is no doubt that being an animal is the act of going beyond it, of crossing every possible barrier, condemning oneself to imbalance, to the expulsion from Eden, to the suffering implied in the impossibility of returning to the initial condition, to placing one's desire in new dialectical conditions between body and world. For this reason, I consider animality a metapredicate impossible to cage within a predicative definition. For the same reason I define the animal condition as a perception of implicit deficiency, which has nothing to do with the endowments one has but concerns one's projection into the world, experimenting with new situations.

Desire dictates the conditions of suffering and pleasure, which are ultimately two sides of the same coin: when one grows, the other increases as well. This perception of deficiency is not necessarily a representation, nor does it require a process of action, explanation or declarative emergence. The perception of deficiency is a languor that takes place in different layers on every level of the body's being. This dimension concerns and at the same time transcends the individual, because it affects the entire chain of prospective genealogy, including phylogenetic genealogy. Animals are not only protagonists in defining their biographical structure, but also in determining the phylogenetic course of their species. The animal that interprets its script and invents new things is not a simple "receptacle of genes", as the reductionist canon would put it. Today, with the new evo-devo theories, we are increasingly aware of how the choices made by individuals modify the phylogenetic trajectory of the species, though not as described by the Lamarckian simplification. Much remains to be discovered in terms of epigenesis and niche construction, but the point is that even the term "adaptation" seems outdated. Life is not a simple adaptation to regimes imposed from the outside. Life works like an artist: it does not only reproduce what is already there but it is always creating something new. Life is never stuck within a border but always trenspasses it, goes beyond it, reinvents itself.

5.4 The Sentience Principle

A fundamental aspect, in order to fully understand subjectivity as the participating presence of the individual in the moment, is undoubtedly the sentience principle, which all too often is traced back to the ability to take charge—or to be aware—of the findings concerning pain and sensory afferents. From my point of view, sentience is not only superimposable to states of consciousness, but concerns the entire "feeling" of an animal—a conjugation that cannot be ascribed to a single sphere and that should be traced back to its "being-a-body". We could say, then, that the body is a multilayer of sentience, as it is interrelated to what it encounters in a plural way. The principle that I am going to examine here concerns the multilevel dimension of conjugation that the body establishes in its relationship with the world. Sentience is therefore the vocation or the participatory foundation of the animal condition, its being in a relationship. It follows that to admit animal sentience means to consider the body not as a delimited entity, enclosed in a skin boundary or in a biographical space–time, but as a threshold. Better still, the body is a set of reception thresholds, capable not only of acquiring information from the world but also of attributing value to events and of diachronically connecting the subject's present experience to the past and the future. Feeling, therefore, is not exposure, be it in the sense of Bentham's potentiality[10] or Derrida's vulnerability.[11] It is a protagonism that is realized in the construction of perspectives that the animal carries out in its intersection with external reality.

The character of being-sentient is expressed through multiple predications of animality, among which nociception, in its conscious expression of pain, is particularly important in the consideration of inherent interests that can be attributed to the condition of animality. In fact, an animal cannot be viewed as an inert object or a machine, because it is able to feel pain and to experience feelings in general. This is the reason why, for example, utilitarianism considers animals moral patients, but it is also the reason that leads to a different consideration of animals compared to objects, recognizing that the former have "interests that concern them". All animal welfare proposals, starting from the Brambell commission,[12] have followed the precept of sentience, to the point that the concept of subjectivity often tends to overlap with it. As we have seen, there are at least two other concepts—the ownership principle and the desiring principle—that are indispensable to understand subjectivity. At the same time, restricting the field of sentience to physical suffering prevents us from fully understanding the sensibility of an animal being, which should be traced back to conjugation rather than to exposure.

---

[10] Jeremy Bentham famously noted: "The question is not, can [animals] reason?, nor can they talk? but, can they suffer?".

[11] Derrida (2008). Here Derrida highlights that there is a level of existence that unites us and other animals: the non-equilibrium and fragility-vulnerability of life.

[12] Written in 1965 by the Farm Animal Welfare Council (Fawc), the Brambell Report listed the "five freedoms" that would ensure the "welfare" of farm animals, namely the freedom: (1) from hunger or thirst, (2) from discomfort, (3) from pain, injury or disease, (4) to express (most) normal behaviour, (5) from fear and distress.

Being-sentient has got to do with the continuous encounter between animality and the world, as a process that not only takes into account what is external, but continuously translates what is external into internal content. In this sense, sentience can be associated with a participatory motion of involvement or concertation. Sentience undoubtedly means having access to the world through sensory windows, capable of catapulting one's interiority into the world and of stretching it along a particular field of sensibility. This gives rise to the phenomenological participation of the individual, who feels their presence as an interrelation along a coordinate of protention. On the other hand, what the individual encounters in this process of extroversion gives form to the perspective partiality that characterizes subjectivity. There is therefore an important ontological link between perception and the metapredicate of being-there since, as pointed out by Maurice Merleau-Ponty,[13] perception is openness to the *Lebenswelt*, an act which is implicitly endowed with intentionality. The perceptual perspective can be clearly considered a principle "through" which the individual takes control over the world. However, it is also true that it represents the "ontopoietic engine" that allows it to build an inner world and to make space for itself in the outside world. Perception is therefore primarily a dynamic process of involvement, which aims to overcome the internal–external dichotomy through a distension of the individual in its context and an introjection of the latter into the individual.

Every cell of an organism perceives its field of action, modifying it and modifying itself to seek assonance. Feeling therefore means actively seeking the correct intonation, lowering the dystonic gradients. One perceives the world in view of a correlation, or rather, of a related action that creates the predications not within the individual but on the "connection structures", to quote Gregory Bateson.[14] Here the predicate can never be referred to some essential content but rather to a relational emergence. The cell has thresholds of access to the world, able to watch over the smallest gradient variations (that is, to be sensitive to fluctuations), as well as to amortize the oscillations, (that is, to be sensible towards compensations). But this double-track sensibility, this focusing on some gradients while overshadowing others, is already a form of world

[13] I believe that the question of subjectivity cannot be deciphered through specific predicates. The latter, on the contrary, testify to the particular conjugation of intentionality of the animal body, to take up Merleau-Ponty's good intuition (Merleau-Ponty 2003). Following this line, one defines subjectivity as a metapredicative condition which underlies the predicates and supports or realizes them but does not correspond to them: it is a common and shared dimensionality in animal being, beyond the particular and circumstantial onto/phylogeny states that a particular individual happens to experience in the here-and-now. A careful view of the relationship between nature and life in Merleau-Ponty can be found in Scotti (2015).

[14] Bateson (1980). Here Bateson, with the expression "connecting patterns", has well highlighted how it is not possible to understand the unstable magma of bios by enucleating the living individual from the network of conjugations that express their vital breath. For Bateson, the living world is filled with a series of links that connect «the crab to the lobster and the orchid to the primrose and all the four of them to me?», p. 8. These relations are not fixed, internal and constitutive of the subjects involved, but organize the whole biological world according to connections that fluctuate in time (responding to a "stochastic" process), which Bateson sees as «a dance of interacting parts and only secondarily pegged down by various sortS of physical limits and by those limits which organisms characteristically impose», p. 13.

construction. Being-sentient therefore indicates a protagonism in the emergence of a level of reality that is not already given. And sentience is not a sensibility localized on the plasmalemma of a cell or on the processes of intracellular transcriptase, on the sensory organs or on the endocrine or immune mechanisms; it is not found in the synaptic network, let alone in the metabolic neuromodulation. One is sentient because this is the condition of being a body.

Sentience speaks to us, among other things, of the affective-emotional nature of the animal being, of its continuous attribution of non-representational value to events. Emotions, in this regard, not only mark events, assigning links of feeling between states of the body and circumstances, that is, transforming the world into a collection of conjugations of experience. They also connect the different phases of the genetic chain of the individual. In other words, emotions construct a biography through a network of locks marked by value seals. The individual, in fact, cannot be considered a mere temporal sequence of events that have followed one another, sedimented by stratifications, but must be seen as a network of affective involvements present or mobilized in the here-and-now. In other words, emotions transform memories into present entities that are felt and participated in, just like experience. In so doing, emotions give diachronic consistency to the individual, making it a biographic entity, not a stratification. Emotions thus provide the individual with subjectivity in that they are not fixed in the here-and-now but emerge from it: they are capable of experiencing a here-and-now in that they are not entirely included in it. The distance that Martin Heidegger identifies in human contemplation, as opposed to the animal's stunned state, should be seen in the emotional involvement felt by the subject in the intersection with the world. In fact, it is strongly connected along its genealogical-experiential chain through the emotional glue that makes both experience and heritage present.

We have seen how being-desiring leads to copulative processes that give individual content specifications to the projective act, defining expressive propositionality in a singular way. We have also said that experience, through the process of construction-solution of setbacks, is a fundamentally creative event in the redefinition of knowledge endowments. These two aspects are obviously affected by important consequences in the prospective emergence of animal subjectivity. At this point, I wish to underline that the intersection between the individual and the world is never passive—as photographic film is passive to the incidence of light—but implies an attribution of value, on the basis of an emotional marking that involves the multilayer of feeling. Also in this case, we can notice how the metapredicate of subjectivity concerns partiality, arbitrariness (in a way), and the protagonism of feeling. Subjectivity is the result of the animal's appearance on the proscenium of the world, personally assigning connotations of value to it. In other words, an animal feels the world through its own emotional state, in a process that can recall a concertive act, bringing the body on a particular frequency of feeling. But this feeling is never an objective translation of the contents of the world, nor is it a representation or the attribution of a meaning to an event. Rather, it is the definition of an in-itself on the basis of a subjective dialogical exchange.

The body stretches out into the world so as to participate in it. It is a body that vibrates like a string instrument at the touch of the world. It transforms this touch into a singular sound that brings back to the present not only the ontogenetic experience of the individual, but also the heritage of others that preceded it. In this sense, feeling transforms individual life into a sort of resonance that makes remote things current, as a kind of time machine that fills the fossil heritage of ancestors with affective living flesh. It is a body that stretches out to feel the warmth of the sun, taste the sweet taste of a fruit, the turgidity of its own sexuality, the epidermal thrill of touching the other, the exuberance that runs through its muscles, the proprioceptive tension in its joints and tendons, its heartbeat, the transductions of chemical and electro-magnetic currents that fill the surrounding space. It is a body that feels and transforms the moment into an experience. But the body also transforms feeling into a resonance capable of colouring a moment to come, turning feeling into a flow that brings back and anticipates, and shaping the world on the basis of a sentient perspective.

To be a body means to be referred or refer to something external, i.e. to somatize a need, to introject in a morphopoietic and functional way a previous relationship for or in-view-of a relationship to come. The body is therefore arranged and prepared to be completed through contributions from the outside, which are already presupposed in its structure. Being a body means (i) being non-autonomous and impossible to fully explain, from an ontological point of view, through internal reconnaissance alone. It also means (ii) being non-placeable and inexhaustible in the here-and-now, impossible to box within a synchronous frame. The body ultimately emerges from the here-and-now, unlike a machine that, vice versa, can be placed in a functional here-and-now. What characterises the body is precisely this circumvention of any attempt to make it fully explicable, both from the point of view of self-sufficiency (i), as if something was always missing in the explanation or something else had to be referred to, and from the point of view of spatial and temporal positionality (ii), because each somatic expression belongs to the past and yet is already in the future.

A sentient being, therefore, rather than being concerned with exposure to pain (an aspect which of course I do not wish to deny or dismiss, especially in terms of inherent interests), is a being-in-relationship. In other words, I wish to emphasize the relational and not autarkic nature of subjectivity. Sentience indicates the prevalence of being-in-relationship in animal ontology, its undulatory nature alongside the appearance of a disjunctive or identifying corpuscularity with respect to a presumed background. Sentience is the predisposition of animal subjectivity to seek its own predictive emergence in the encounter. The body brings the context to life, in the same way in which it enlivens past resonances. As my body stretches out, my ancestors come to life like ghosts, along an increasingly broad and remote *basso continuo* that makes it possible to interact with other animals, even those that are very distant from the taxonomic point of view. As Charles Darwin had already intuited, emotional expression allows for a trans-specific dialogue (1872). In sentience, belonging to the animal community becomes more evident than in a thousand other performative homologies. At the same time, sentience is an opening towards heteronomy, that is, the willingness to let oneself be organized by the world.

With respect to point "i", one can say that the evolution of the body, in all its phylo/onto-genetic temporal phases, is a predisposition to apparent deficiencies, which it would be more appropriate to call redundancies aimed at complementation.

The neurobiological structure, for example—but one can also refer to the foliage of a tree or the distribution of a population—does not follow the logic of morphopoietic autonomy, but that of the ability to acquire information about the external context. The structure of the synaptic network as well as the development of foliage follow fractal directives of redundancy, where the repeating morphological unity allows for an introjection of organizational formation within it, so that it does not betray its own internal logic while being shaped by the environment. We could then say, almost symmetrically, that an oak reflects the world through its foliage and that the world reflects itself through the oak. Mirroring is therefore an interpretation and a representation that brings novelty. The structure of the body is far from being self-referred autopoiesis, since the body is always normalized from the outside (heteronomy). Likewise, it is far from being simply modeled on the environment, because the body always imposes a precise fractal structure.

As to point "ii", it is important to understand the body as a becoming entity that creates the present and does not remain in a condition of explanatory isochrony. The past echoes in the body, not like a fossil testifying to a past passage, but rather as a beating heart that continues to claim its presence and is the engine of this presence. At the same time, it is impossible to understand the act of presence through the enucleation of instantaneousness, because presence in the moment is always a projection into something to come and yet already implicit: the future is already implied in the present, as if it had already happened. Its diachronic conformation, with a plurality of intentional coordinates, makes the somatic dimension elusive and ambiguous with respect to any attempt at a machinomorphic explanation or mathematization. This has introduced the need to make use of several causal/explanatory matrices—for example Ernst Mayr's dichotomy between proximate and remote causation and the four questions of Niko Tinbergen's ethology—since each highlights a single phenomenological position of the body. The problem is that, no matter how many causal devices we add to our explanatory catalogue, we will never manage to grasp the whole process of a body in its entirety, but only some elements of its intersection with the here-and-now. One could then say that the body has a here-and-now *because* it is not part of it.

The body seems to escape causal determination, folding in the creases of time and hiding in the fractal illusion of internal/external dichotomies. By lending itself to external completion and engaging in a game of references—i.e. by continuously referring to something missing, something that no longer exists or is not yet there—the body obtains its own self-ownership and is emancipated from the nomothetic. In its predicative process, the body does not present a causality that can be said to be exhaustive: it cannot be compressed into a formula, neither in a linear nor in a recursive sense. Whence a surplus that renders every explanation partial by only focusing on one of several possible explanatory levels. The sentience principle, as affectivity and involvement that transcends the moment in its causal networks and in manifesting a being-in-relationship, is the most interesting and most difficult aspect to investigate, especially while sticking to the conception of *res extensa*, i.e. an entity that is mathematizable through the Galilean-Cartesian synthesis.

There is a principle in Darwinian thought that more than any other should prompt reflection, not only for its counter-intuitive character, but above all for the ontological consequences it opens up, which in some respects are contralateral to the deterministic assumptions of 20th-century neo-Darwinism and to the claim that the history of life on Earth can be compressed into an algorithm. I am talking about the morphopoietic foundation: the view that a given organ has not developed *for* its function, but that its function is only the consequence of the event that brought it to light. This thought, which views the function as a conjugation of the organ and not as its morphopoietic cause, carries many consequences, including the principle of cooptability of every character, i.e. the functional range of the predicate. In other words, the emergence of a predicate does not *ipso facto* dictate a function, but paves the way to a possible conjugative space. This means that an endowment is not binding but, on the contrary, allows for the individual's protagonism because it sets up a precise conjugative perspective. Consider the hand: it is not bound to a specific function, not even the one that sanctioned its adaptive fitness—and this is demonstrated by the fact that, starting from a prehensile function, today it has taken on a performative multiplicity, from gesturing to typing. Likewise, any other character that enters the stage of life opens up a range of sensibilities as well as performativity. Every living being has margins of functional protagonism, that is, of ownership over its own endowments, but also of protagonism in feeling, in stretching out towards the world.

After all, it is precisely this protagonism that makes the animal a subject and not a puppet moved by threads. But there is a further aspect behind this principle of functional succession or conjugability of the predicate, a consequence that often escapes those who pursue the determinist view or an essentialist conception of the living. We tend to forget about sentience, i.e., the hedonic judge that regulates the expression of every animal. The mistake that is often made is to think that the expression of a predicate is bound to the performative fitness that caused its affirmation within a given population. To take up Ernst Mayr's dichotomy (1961), in so doing we mistake the remote cause, the judge of replicative compatibility that calibrates the value of a character on the reproductive dimension it enables, with the proximate cause, which is always linked to the singular creativity of the subject and to the hedonic value of the expression itself. The body does not follow the dictates of fitness, which instead is always a long-term matter and must be considered a consequence, not the engine, of expression.

The real judge of the expression is the sentience principle that, through pleasure or discomfort, orients the choices made by the animal. To clarify: cats, as a species, present a predatory character because this character has been positively evaluated in terms of fitness, that is, it has given more reproductive chances to those who owned it. But *that particular cat*, as an individual protagonist of its own here-and-now, expresses its predatory inspiration in any possible conjugation, even in the form of a game, simply because doing so gives it pleasure. The conjugability of predicates makes the animal the subject of its choices on the basis of sentience feedback and thereby allows the individual to be the protagonist of its own here-and-now. The body ultimately connects the subject to the world through coordinates of feeling that do not bind it to the past, but transform the past into fields of creative possibilities.

The body is therefore a composite set of expressive protagonisms that have been realized in different temporal phases, a concert of echolalias that claim their conjugative presence in the sensory dimension of the here-and-now. The body inherently attends the ceremony of the moment, as it is part of it. The body is a reference to something affective that is not only marked but undergoes a real ontological introjection. The body is therefore a potential feeling, but a feeling that does not simply monitor or dominate external reality, but that rather realizes a presupposed relationship in a singular way, through affective protagonism. Recalling Brentano, but extending his view to all somatic dimensions, we can then speak of an intentional plurality of the body, which does not exclusively concern the multilayer of somatic geography, but also the diachronic dimension. The body therefore does not resemble a workshop or a machine, made available by nature or a *res cogitans* capable of transcending its extended condition through intentional flights. Through the sentience principle, the body subsumes a multiverse of non-enumerable and non-explicable intentions.

The body is an ever-changing entity, because its copulative engine continuously creates new reception thresholds: its very being in the moment in a singular way, its being there, inevitably builds new spaces of complementation that presuppose an external contribution. It is then evident that *Dasein* is not revealed in distance but in the creation of contexts of hospitality, mooring points, internalisation of needs. The process of mirroring, trivialised in the concept of adaptation, is in fact more of a reference or complement to an external presence, from the union of which a predicative presence emerges. The germinative structure of a tree is an implicit appeal to a given conformation of reality, so rather than adaptation we should speak of reference or relational poetics. In this sense, mirroring has a copulative structure or an implicative connection to the outside: the wings of a bird imply thermals, just like the gills of a fish imply oxygen dissolved in water. Therefore, the body does not have its own autonomy and does not develop along self-centered morphopoieses, but just the opposite: the body has a conjunctive structure—it is an "if" or a "that". Whence its being predicative as intentionality.

The body represents the place where the relationship with the world is realized, along with the dimension that defines the "epiphanic level" of otherness,[15] the complicity inherent in the process of perception that not only unveils but also

[15]Thus, in the relationship with non-human othernesses, the human being discovers the highway of their own desiring being, since otherness incardinates in an exemplary way the different forms-opportunities of ontological expansion. The flight of a bird thus becomes the inspiration/announcement of a possible existential dimension "within reach", teaching the human being "that one can fly" and only later how to do it. In that principle of ontological dimensional opening—which we can define as "ontopoiesis"—lies the conjugation of desire that transforms the flight of a bird from "other-from-oneself" into "other-with-oneself" or "other-in-oneself". As a result, that existential action is not an external phenomenon observable with detachment, but rather the "epiphany" of a condition of desire ("*de-sidera*"), that is, something distant, missing, longed for and therefore achievable. On the other hand, such an epiphanic process would not have been possible without a threshold of encounter with otherness, i.e., if birds were completely other from human beings and therefore devoid of a different perspective or so distant or reifiable as not to trigger the identification process in human beings (Marchesini 2017).

reveals the entity. It does so through peculiar references: specific characteristics of the somatic condition that could be defined "reception thresholds",[16] i.e. free ports of hospitality. It is there that the magic of donation, both ontological and epistemological, is realized, igniting some predicates and overshadowing others. The body is therefore a multi-level conjugation *with* other entities, and at the same time it is intentionality referred *to* them. In authors like Bergson and Merleau-Ponty, the body presents itself as the "chiasmatic centre" of a becoming that mixes together time and space, structure and function, identity and otherness, past and future. The body therefore is a plurality of coexisting conditions: it is the "how" of the manifestation of mirroring events introjected at different times. Every dissertation on the body in the humanistic tradition, instead, tends to propose a unidimensional image of it, transforming the somatization of experience into a crystallized sediment instead of energy to be spent or a virtuality to be actualized.

The transformation of the body into a *res extensa* implies that the somatic dimension is fully blind to the world. The body is regarded as a mere chain of triggers that fall like dominoes without bringing anything new to the scene. Flattening onto a single level of performance—which also means zeroing all meaning—the body becomes a layer of intertwined elements, a fabric which stretches out on the same phenomenological level as experience. The body, instead, involves several "intentional levels", multiple levels of reference or reception that are constantly combined with the world, giving rise to new configurations of reality. Translating the body into *res extensa* means ignoring the multiplicity of its resonances and the emerging virtuality of its predicates. The humanistic body is poured onto the single level of the histological slide, so as to reach the conclusion, already implicit in the premises, of a dead somaticity.

In reality, the body realizes the principle of "coexistence" of different times as described by Merleau-Ponty. One could define it as a flow of instances diachronically located along matrices that can only apparently be attributed to the genetic, epigenetic and environmental dimension, and which do not support the ontogenetic diarchy. In fact, we must not consider these morphopoietic legacies as crystallized structures, informed by deterministic or economically oriented elements. Diachronic coexistence can be thought of as a creative energy or a virtual condition that, in the multiplicity that it implies, allows the body to embrace a continuous process of ontological singularity through the emergence of possible worlds. Here one can see the common thread that binds authors such as Henri Bergson, Maurice Merleau-Ponty, and Francisco Varela. And it is precisely by seeing the body not as a scheme or

---

[16] Cf. Marchesini (1997), *Il concetto di soglia* (*The concept of threshold*). The term "threshold" is ambivalent, but for this very reason it has a very useful heuristic function to understand how the human is the manifestation of the non-human sign in humans. I have used the concept of threshold to define precisely this explanatory inversion that sees human identity emerge in allowing for the hybrid effect and not in placing an insurmountable *limes* between human and non-human. The threshold represents a point of conjunction between different domains, able: (a) to circumscribe them or to make them possible because they are interfaced or not separated; (b) to give rise to a space that becomes such only in the very act of crossing it, or rather to overcome it, since the threshold is dialogical; (c) to define a limit-transition that by definition is boundless or not confined.

an informed level, but as an opening to what is new, to something unpredictable that escapes both determinism and the principle of causality, that the principle of sentience is established.

The body dialogues with the world, it is positioned on a peripatetic level of space–time, it chats with everything that surrounds it, it welcomes othernesses into relational occurrences and introjects them to start subsequent dialogues. Of course, I am talking about a dialogical dimension that is not only linguistic, but also involves the intersection of meaning: meaning has got to do with sentience. It is neither attributed nor assumed as it is by the thing itself, but is the product of the dialogical process, that is, of the double flow of reciprocal donation-hospitality. The virtual reality of the body—to quote Bergson—is therefore a sort of poietic redundancy or conjugative potential that paves the way to the need for an external reference to find a transitory configuration. Redundancy is therefore not a shortcoming or lack, but an opening to hetero-organisation. The body is therefore autopoietic in heteronomy. Its virtual nature, that is, its subsumption of an infinite multiplicity of actualizations, realizes/makes possible: (i) its dimensional singularity, (ii) its mirroring of the world, (iii) its continuous infidelity to the identity principle. The body has an identity status only if it is "fixed", as Deleuze would say.[17]

The body is always in a state of singularity—i.e., of emergence—with respect to its contents and of unrepeatability with respect to its condition. It is singular in all its elements, even in the intimate physiology of its genomic string. It is never a mere translation, an algorithm or a compression of data: the body is not the result of a project deposited by phylogeny or the sedimentary outcome of some organizational information extracted from the environment. The body is always on the brink of emergence. Every experience is something profoundly different from mere acquisition or subjective construction. The body reflects the world through the previous series of mirrorings. In so doing, it gives life to a sort of diachronic fractal. Let me provide an example. The leaf of an oak tree reflects a set of adaptive correlations: it is therefore a reflecting poetics. Each oak uses its own reflecting leaves—its own poetics—to reflect, in the singularity of its foliage, the environment in which it grows. But this is not a faithful reproduction of the surrounding world: rather, it is a subjective representation of the context carried out through heritage information, which is also represented. Each predicate is projected into the world *to* dialogue and is the result *of* previous dialogues: it expects a copula to be realized.

Sentience speaks of a living body, which does not only react, which is not moved around by the world like a dead leaf by the wind, but which attributes value to the context, and which in the first instance creates a biography in time… who knows! If traumas are transmitted epigenetically, this means that sentience regulates more than individual life. On the other hand, sentience speaks of subjectivity as a continuous becoming, as implicit self-overcoming. This takes place precisely in the frenetic succession of affective conjugations that turn the present not into the final result of an arithmetic of the past, but into a network that allows or leads the individual to continuously oscillate between different temporal dimensions. Sentience, therefore,

[17]G. Deleuze, *Lectures on Spinoza*, available at https://www.webdeleuze.com/cours/spinoza.

is not a simple conjugation that presumes a dichotomy, but a plural relation that invalidates the claim of any dichotomous explanation.

The body is never a fact, a *res extensa* that can be nailed on the space–time axes and can be mathematically measured; even its here-and-now, as Merleau-Ponty suggests, is a pseudo-present. A dead body is not a crystal, because it falls into immediate degradation and this is a living process. The body rejects the identity principle, it is unfaithful because of its founding character, it is for-the-other, it is in becoming, but its state of infidelity does not mean that it is nothing. This, in my opinion, is the legacy that the French philosopher left us as a workshop for future reflections: the overcoming of opposing categories which, through the principle of noncontradiction, have removed the possibility of fully developing a philosophy of nature. The condition of the living does not pre-exist its continuous making. As Bergson (1988) would say: the consequent cannot be explained by the antecedent. The reasons for this, in my opinion, lie in the fundamental aspects of sentience: the singularity of the condition of feeling, the mirroring or heteronomic character of feeling, the infidelity of feeling, and the experiential progression of feeling. The body *is* because it always presupposes an external contribution, it is a "matrix of relations" with the world, and it is precisely from these relations that predicates emerge, so that they can never be inferred through internal observation. Without reference to implicit heteronomy, to a multi-layered intentionality, it cannot be comprehended.

Merleau-Ponty emphasizes the inseparability between being and phenomenon, but this inseparability is based on the implicit partiality of the axis of feeling. If feeling were nothing more than a view of the world, one would expect something like an objective view, albeit oriented by the function of praxis. Yet this is not the case. If what Lorenz suggested is true, namely that a-prioris are phylogenetic a-posterioris, it becomes evident that feeling is the protagonist, which on the one hand denies the Kantian disjunction, but on the other hand only reports the constraints encountered in the perspective predisposition. Feeling is therefore a precise bond contracted with the entity: in other words, the perceived cannot be foreign to the perceiver. Therefore, the body cannot afford a general overview, a "bird's eye" perspective; so, perception is more likely to be an epiphany. But, on closer inspection, the epiphany is already a recourse to a time that escapes the here-and-now. Epiphany is both a resonance and a projection. Drawing a transversal line across very different authors, such as Darwin, Proust, Bergson, Lorenz, Piaget and Bateson, we find that the sentient body contains the Husserlian concept of institution/donation (*Stiftung*) which is the mark of chronic implementation. The body is the place where time is recorded, or accumulated, for the dialogues to come.

5.5 To Conclude

In order to talk about animal subjectivity, it is not enough to follow Descartes who, like an illusionist, shifts the attention to the dichotomous interplay of opposites. It is not enough to resurrect some *res cogitans*, or however you want to call it. It is necessary, on the contrary, to overcome the dualist conception and to reconsider the concept of *res extensa* as an exhaustive paradigm. It is not enough to hypothesize animal consciousness, if the underlying level remains attached to an explanation based on automatisms and reactivity. Mechanism can be considered as a full identification of the individual with the functions potentially available and inherent in it, i.e. as a full overlap between the two terms. Overcoming mechanisms therefore means addressing the individual as an entity that exceeds its endowments and has full protagonism. The de-subjectivization of animality has been fully expressed by Martin Heidegger in the concept of the animal as "poor in world" (1998),[18] i.e. totally immersed in its functions, and in this sense unable to emerge from its stunned state. Subjectivity is therefore the admission of an animal *Dasein*, i.e. the intrinsic ability of the animal condition to exceed its functions, in a conception that has nothing to do with the judgement of the typology or of the predicative gradient of that function. It is not the complexity of the function that proves the character of subjectivity, but rather the capacity of the animal being to reinvent its presence in the world at any time.

This, in short, is Dennett's true "dangerous idea" (1995)—a reminder that is difficult to digest. It is as if human beings rejected the idea of subjective otherness and had to take refuge in determinism due to gnoseological necessity, as the latter is the only condition that allows them to maintain control. The animal-machine is therefore de-subjectivized, transformed into a set of mechanisms which, like switches, are directly and exhaustively responsible for the functions expressed. However, there is no doubt that even under the coercive rigour of the *res extensa*, the animal seems to escape from the autopsy table and exceed its limits, so that the mechanistic explanation has to resort to tautological artifacts—real explicative epicycles—as it fails to accomplish the main task it had set itself. It is no coincidence that both evolutionists and ethologists have had to appeal to diachronic explanatory principles—such as the concept of remote causality, teleonomy, motivation, etc.—to explain not so much the "how" but rather the "why" of different behaviours.

On the other hand, when we see an animal bend external reality to its own needs, adjusting its species-belonging to the emergence of its individual singularity, charging the events with arbitrary semantic values, creating new solutions capable of solving and managing novelty, we cannot fail to witness an overall condition that has little to do with that of a machine. At most, one can explain a behavioural endowment—be it a sequence of associative structures, a sort of processing utility, a complex set of possible connections to form hidden entities—as a tool called to perform a particular cognitive-behavioural function. However, there is no doubt that

---

[18]In this regard the German philosopher wrote: «For the animal is related to his circle of food, prey, and sex in a way essentially different from the way the stone is related to the earth upon which it lies. In those living things characterised as plant or animal we find the peculiar arousal of excitability, by which the living being is "excited", i.e., stirred to an emerging into a circle of stimulability on the basis of which its drawers other living things into the circle of its activity No excitability or stimulability of plants and animals ever brings them into the free in such a way that what is excited could ever let the exciting "be" what it is even merely as exciting…Plant and animal are suspended in something outside of themselves without ever being able to see either the outside or the inside, i.e., to have it stand as an aspect unconcealed in the free of Being.»

the individual uses their endowments as tools. The animal owns the endowments—it uses *them*, they do not use *it*—and, consequently, it manifests sovereignty over them.

Animality is primarily existence, a very complex term that one could define, at least in part, with the following characteristics. One is (i) the ability to position oneself in a state that previously did not exist and could not be deduced from previous ones: animality means not being contained in the already given world but continuously creating worlds. Another characteristic is (ii) the property of presenting a *quid* in addition to one's functions or of exceeding one's own endowments, i.e. not being fully definable by one's predicates. Animality means (iii) not being completely immersed in one's here-and-now, i.e. owning a present by being diachronically positioned above it, or, again, being inexplicable by referring exclusively to the forces acting in the here-and-now. Animality also means (iv) owning one's endowments and using them in complete freedom (flexibility, co-optation, redefinition), i.e. having expressive sovereignty. Finally, it means (v) owning inherent expressive motives, which I have defined as the principles of desire and sentience, and on the basis of these, having interests.

Also with reference to the term "subjective", used to define specific traits such as singularity, partiality, arbitrariness—in the interface, in the orientation, in preference, in judgment, in motivation, in representation, in decision—it is indispensable to outline not only its different characteristics, but also and above all the assumptions that make them possible. From my point of view, conscious access is nothing more than one of the many expressions of animal subjectivity: it is neither the *conditio sine qua non* nor the most striking expression of the individual's subjective being. The unconscious—if it still makes sense to practice this dichotomy—has proved to be a powerful motor of subjectivity, much freer in the emergence of singularity than the seclusion and censorship of the explanation. Subjectivity emerges from the whole system, in its overall and dislocated expression, and in this sense I wish to avoid any conception that appeals to transcendent entities, however they may be defined or presupposed.

Therefore, when I speak of the protagonism of the subject—as the owner of their endowments and the bearer of inherent interests—I do not refer to some control room that decides how to use the endowments or to a level subsuming the desiring and sentient principle. Protagonism is rather the result of three things. First, there is (i) the systemic character of the body called to express its state in the here-and-now. Then, there is (ii) the requirements of the endowments which, rejecting the mechanistic conception, cannot be assimilated to automatisms;. Finally, there is (iii) being-a-body, i.e. implicitly conjugated and copulated towards the world. On the other hand, I would like to underline that the Darwinian revolution, in its ontological assumptions—the most obvious example being the scarcity of species-specific behaviours in the ethogram of a species—does not allow us, in complete intellectual honesty, to maintain the "human vs. animal" dichotomy, which by transforming animality into a counterterm has depicted the animal condition as a dimension of being that doesn't directly affect us.

Subjectivity therefore does not concern species peculiarities or expressive gradients—what that particular animal is able to do—but rather the animal condition and, consequently, the model adopted to explain the dimension of animal-being. For this reason, I think it is misleading to seek subjectivity in the complexity of predicates, since each predicate can be traced back to the chosen model through explanatory approximations. Only in this way is it possible to speak of animal subjectivity without getting lost in the description of the individual forms of subjectivity, the analysis of which is inevitably a work in progress. Species characteristics are important to define a style of presence, which can be described as a range of positional possibilities or a level of practicability—and this is the complex task of ethological analysis—but are not necessary to understand the paradigmatic sense of subjectivity. Subjectivity is a metapredicate, just as the mechanistic explanation of animal behaviour is metapredicative, i.e. it relates to the explanatory model used to account for the general condition, not a particular function. When one talks about a mechanistic metapredication, one means that, once we have chosen the explanatory model—the machine—any function, no matter the type or gradient of complexity, will be interpreted in a mechanistic way. The metapredicate evolves over time not on the basis of ethological descriptions but in accordance with the evolution of the model itself, as capable of subsuming the functions. The "machine" model has changed along with the historical technological metamorphoses: first there was a hydraulic explanation, then a thermodynamic one, then a cybernetic one and today a computer one. All were used in their time to account for animal behaviour: the type of machine changes but the mechanistic explanation stays the same.

Subjectivity is instead the recognition of a space of action for the individual who, owning their endowments and possessing inherent integrity, is called to the creative and participatory level. It is the admission of the animal's existentiality, of its being-there in the full sense as a "creator-of-worlds", since animal being implies not only the production of solutions responding to the singularity of reality, but also the emergence (conception) of the setback itself within the fabric of reality. The animal being must have a positionality in the here-and-now precisely because it transcends the here-and-now, because it is involved in a gap that does not allow one to explain the animal's expression by exclusively referring to the causal matrices acting at a given moment. To position oneself in a here-and-now means to also own an elsewhere, so that one's perspective—matured through the legacies of phylogeny and ontogeny—is subjected to a continuous process of renegotiation of meanings. The focus of interest and specific objective of this work is to show that animality is a condition of full presence in the here-and-now precisely because it emerges from the total immersion in the here-and-now. It is an existentiality that, in order to be able to dwell in the fluctuation of reality and in the singularity of the situations it has to address, must constantly be present in the immediacy of its intersection with reality.

References

Bateson, Gregory. 1980. *Mind and Nature: A Necessary Unity*. New York, NY: E. P. Dutton.

Darwin, Charles. 1872. *The Expression of the Emotions in Man and Animals*. London: John Murray.

Dennett, Daniel. 1995. *Darwin's Dangerous Idea: Evolution and the Meanings of Life*. New York, NY: Simon & Schuster.

Derrida, Jacques. 2008. *The Animal That Therefore I Am*. New York, NY: Fordham University Press.

Gould, Stephen Jay, and Vrba Elisabeth. 1982. Exaptation-A Missing Term in the Science of Form. *Paleobiology* 8 (1): 4–15.

Griffin, Donald. 1992. *Animal Minds: Beyond Cognition to Consciousness*. Chicago, IL: University of Chicago Press.

Heidegger, Martin. 1998. *Parmenides*. Indianapolis, IN: Indiana University Press.

Marchesini, Roberto. 1997. *Il concetto di soglia: Per una critica all'antropocentrismo*, Roma, Theoria.

Marchesini, Roberto. 2016a. *Alterità*. Mucchi: Modena.

Marchesini, Roberto. 2016b. *Etologia Filosofica: Alla ricerca della soggettività animale*. Milan: Mimesis.

Marchesini, Roberto. 2017. *Emancipazione dell'animalità*. Milan: Mimesis.

Marchesini, Roberto. 2018. *Over the Human: Post-Humanism and the Concept of Animal Epiphany*. Springer.

Mayr, Ernst. 1961. *Cause and Effect in Biology*. *Science* 134 (3489): 1501–1506.

Mayr, Ernst. 1982. *The Growth of Biological Thought: Diversity, Evolution, and Inheritance*. Harvard, Harvard University Press.

Merleau-Ponty, Maurice. 2003. *Nature, Course Notes from the Collège De France 1956–1960*. Evanston: Northwestern University Press.

Michelini, Francesca. 2011. *Il vivente e la mancanza. Scritti sulla teleologia*. Milan: Mimesis.

Michelini, Francesca, and Davies Jonathan (ed.). 2013. *Frontiere della biologia: Prospettive filosofiche sulle scienze della vita*. Milan: Mimesis.

Minelli, Alessandro. 2009. *Forme del divenire: Evo-Devo: la biologia evoluzionistica dello sviluppo*. Milan: RCS.

Scotti, Alessandra. 2015. *Il mondo del silenzio: natura e vita in Maurice Merleau-Ponty*. Pisa: ETS, Pisa.

Tinbergen, Nikko. 1991. *The Study of Instinct*. Oxford: Oxford University Press.

Chapter 6
Intus-Legere: Knowledge as an Actualization Process

Roberto Marchesini

Abstract Despite its many possible interpretations, the Western tradition has restricted the act of knowing to disjunction from bodiliness, in a sort of detachment and contemplation of the world ascribable to the *res cogitans* and to the exclusion of mere fruition. The observer becomes an entity *other* than external reality, looking at the latter from the outside (albeit within their own perspective). Knowledge, as a neutral and objectivizing act, as something neither participating in, nor emerging from, the relational predicate, has marked a sort of divorce between knowing and known: a dichotomy that is reflected in a fractal way and itself produces other dichotomies. Far from being interpreted as a particular and partial gaze on the world, the human perspective has acquired universal connotations: the perceived is suspended, and the perception and connotation of reality are taken to be separate things. This *consecutio* goes from the Cartesian doubt to the Kantian critique; then Konrad Lorenz operated a first conjunction by connecting the aprioris to the adaptive processes and making them phylogenetic aposterioris. Now we should ask ourselves: (1) whether we can consider the epistemological perspective as a state of approximation to reality, which has an ultimate configuration even though the latter is not necessarily knowable; (2) whether, on the contrary, we should accept a principle of absolute relativism to deny any conformation to reality and theorize the latter's non-consistency, along with the total perspectival creativity-freedom of the subject. There is also another option, though: perhaps (3) we can chose a third path, one that presumes that reality per se is unknowable, not for a lack of conformability but for its virtuality, and the only observed datum acceptable as a starting point is the resistance to the actualization process. In this chapter, I shall try to elaborate on this third option, referring to Lorenz's view on the adherence of cognitive tools to the configuration of reality. I will also try to understand how to conciliate an interpretation based on the animal being's subjectivity, full of perspective protagonism in any interaction with the world, with an epistemological framework that accounts for the phylogenetic conjugation as a given dimension. This interpretation acknowledges specificity but also takes into consideration the homological, analogic and universal factors. I believe that Lorenz's thought, and evolutionary epistemology in general, have left us a legacy we cannot ignore. Instead, we should avoid Kant's disjunctive parallelism; indeed, in light of animal subjectivity, it is evident that we need to revise both the concept of perspectival approximation and that of perspectival passivity.

R. Marchesini and M. Celentano, *Critical Ethology and Post-Anthropocentric Ethics*, Numanities - Arts and Humanities in Progress 16,
https://doi.org/10.1007/978-3-030-74203-4_6

6.1 Premise

When we analyze the plurality of different species and their different ways of interacting with the external world, but also when we explore our individual image of the context going beyond the surface of the already known—hence *intus-legere* as an act of searching in depth and beyond appearances—it is essential to start from some preliminary considerations. First of all, we must ask ourselves if we can still understand the intellectual act as a disjunction from the body, that is, an event that takes place by virtue of transcendence or distance from feeling and acting, from desiring and obtaining. And we must wonder whether instead, following Nietzsche, knowledge should be seen as one of the many expressions-needs of being-a-body,[1] which can be conjugated within the correlative canon of the evolutionary process. The rational Self that knows by distancing itself from contact with or fruition of the object of its knowledge, contemplating it with detachment while not being somatically involved in it, is one of the most rooted prejudices of our culture. This view prevents us from connecting knowledge to the multilevel of feeling, so that we are led to consider the intellectual act as a sort of emancipation from the body. This disjunction, in the de-somatization of the *res cogitans* that distances feeling from knowledge, separates contact and involvement from *intus-legere*, and sees the innate as stunning and passions as misleading. It transforms, by attribution, animality into opacity.

This disjunction has transformed the epoché into a sort of negation of the conjunction between feeling and the structure of reality. Secondly, it is therefore fundamental to ask oneself whether knowledge travels alongside reality—by virtue of reconstructive phylogenetic endowments, obviously not only of a perceptive order, but also of a metacognitive and practical-gnostic one—or whether, as evolutionary epistemology suggests, the endowments of knowledge, in their correlative partiality, already give us information on reality. Also in this case it is a question of recognizing the reasons of the body and the epistemic productivity of phylogeny. We all agree, in fact, that the evolutionary process is a morphopoietic laboratory, capable of developing structural and functional solutions able to give rise to context-referenced performativity. What, instead, is not always emphasized is the context-referenced epistemic value produced by the phylogenetic atelier. The evolution of the living is not only a process of building performativities—which, *contra* the essentialist vision, "emerge" (the wing is an emergence and not the realization of an idea). Evolution is also a laboratory of epistemologies, which also "emerge" and are able to actualize a dimension within a field of possibilities. This indeed is the case with the wing, which is only the organization of a new image of reality. With respect to this, one could ask if the Cartesian doubt should not annihilate the acknowledgement of the intersection and only favour criticism, entrusting a particular "validity domain" to the particular image obtained.

[1] In "The Bestowing Virtue", Nietzsche wrote: "In knowledge doth the body purify itself; attempting with knowledge it exalteth itself; to the discerners all impulses sanctify themselves; to the exalted the soul becometh joyful" (Nietzsche 1917: 81, modified translation).

But what does the concept of "validity domain" mean? Can we consider epistemology as a part of this concept? I think so, because knowledge entails a certain level of interrogation, a way to probe specific resistances of reality to test its organizational opportunities. Sensory and cognitive organs, predisposed by phylogeny, are undoubtedly the result of a dialogical-interrogation process with a precise context-world, based on specific cognitive interests. Take the visual sensitivity of the human being, able to disjoin two electromagnetic bands (such as green and yellow) that are very close from the point of view of physical feedback, to the point of transforming the finding into a chromatic and not tonal difference. This tells the story of an animal that, in its phylogenetic path, has spent a lot of time among trees (hence the sensitivity to a given light spectrum), in the daytime (hence the lack of hair), eating fruit (hence the prototypical spherical shapes and yellow-red colours). The same can be said about our cognitive characteristics of extraction of iconic structures, typical of omnivorous gatherers, and about our sillegic orientative motivations, the same that lead us to collect flowers, shells, mushrooms, or stones when we are in their respective environments. The third consideration is whether the knowledge that comes from this epistemic dimension, elaborated by the phylogenetic craftsman, should in any case be considered as an "approximation to reality", as Lorenz's considerations seem to suggest. And we should wonder whether instead it is more productive to consider reality not as a completed form to be discovered, but as a virtual field—that is, a set of potentialities that the given epistemic intersection actualizes in a given way.

The fourth point is, in a sense, the consequence of the questions I have mentioned, even if it poses a new question, namely whether knowledge can or cannot be framed within a directionality. If it can, there are again two options. Either this direction can aspire to give us a universal image of reality, albeit through long processes of actualization—that is, by measuring the resistance of reality—or instead epistemic pluralism can be subsumed within a single canon, on pain of losing its morphopoietic richness. The first question is about whether pluralism translates into a sort of absolute relativism or a relativism mitigated in the dimensional definition, or whether instead we can glimpse a direction aimed at widening and deepening our view of the world. With the second question, I ask whether this path will lead us to have a single image, definitive and final, of reality. To explain my take on this, which I have developed over the last ten years in my research on "levels of reality", I believe that knowledge has a direction in widening the field and in listing laws capable of subsuming events. However, as a means of correlation and configuration of reality, knowledge cannot claim to exhaust all that is possible in a single image, because knowledge is always related to a style of involvement in the individual's intersection with the context. Claiming the opposite would be like demanding a performatively exhaustive organ.

The fifth reflection, which I would like to add in the introduction to this work, concerns the role of the individual and, more specifically, of the here-and-now of feeling and being there, understood as active protagonism in the intersection and in the act of knowing. In fact, knowledge undoubtedly has the ability to go beyond the already given, to seek new configurations among the folds of reality, and to bring out new answers in a meta-cognitively innovative way, by virtue of the singularity of interrogation that the individual lives in its protagonism. The wing, to take up the example of an anatomical functionality forged by phylogeny, is an entity at the disposal of the individual, not a mechanical structure that directly imposes its performances. The wing is a conjugation of the body, it is operative and sensitive flesh which adapts to the stimuli of the here-and-now—it is a morphopoietic dimension, not a simple already-given structure. The individual has an epistemic dimension not to limit its access to the world, but to start from a collection of useful materials in order to give life to new contents. The epistemic protagonism of the individual therefore does not prescind from the phylogenetic heritage, but the latter only establishes the starting coordinates—not a finished epistemic product. The act of *intus-legere*, i.e. the cognitive protagonism of the individual, is therefore not an accident or a surplus with respect to the species-specific epistemic dimension, but the very meaning of the latter.

Before addressing the various topics related to this important part of the singular and subjective life of every animal, namely its ability to play an epistemically leading role, I want to clarify these five points with further considerations that will then serve me when dealing with the different levels of intelligence. The first question I have highlighted emphasizes the indissoluble relationship between somatic dimension and knowledge, not necessarily in terms of limitations, but rather in pointing out that the act of knowing is a questioning process that, as such, arises from a perceived lack. Knowledge comes with a tension that cannot in any way be de-somatized, but that arises from a languor of the body that has its first *movens* in the motivational and emotional coordinates. Often we speak of knowledge as an end in itself for the human being, because perhaps it cannot be clearly traced back to or, in any case, induced by any immediate applicative translation. The mistake lies in attributing to a de-somatized dimension all that we cannot trace back to some practical application or to a confirmation in terms of physical gratification. On the contrary, exploratory pleasure as such, or the social recognition deriving from a discovery or a professional title, also represent an involvement of the body.

On closer inspection, there is no overlap between the motivational drive and a necessary-implicit practical application or the fulfillment of a need. As we know, intrinsic motivations like "mental languors"—the propensities of our species include the exploratory one, the epimeletic one, the syllegic one, the imitative one, the social one, etc.—have developed in the human being along its phylogeny because they bring fitness. In other words, they favour the satisfaction of particular needs (for example, being led to explore and collect facilitates the procurement of food). On the other hand, they act on the subject in a way that is independent of its needs, for example in playful activities. When a scholar is passionate about a line of research, they do not necessarily pursue a practical objective, but this does not mean that their cognitive tension is separate from their bodiliness: they are following their motivational coordinates, in one or more dimensions, participating in the act of research from within their body, primarily with the active contribution of their emotions. Knowing is therefore an expression of all the dimensions of the body.

The second question poses the problem of the multiplicity of points of view on reality, which can be referred to different aspects. A first point is that (i) each species lives in "different contexts", where reality acquires a specific configuration, so the individual is not phylogenetically confronted with the overall field of virtuality but with a precise conformation of reality that manifests its own opportunities and resistances. Secondly, (ii) each species can also be considered a particular "life project", which therefore seeks opportunities and is confronted with specific and never absolute resistances, developing a gaze on the world related to its existential project. This means that we have to start from the epistemic dimension typical of each species, which therefore not only brings out opportunities and resistances, but also builds epistemic directions that are often epistemic artifacts.

This "epistemic artifact" effect is also found in the human being, in the so-called "naive physics"[2] since, for reasons of adaptive efficiency-effectiveness, each species makes use of (heuristic or biased) shortcuts that have value in replicative and not absolute computation, zeroing or obscuring the conformations or findings that have no meaning or have proven irrelevant in terms of fitness. The questioning of appearances—i.e. criticism—therefore has an undoubted epistemological value, not because there is no correlation between epistemology and reality, but because every epistemological dimension, as well as every theory, has a domain of validity. If we consider the structure of a wing, we undoubtedly see findings about how to fly, and therefore how to use the air currents to one's advantage; a fin speaks of how to slide in the water, that is, how to use the viscosity gradients for propulsion and, on the contrary, to lower its resistance.

So what does the shape of a wing or a fin tell us? This is the third question I would like to ask. Is a wing an approximation of air and a fin an approximation of water, or are they something else? I think Konrad Lorenz is right when he points out that phylogenetic endowments—and the so-called a-prioris in the epistemic act—are connected with the dictation of reality, because they are forged by fitness as a result of correlative results. However, this does not mean that they should be considered approximations to the conformation of reality. In other words, the structure of a wing summarizes the particular performative conformation of animals who want to adopt a given lifestyle in the air. It is the conjugation or actualization of a life project, i.e. flight, which is confronted with the opportunities and resistances of air. Flight is the emergence of a level of reality that is realized through specific questions posed to the context.

The fourth consideration concerns epistemic plurality as a necessary and inevitable condition that neither denies an increasingly omni-comprehensive view of

---

[2]The studies inaugurated by Gestalt psychology show that the human being possesses a particular perceptive cognition, beyond its sensorial interfaces, such as the amodal completion of objects, the construction of illusory margins, and pregnancy. In the same way, the cognitive sciences have highlighted certain interpretative or heuristic tendencies, useful to solve in a hasty or approximate way certain problems, but equally fallible or leading to distortions, defined precisely by the term "bias". As a whole, we can speak of a phylogenetic epistemology that is characterized by interpretative or resolvent methods that should not be considered absolute but adaptive specializations, and therefore relative and endowed with a point of view strongly centered on the characteristics of the species. Also defined as "naive physics" and described by Gaston Bachelard as an "epistemological obstacle", this epistemology gives rise to what I call "anthropocentric perspective", which is not relevant culturally but in terms of phylogenetic heritage. For the notion of "naive physics" Bozzi (1998).

reality, nor brings out some ultimate configuration: it rather highlights the resistances that we gradually encounter when we broaden our view of the world. On the other hand, knowing always implies asking a question, and it is not certain that there is some final question capable of bringing out all the resistances of reality—personally I don't think so, but I am inclined to suspend my judgement on this point. I believe that scientific knowledge is able to give us increasingly accurate information on these resistances, through progressive amendment of the prejudices of our epistemic dimension. Therefore, I do not agree with the Kuhnian paradigmatic disjunction (1962), because I believe that every theory is deeply linked to the previous ones, which is why a direction is more likely than another. However, every way of investigating reality brings out levels that cannot be subsumed by one another; it's as if reality were a skyscraper where each level shows us different aspects of reality, without there being a level capable of summing up all the others. Each level rests on the one that precedes it, but each level has its own epistemic legitimacy, as it shows us a particular view on the world that has its own relevance.

This leads me to the fifth consideration: the epistemic protagonism of the individual and the feedback that this protagonism has on the morphopoietic processes of the living. We are used to considering evolution as a process that relieves the individual of any responsibility for its phylogenetic trajectory, despite the considerations made by Darwin, Lorenz, Waddington, Piaget, and Baldwin. The interpretation that has been affirmed in the neo-Darwinian synthesis, in fact, regards phylogeny as the product of two factors: (i) the randomness of the mutations that produce morphological variants; (ii) the selection made by the environment on the basis of the replicative potential. However, thanks to the theory of niche construction, migrations, and epigenetic transmission, we know that—through its own action, and consequently also through the discoveries it makes and the solutions it adopts in its behaviour—the individual is able to determine some "shifts in the selective pressures". Consequently, even if indirectly, it can modify the fitness parameters within the population itself and therefore the morphopoietic trajectory that in the long run will eventually codify the species canon.[3]

One must also consider the protagonism of the individual in the act of knowing: the subject starts from its own species-specific epistemic dimension, but does not remain trapped in it, and apprehends new information and strategies also through a creative singularity. It is not only the human being that is able to go beyond its specific epistemic legacy: it is part of the animal condition in general to continuously

[3]I have had the opportunity to explore in depth the subject of the "shifting of selective pressures" in the book *Post-human. Verso nuovi modelli di esistenza* where I wrote: "Every shift in evolutionary pressure, achieved through technological mediation, actually inscribes that technology in the genetic heritage of the species. Technology therefore becomes a sort of performative externalization that extends man's operational domination over external reality, not impoverishing the species but enriching it through its needs", p. 32. The example that I used in support of my theory was that of antibiotics: the introduction of that technology within the sphere of human identity has made it possible to shift the selective pressure from man to the man/antibiotic combination: "Man is no longer selected for his ability to produce antibiotics, on the contrary it is the antibiotic that is selected for its ability to offer man antibiotic properties" (2002: 31).

explore or interpret one's intersection with the world, to seek new opportunities of configuration of reality. The act of *intus-legere* is precisely this ability to read between the lines, bringing out new images of the world-context, so that the epistemic heritage should be considered a sort of epistemological handbook that does not exhaust the cognitive possibilities of the individual. Therefore, it is necessary to put the individual back at the centre of the cognitive act, recognizing the animal condition as a metapredicative dimension of this potentiality, through a fundamental reconsideration of what it means to be an animal in the intellectual process.

6.2 Knowledge as an Expression of Being-a-Body

Interpreting animality as a condition that concerns us in the cognitive processes means starting from some basic assumptions. First, we must: (i) reconnect the intellectual process with the dimension of bodiliness in all its aspects, because knowledge is dislocated on many levels and realized through all the somatic components. We must abandon the Cartesian dichotomy to recognize the somato-emergential level of epistemology. Secondly, we must (ii) admit an epistemic plurality, which is not essentialistic but based on homological, analogical and, in some respects universal similarities, just as happens with other physiological functions, avoiding the hierarchization of entities in the name of correlative specialization. Finally, we must (iii) recognize the epistemic protagonism of animal subjectivity, capable of inventing its own perspective on the world in a singular way and of intervening in its evolutionary trajectory; i.e. we must view the epistemic act as one the causal factors of morphopoiesis. The Cartesian disjunction that described the act of knowing as a departure and emancipation from the bodiliness of the *res extensa* is a humanistic leitmotif. This view goes as far as the Heideggerian notion of the stunned state or opacity of the animal in its intersection with the world, and transforms animality into a condition of contralaterality. Today, after Darwin and after Lorenz, it is no longer possible to accept this perspective.

Animality and the body are thus at the centre of the epistemological debate, in a logic of connection between a-prioris and the resistances of the world, acting as a mirror or a posteriori of the demands that have calibrated their characteristics. The somatic dimension of epistemology leads one to consider the centrality of animal-being as the starting point for any reflection on cognitive processes. In particular, one ought to recognize the phylogenetic role in the epistemic definition of the individual, but also, recursively, one should acknowledge the morphopoietic value in the epistemic singularity of the individual, which affects the phylogenetic trajectory itself. In other words, the point is to place animal subjectivity back at the centre both of existential singularity and of the phylogenetic process that transforms every trajectory into a life project, where the latter is guided by an intrinsic teleology, at least as a cofactor in the ontopoietic design process. Placing the body at the centre means admitting a multi-level epistemology that goes beyond the characters predefined by the heritage, because it contemplates the overall ontogenetic evolutions of the somatic dimension and also the here-and-now of the body itself. Therefore there is a close correlation, notable at multiple levels of recursion, between heritage and context, in a yin-yang dialectic of inclusions for which epistemology is a relationship.

On the other hand, admitting a connection between inherited epistemic dispositions and the conformation of reality can give rise to some misunderstandings. One of those is (i) the belief that the heritage epistemic dimension is a predetermination that does not allow for any subjectivity or singularity in the epistemic protagonism of the individual. Another mistake is (ii) to consider different species-specific epistemologies as non-overlapping entities, whence Thomas Nagel's observations (1974).[4] Or else (iii) this connection between the phylogenetic investigative apparatuses and the world is viewed as an approximation to the configuration of reality. In short, we must take up the considerations made by von Uexküll and Lorenz, revisiting them in the light of animal subjectivity. Thus: (1) the *Umwelt* is not a bubble that encloses the individual within an epistemic dimension, but an elastic mat that allows for the emergence of an epistemic singularity; (2) the different *Umwelten* are not separate monads, as posited by the essentialist conception of the living, but present wide epistemic overlaps due to homologies, analogies and universals; (3) the particular epistemic intersection of the individual, in its phylogenetic and experiential determinations, is not an approximation to reality but an actualization of reality, so that the dictate of reality can only be partially deduced from the epistemic configuration.

The body is therefore a singularity in relationship, a predication that emerges in the connection event thanks to "connecting structures", to put it with Bateson, which nevertheless are works in progress and not finished structures. In other words, these structures exist in relations that are never impermeable, as each relationship modifies them by virtue of a permeability that is already taken into account. Put differently, the animal is not a machine created by an imaginary craftsman, be it a demiurge who converts ideas, a creator who realizes them *ex novo* or natural selection that brings them out from randomness. The animal is an entity that is made by expressing itself. Placing the body back at the centre means shifting from a merely informative view of knowledge to a relational and morphopoietic interpretation. In fact, there is no such thing as a cognition that is not also a metacognitive enhancement of all the components (such as memory or motivation) that are exercised in the very act of knowledge. We never simply learn something—we transform ourselves in our epistemic conjugation with that something. The knowing body is a growing entity, in a continuous internal metamorphosis directed by the cognitive processes. And it is illusory to think that knowledge takes place through perceptual and processing channels, i.e. that it only has neurobiological value. Knowledge is achieved through modifications of the endocrine, metabolic and immune systems, just to give a few examples, and is already implicit in the conformation of the cell itself. The cognitive impetus is not supported by an element that transcends the body, but by the somatic reasons of the body itself and by the animal dimension as a flywheel (Rosenblum 2010).

---

[4]For him, the subjective experience that non-human animals have of their world is inaccessible to us (Nagel 1974).

The body summarizes several stories, i.e. the different stages of ontogeny and of phylogeny—which is why many of our views still speak of a confused tetrapod on the edge of the mainland. Whence our embryonic palpitations that are not yet split from the mother's body, the fetal experiences of dreams that precede experience, the first jolts of the process of attachment and the journey along the path of our relationship with the world. However, the various stages of phylogeny and ontogeny are not the result of stratification, which still gives rise to the idea of a reptilian backdrop, but of subsequent reconfigurations of the past, so that the phylogeny of the past is no longer ancestral. The body clearly shows that the conception of animal inertia, which we have translated into the behaviourist view of ontogenesis and into the neo-Darwinist view of phylogeny, is totally wrong. The animal actively intersects the world, continuously reconfiguring its access threshold to external reality through a plurality of somatic dimensions, so that knowledge is never positional on a single somatic level, but is the emergence of a multilayer where each one contributes to shape, or rather to realize, a given configuration of reality. Therefore we cannot speak of approximation, because epistemic subjectivity is not a process of approximation to an ultimate form, but the construction of a form on the basis of a life project, exploring the opportunities and resistance of the context.

Animality thus becomes the core of a reflection that questions the passivity of the living in the face of the morphopoietic—and therefore also epistemic—forces that claim to have the environment as their only configurative agent (whence the concept of approximation). Instead, a conception based on the somatic subjectivity of the animal, which evidently translates into the protagonism of *intus-legere*, speaks of an animal that actively constructs its own epistemic configurations during phylogeny. Despite being endowed with an a priori epistemic dimensionality, the animal does not blindly reiterate an intersection with the world that is already predisposed, but uses its apparatuses of world image to give life to its own epistemic singularity. In other words, we need an epistemological view that is evolutionary and not only evolutionistic: that is, a view that considers the individual as a "work in progress" from an epistemic standpoint.

The body summarized in the phylo/onto-genetic process is not a sediment but an entity that reshuffles and subsumes a singular project. The phylogenetic handbook is therefore to be considered as "construction material", not a backdrop that definitively predetermines the intersection. In the same way, it is indispensable to recognize that this protagonism does not have a role that can be circumscribed to the being-there of the individual, but rather is itself a morphopoietic driving force in phylogeny. In other words, we need to shift from a view of the animal as passive, forged by the world through selection and enclosed in its epistemic bubble, to a view of the animal as endowed with double epistemic protagonism, both in the construction of a singular perspective on the world and in influencing the fitness dynamics by shifting selective pressures as a result of every epistemic emergence, be it due to discovery or creation.

But let's take a step back. Admitting animal subjectivity mainly means overcoming the prejudice of non-existence and non-presence without this turning the non-human into an entity that can be placed in the past, already completed in the process of intersection with its own context. In other words, it is necessary to question the humanistic tradition that considers non-human otherness as an entity incapable of rippling the surface of the already given. To use Heidegger's metaphor, we must therefore overcome the prejudice of epistemic opacity, which leads to the belief that the animal is stunned in the already given and "poor in world". It is therefore essential to acknowledge the principle of *intus-legere* in animality, that is, the ability to engrave the epidermis of appearance in order to construct one's own vision of the world, being present in one's own here-and-now. I think that this point is essential, not only to understand non-human animality, but also to explore the very concept of *intus-legere*, by discussing the disjunctive and computational assumption that today seems to prevail in the interpretation of the intellectual act. If we start from a Darwinian conception and take into consideration neurobiological and ethological findings, it is very difficult to accept a discontinuist view, even if we admit that some human aspects—such as language, writing, *techne*, science itself—may have favoured the dimensional widening of the epistemic sphere.

In my opinion, an evolutionary view, which aims at analysing the performative—and therefore also epistemic—context of different species, should treat the act of knowing in the same way as all the other functions or endowments of the animal being. Evolutionary continuism leads us to consider the great animal family as a set of presences characterized by predicates that are different but still shared, as part of a common heritage or as the result of adaptive convergences. This paradigm speaks of similarities and differences, but in a logic that does not contemplate gaps or discontinuity. Phylogenetic processes are based on the somatic correlation to a given environment/lifestyle, i.e. they are specialization events, which therefore cannot be framed in hierarchical terms, but only in terms of performative specificity. The latter, however, does not prescind from the common heritage, i.e. from the fact that many predicates of a species are not exclusive but shared because they emerged in a common ancestor.

Specialization is therefore a correlation that somehow reflects three aspects. First (i) a history of sharing by succession of predicative emergences, so that the similarity-difference operator does not grant anything to disjunctive essentialism or to the hierarchy of entities. Secondly (ii) the specific life-context on the basis of the relationship between the life project of a given species, and therefore performativities requiring given epistemic coordinates, and the resistance-opportunities presented by that context. Finally, specialization reflects (iii) the type of reproduction—for example the "r" or "K" strategies[5]—which also defines the ontogenetic adaptation needs of the species and therefore its virtuality gradient in the phenotypic conversion. In the face of all this, we must therefore expect that, just as there are many somatic morpho-functional organizations, there is a perspectival plurality of intersection typical of each species in the construction of an image of the world. On the other hand, the greatest mark of the phylogenetic success of a character is the

[5]The "r" reproduction strategy is found in all insects, almost all amphibians, reptiles and some bird species. Except for some important exceptions, these species give birth to a great quantity of offspring which is not cared for by the parents. The "K" reproduction strategy, on the contrary, is found in many bird species and in mammals, which have few cubs that are cared for by one or both parents.

principle of fitness, or replication, so it makes no sense to think that knowledge is detached from somatic involvement.

To this day, a profound dichotomy still separates our species from all the others. This is precisely because of our denial of the subjectivity of the non-human, which is reduced to a *res extensa* governed by mechanisms in a deterministic way, no matter how complex it is. Instead, we believe that the human being is free and to some extent independent of biological laws. One hundred and sixty years after the publication of *The Origin of Species* (1859), we must recognize that Darwinian thought is struggling to permeate the upper salons of our culture, so that the view of other species as categorically opposed to the human, who claims to carve out its identity in opposition to non-human beings, is a philosophical leitmotif that is still difficult to dismiss. Ernst Mayr rightly points out that Darwinism is on a collision course not only with the monotheistic tradition of the West but also with the three pillars of Western philosophy, namely Plato's essentialist conception of entities, Aristotle's hierarchical vision of categories, and above all stoicism and its interpretation of the human (Mayer 1997). When, in his essay *The Expression of Emotions in Man and Animals* (1872), Darwin draws an interspecific comparison of behavioural motives, he challenges not only the Cartesian idea that compares the animal to a machine, but a whole humanistic vision that claims to see man as the expression of a dimension that is opposed to and distant from all other species.

Indeed, the Cartesian interpretation does not only separate the human being from other species, but also indelibly separates the mind from the body. In this sense, placing animality at the centre as an "epistemic principle" means placing the body back at the centre of cognitive processes. Thus, doubt, imagination, reference, desire—just to give a few examples—can no longer be seen as de-somatized or ecstatic acts that can be attributed to: (i) an entity that has nothing to do with the condition of being-a-body; (ii) a process of delaying, suspending or aligning the body's reasons. This means overcoming the interpretation based on abstraction and computation that has always characterized the humanistic tradition. The latter, in pursuing the Promethean myth in the genealogical attribution of the human being, has transformed man into a kind of trickster who tries to circumvent the Epimethean principle through the intellectual process. The dualism of the founding myth of the two Titans is the root of the dichotomous thought that has de-somatized the act of knowledge. Also, this prejudice makes it impossible to understand the motives of *intus-legere*, because it makes it look as though the cognitive propensity of the human being were detached from human nature. This is why, in pedagogy and especially in teaching, we do not take into consideration children's intrinsic motivations, but we either appeal to a deontological factor or to a supererogatory one, at most dispensing compensations that, however, have nothing to do with the cognitive act itself. The results, incidentally, are under everyone's eyes, because a child learns only if involved, i.e. if interested and eager.

But, then, there is a question that needs to be answered: what is the act of knowing and, more generally, what is intelligence? In the 17th century, Thomas Hobbes wrote: "When man reasoneth, he does nothing else but conceive a sum total, from addition of parcels" (1909); if we followed him, we would have to answer that intelligence amounts to "computational skills". In reality, this is a question that has probably never been fully answered, because we have often taken human rationality as a model, trying to construct a computational entity that simulates its performativity. Reproducing thought through an algorithm means completing the reductionist parable started with Galileo and followed by Descartes' analytic geometry. What these thinkers did was try to "mechanize" reasoning—an aspiration that would then be taken up by Leibniz, an advocate of the need to describe the world by means of symbols, according to his symbolic logic. Indeed, as early as 1679, Leibniz imagined a calculating machine whose opening/closing principles—albeit in the form of holes for marbles—were the same as those used today in electronic microprocessors. Leibniz's thoughts would thus resonate in Charles Babbage's computer projects, in George Boole's logic and even in Alan Turing's machine.

But it was not only reasoning that was subjected to the logic of the great god of calculation. Science fiction, as it was delivered to us by authors such as Isaac Asimov or Philip Dick, has often presented a dystopian panorama with androids and cyborgs able to replicate emotions, desires and memories, because even the most intimate aspects of feeling were assimilated to a computational order. These visions reveal a profoundly anthropocentric way of understanding intelligence, based on the Promethean principle of de-somatization, that is, of expunging the Epimethean principle of being-a-body from epistemic performativity, adopting an interpretation free from the biological dimension of *intus-legere*. However, knowledge is born from a tension between the state of the subject, whose here-and-now is a perpetual oscillation between languor and gratification, and the objective resources they encounter based on the evidence coming from the epistemic intersection. It is inevitable, then, to consider *intus-legere* on the basis of the somatic dissatisfaction experienced between the oscillating state that projects the individual into a desire—i.e. a condition that is either not present or not achievable—and the cognitive resources that it possesses. Knowing is therefore the attempt to bridge this gap and soothe this languor. We could say then that an animal is brought to knowledge by the problems posed by its desire.

Now, the world is not made up of objective problems that the individual must solve, but of subjective goals that the individual wants to achieve. The concept of *intus-legere*, that is, of "reading-into" from which the term "intelligence" derives, therefore indicates the act of knowing how to look in depth and beyond appearances, in a condition of cognitive languor that involves the whole somatic dimension of the individual. It is in the body that we find the desire sustaining both the effort and the risk involved in sailing beyond the already given, beyond the clear appearance provided by one's epistemic condition. Deprived of desire, the body, in all its aspects, would not be led to set sail towards unknown lands: it would have no reason to do so. The act of knowing is the result of a tense relationship with the world that, even before being a *solution to a problem*, is immersion in a problematic condition that becomes a breeding ground for the *creation of a problem*. Computational machines are able to solve problems but, in fact, they do not *have* problems, because they lack the desiring tension that motivates the posing of a problem. A computer may be able to solve even complex problems and perhaps in a performatively superior way compared to a human being, but it solves the problems that *we* pose to it: the machine is devoid of desires and therefore has no tensional immersion in the problematic condition that brings out problems.

On the contrary, each animal is constantly immersed in problematicity, for one simple reason: because it desires. Desire is the result of one's relational history with the world; one could say that desire arises from the series of detachments that characterise one's phylo/ontogenetic emergence. In ontogeny, for example, a child constructs his own profile in the relationship with his mother, but then separates from it, retaining the emotional desire of conjugation within himself. In the same way, any species-specific character comes from dependencies introduced along the phylogenetic path, such as a predator's need to chase after a body in motion. An animal being emerges by recreating introjected conjugations that express themselves in desire, as if its heterotrophy extended beyond metabolism and characterized all the dimensions of animality, including cognitive ones.

I argue that the animal is structured by heteronomy because I believe that this aspect of relational dependence is the foundation of animality. In this sense, the languor that makes the animal oscillating and peripatetic underlies the need to go beyond, i.e. to overcome oneself as a tension that is discharged into epistemic singularity. To understand the act of penetrating reality, it is therefore necessary to focus on this tension that induces the subject to re-organize the structure of reality according to a different matrix for the conjugation of entities. We could say, then, that *intus-legere* means flexing reality according to one's own scheme of action, reconfiguring the given level of reality just as a perceptive Gestalt creates a morphology through an internal scheme for the conjugation of findings and the completion of the perceived. *Intus-legere* means, therefore, to bring out a new level of reality, to discover a possible configuration of the real—an opportunity hidden between the lines of appearance.

The concept of *intus-legere* involves going deep into the reading of reality, searching between the lines, not stopping at the already given—i.e. on previously structured reality layers—but building new organizations with respect to the virtual condition of reality. *Intus-legere*, therefore, is not only the act of extracting something that already exists, even if encrypted between the superficial meshes of appearance, but also the act of bringing out what *could* be, assigning a particular form to virtuality. An animal never dwells on the surface of the given: it connects different elements in the great sea of givenness, and prunes it singularly in order to discover—or rather "to bring out"—opportunities that are not only unknown but come only after the animal's epistemic act. An animal therefore understands reality by assigning it a form that does not precede the epistemological act: it does so by bringing out new organizations of reality, i.e. by carrying out processes of singular actualization. Well, to do this it is necessary to have an internal motivational system, i.e., inherences. Indeed, the animal "goes deep" into reality, which is a metaphor to say that it does not settle for appearances, because it desires. Now the question is this: can we consider a non-desiring entity to be intelligent? Is intelligence a computational skill or, vice versa, a creative organization of reality by virtue of a desiring condition?

I think that intelligence, as a dialogic-emergent act which treats reality as a range of possibilities to be discovered or probed, is never achieved without a motive, i.e. it is not the result of an objective process, but the outcome of an innermost desire. In other words, we think because we desire, not the other way around: it is desire that pushes us far beyond the veil of appearance to seek new opportunities, it is desire that transforms the world into a territory dotted with problems to be solved. This means that without desire it is unthinkable for any problem to emerge. Desire is also connected to feeling, to the pleasure felt in involvement, to the deeply somatic sensation of *thaumàzein*, where the epistemic dimension slips away from the coordinates of the already given and shows us an unexpected and unimagined continent. We cannot ignore this emotional participation in the cognitive process, which also feeds on the adrenaline and endorphin thrill caused by the vertigo of exploring unknown lands. Knowledge is an aesthetic act, nourished by beauty, by the serotonin effect of a cliff, and by the adrenaline sublime of infinity. *Intus-legere* takes place in the body, it is a conjugation of the reasons of the body sustained by feeling and by desire, which are the foundations of our animality.

Animals are endowed with intellectual abilities, that is, they all practice *intus-legere*, albeit in their plural way of knowledge, because they all desire. In other words, they have a dispositional system characterized by emotions and motivations that solicit the projection of an objective and, consequently, put them in a condition of constant existential problematicity. All species, including humans, conjugate their being in the world through motivational structures (chasing, collecting, searching, exploring) and states of feeling (fear, joy, thrill, excitement, curiosity). They define a tensional condition in their relationship with the world, which one could simplify as objectives to aim for, objectives which allow animals to experiment with creativity. It is in this emergent and intellectual act that the individual discovers new causal links, new temporal correlations, new categories. Going deep means putting things together in an unexpected way, and this is what the intellectual act amounts to: the subject has objectives because it is a body in continuous redefinition and in a relationship with the world. In order to understand the act of *intus-legere*, it is essential to enter the workshop of life, so as to clarify the behavioural mechanisms underlying activities such as the formulation of objectives, the evaluation of opportunities, involvement in the here-and-now, judgement of value, the relationship with otherness, curiosity and so on... i.e., the set of properties that allow one to widen one's epistemic space.

When one builds an entity with computational skills but without desires and calls it intelligent, one is actually operating a semantic-conceptual stretch with respect to the principle of *intus-legere*. My impression is that we want to build a house starting from the roof, with the result of creating a big calculator, yet unable to really read reality, let alone go deep into it, beyond appearances. To get closer to intelligence, conversely, we will need to better understand the reasons behind desire, before venturing into the continent of thought. Perhaps we should dismiss the disjunctive and computational approach typical of humanism and build a sort of "ethology of robotics" that starts from the simplest motivational systems to understand how to create an artificial entity capable of desiring in its relationship with the world. At this point, it is important to note this: only a machine that really desires is able to *intus-legere*, that is, to go deep into reality and build problems to be solved. But a desiring machine would inevitably take ownership of itself: it could no longer be defined as an instrument at the disposal of the human being, because it would be endowed with inherences that require forms of consent and negation, not absolute power.

6.3 The Somatic Multilayer of Knowledge

What do we normally mean by intelligence? In general, in order to talk about intelligent performance, we presuppose some functions or performativities in decoding the interface landscape. These include: (i) the ability to learn or to modify the performativity on the basis of one's experiences; (ii) the decision-making principle, i.e. ownership of the choice between one or more possible options on the basis of given criteria; (iii) competence in solving or overcoming problems posed by the context. They also include: (iv) the inductive or deductive inferential capacity, i.e. the ability to draw logical conclusions or laws from information available or in the process of being acquired; and (v) the ability to construct possible alternatives on the basis of probability indexes through operational biases. Finally, more such functions are: (vi) the flexibility to deal with situations of potential singularity, exploiting conditions of uncertainty to invent new operators; (vii) readiness to extract useful data in the intersection with the context, by means of logical operations of distinction and categorisation, but also through the emergence of orientation keys; and (viii) the ability to use concept maps to create new representative endowments.

Obviously these few examples do not claim to define the main chapters of intellectual activity, nor do they aspire to any degree of exhaustiveness, but it is clear that when we speak of artificial intelligence most of the time we focus on these performative guidelines (Bostrom 2014).[6] The tendency to favour the resolvent and computational character has made it possible for some performances—such as playing chess or demonstrating theorems—to be considered as leading intellectual skills when, on the contrary, they have proved to be easy to reach, even with programs that are, after all, quite simple—unlike others, such as orientation and decision-making. On the other hand, the approach to intelligence remains stuck in the field of performance, that is, in the evaluation of the ability and level of competence in carrying out a pre-established task. Also from the point of view of executive ownership, what is taken into consideration is the level of operational autonomy—the ability to conduct, continue or complete a pre-established task without guidance from a human operator—but not creative autonomy or operational involvement. When we then tried to obtain expert performances, capable of extracting objective-referenced information or finalised procedures, we did not at all approach the act of *intus-legere*, i.e. the act of flexing reality according to unprecedented coordinates. Rather, we operated through solutions that one could define as the exact opposite, because they are based on heuristics obtained from human knowledge.

---

[6]Bostrom shows that once we reach a level of intelligence comparable to the human one, it will only take a small step for machines to be equipped with a "superintelligence" that will be unattainable to us.

It is true that what we call artificial intelligence can easily outperform humans in some processing functions. However, AI does not deviate in any way from the preconditions that govern all machines: it is a set of features built to offer services for the use and consumption of the human being, but it has nothing in common with the intelligence of a living being, namely, the ability to build perspective singularities on the world. It is not easy to identify and define the property that allows a living entity to *intus-legere*—unlike a machine, which does nothing but calculate and simulate. Perhaps we will never be able to do so coherently and completely, precisely because we inevitably evaluate the process from within. However, we can at least say that somatic involvement is essential to it, because one does not learn in order to solve problems, demonstrate theorems or win a game of chess: each of these actions, in life, hides a deep volition, which is not implicit in its performance but transforms the context into a problem and only then seeks a solution. When we analyze an animal's intersection with external reality, we find that the act of *intus-legere* is a very common and, if you like, banal phenomenon, in its simple ordinariness. Even if (or precisely because) it produces maladaptive solutions, this involvement yields creativity because it can reconfigure the already given.

While a machine has horizontal functional levels and isochronic states, this is never the case with an animal. The reason for this is very simple, though extremely complex to analyze: an animal has a body, that is, a multi-layer intersection with the world. The body does not just investigate the context with sensors linked to a data processor, but grows by supporting or reflecting its various levels of intersection with it and creating interactive-emergent events between these different levels. For example, immunity knowledge mirrors the antigenic catalogue experienced through the relative lymphocyte population, but this level of intersection does not stop here because, through the cytokine medium, it intervenes in the neurobiological intersection with the world. And it is not only a matter of complexity, but of a different ontological principle, already in force in the simplicity of a cell.

The tendency to confuse intelligence and computation arises, as I said, from a claim that originated mainly from Descartes, but that had distant roots and is probably the most likely or most intuitive outcome of our way of perceiving the intellect as de-somatized. In my opinion, the de-somatization of the intellectual act stems from a number of reasons that should not be underestimated. First, there is (i) the fear of death or the need to desperately attach oneself to an existence that goes beyond transience (this, in this sense, justifies an immaterial presence that inhabits and drives the body). Secondly, there is (ii) the evident fact that the mind has a different degenerative process compared to the body, i.e. that a decrepit body can correspond to a young mind, especially in those who have exercised their intellect during their lifetime—including those who debated such matters and created a tradition about them. Another reason is (iii) the impression that, as we age and distance ourselves from the somatic impulses of youth, we become wiser and more rational, that is, there is a sort of purification of the mind. Then, we tend to have (iv) the sensation that the body, with its needs and impulses, diverts us from objective evaluations, i.e. that the somatic dimension, immersing us in impulse and fruition, obscures the mind and is a sort of prison for rationality and contemplation, so that in order to think clearly we need to distance ourselves from the body.

And there are also other reasons why we tend to de-somatize the intellect. There is (v) the idea that the explorative and perlustrative tendency, which is a phylogenetic characteristic of omnivores, has instead an ecstatic connotation and can be ascribed to a primordial nostalgia towards an ethereal condition lost in the fall into the telluric, because it is a "de-sidera" or lack of the existential hyperuranial dimension. We should also remember (vi) the prejudice that the ability to doubt and the practice of counter-intuition, as well as the use of tools, techniques and theories, are Promethean products (i.e., a way to go against nature) and not Epimethean (following the imitative and introjective nature that characterizes primates, a phylogenetic connotation emphasized in the human being). Finally, there is (vii) the need to mark a distance between the human and the other species, a need which is also the result of a perspective distortion—much like seeing all non-Hellenic populations as barbarians—but which is justified in the progressively growing need to exploit other species.

If on the one hand I understand the reasons for the de-somatization of the mentalistic dimension, on the other hand it is necessary to acknowledge the misleading results of an immanent view of the *res cogitans*—results that still accompany us today. The dualist paradigm, in fact, creates two different and disjointed domains, and tries to extract the entity or phenomenon through "oppositional enucleation", that is, by dichotomous operations. This means that in order to shed light on something it is necessary to identify a contraposition: what-is, and presents particular predicates, as opposed to what-is-not and lacks these predicates. Quantum physics is often difficult precisely because it breaks this claim of a supposed still-image for which an entity emerges by detaching itself from a background. The *res cogitans*, in fact, can take on different dualist conjugations even in a non transcendent view, for example in the idea that subjectivity is consciousness (i.e. an entity detached from the reasons of the body, called to oversee its processes), or that it is an undefined *res informatica* running in the embodied hardware of the brain. These perspectives also follow the tradition of de-somatization: they do not consider the intellectual act as the result of the dimensional plurality of being-a-body, i.e. a "multi-layer of intersection" with external reality, but seek *intus-legere* in something external to the body, dichotomously seen as opposite to it. Somatizing the intellectual act does not mean, however, assuming a sort of horizontal intersection, but on the contrary admitting multiple levels of somatic relationship between the individual and the world, each with its own cognitive specificity and equally able to interact with other levels or to make other levels possible through emergent qualities.

Somatizing knowledge means recognizing this plurality of the body in giving rise to a multiplicity of landscapes, which are interrelated in providing us with emergent levels of knowledge. If we take consciousness, for example, we immediately realize that it rests on other pre-conscious, unconscious or not necessarily neurobiological dimensions; it does not arise out of nothing, as a disjointed entity. Somatizing knowledge therefore means two things. It means (1) recognising this "dimensional multiplicity", where each interface layer determines precise fields of actualisation, that is, of organisation of reality in order to give life to orientative, evaluative and operative meanings. And it also means (2) admitting the "non-prior", that is, present, referred, situated and non-dichotomous/disjoined character of every level of knowledge, by virtue of the predicative relationship (the predicate is born of the very act of being-in-relationship) and not of a supposed predicative essence. Every piece of knowledge is therefore a form of growth and metamorphosis of the level in question and, consequently, of the multilayer. It is not a simple elaboration. Moreover, no level can exist without the underlying one, which entails relational predications giving rise to qualities that supervene on the characters of the level itself. In other words, it is impossible for consciousness to emerge from nothing, for the unconscious not to be situated on precise metabolic dynamics, and so on. Consciousness is therefore not a separate entity, a software that runs on somatic hardware: it is a state of the body, one of the many possible presentations of bodily knowledge.

At this point, I would like to underline that most of our views on intelligence are still strongly connected to this dichotomous-disjunctive vision, not allowing us to fully understand the plurality of cognitive processes *in* the body and *between* different bodies. First of all, intelligence is seen as a universal parameter, unlike all other expressions of the body, removing it, in fact, from the Darwinian paradigm of specialization. Banally, we would never ask ourselves if a bat is more locomotor than a dolphin, if a bear is more endocrine than a wolf, if a pig is more gastrointestinal than a cow, or if a cat is more immune than a rat, yet there are countless studies and articles that claim to identify the most intelligent animal. This implicit denial of the adaptive character, and therefore of the specialization, of the intellectual act, this need to consider knowledge as universal and not in a logic of plurality, is certainly the fruit of de-somatization, that is, of the habit of not considering *intus-legere* as an epimeletic dimension. In this sense, even when we refer to other species, almost forgetting the humanistic imperative, we still continue to apply anthropocentric prejudices, seeking in the heterospecific not their peculiarities but some kind of approximation to us, the holders of the cognitive universal.

A second consequence, also referable to this disincarnated view of knowledge, is the idea that the mind can be transferred, exactly as if it were a digital sequence, from one container to another. In other words, there is the idea that one could potentially implement the so-called "mind up/down-loading" that has been so popular in science fiction, starting from Philip Dick's narratives. This sort of techno-mediated metempsychosis, in the mirages of the new digital dualism, animates a thousand projections. A first example is (i) the grafting of false memories or the compression of mental time and the possibility of living months of experience in the space of a few seconds. Then there is (ii) the creation of back-up copies that are always available, capable of giving us a second chance or perhaps of turning into slaves to be inserted into computer supports at our service. Another example is (iii) the transfer of mental content into strings of compressible data for interstellar journeys, with potential time travel if moving faster than the speed of light. Then we have (iv) teleportation *à la* Star Trek, so that it is no longer necessary to carry the body, because one can reconfigure it at any time and place; as well as (v) the source of eternal youth and immortality by means of clonation, with the risk of potential duplications of one's Self. Another instance is (vi) the emergence of new transcendences, where the true death of the individual only occurs if the corresponding file is damaged, since the body, provided it has any meaning at all, is only a container that can be reconstructed. Then there is (vii) the possibility of living within the Internet and giving rise to super-organisms, where several individuals can merge together or enter into non-human bodies; a finally (viii) the development of new risks and the consequent need of measures to avoid the potential hacking of the mind.

But is this really the case? I find these narrative fantasies extremely amusing, especially in terms of psychological analysis; in them, in fact, we see a liquid identity pushed to such an extent that the very concept of individual becomes rather blurry, leaving Pirandello's perplexities on a remote background. These are transrealist experiments, interesting because they bring to the fore doubts and fears that we are already feeling in the current excess of virtual experiences and in the immersion in a reality which, as underlined by Zygmunt Bauman (2000), is becoming more and more fluid and therefore open to possibilities. However, from the point of view of the intellectual and intentional presence of the subject, i.e. of what the mental identity of the individual actually means, these projections differ little from the humanistic tradition of Cartesian dualism. They also pave the way for hypothesizing a bodiless or artificial intelligence which, in my opinion, totally misses the target of *intus-legere*, i.e. that particular propension-ability to penetrate the epidermis of reality by creating a problem, before finding a solution. The anthropocentric absurdity lies in recognizing intelligence in a computational machine, while denying it to other animals, still considered "reactive machines".

If, on the contrary, we consider intelligence as the ability of the body to construct its own reality dimension through actualization processes, things will appear completely different, for the following reasons. (i) No entity knows the world through a disjunction or along a well-defined boundary of "understanding": knowledge, along the various somatic layers, is a process of heteronomic growth, exactly like the development of a tree's foliage along the evolutionary matrix offered by light. So, the resulting form is a reflection of the context and the latter is an interpretation operated by the heritage contents of the tree itself. (ii) We cannot define sensory accesses as simple channels that convey objective information present in the world, since each sensory organ can carry out several intersections with the world and always works systemically with the whole body and with the other sensory accesses. So, each experience modifies the proximal plane of intersection, and therefore knowledge is not the simple processing of objectively given inputs. (iii) There is no cognition that is not also metacognition because, even if knowledge can recall a performance or an elaboration process, in reality it is also always an exercise, that is, a growth differential, of all the cognitive meta-components called to carry out that activity, be they the type of attention, the motivational structures, memory in its various forms, or the internal organization of the system. Finally, (iv) every cognitive process implies an effort that must be compatible with the motivational and emotional involvement felt by the individual in its here-and-now, that is, by the whole state of the body, since knowledge is always born from a somatic projection, even when it seems to be an end in itself.

Therefore, the body is not the instrument of knowledge, but knowledge is the expression of the body. We experience our bodiliness within a precise dimension, even if our heritage is not a bubble that imprisons us but an elastic mat that allows us to jump onto new levels of intersection with reality through multiple paths, for example with the help of new technologies, theories, techniques, or hybridizations with other species. By observing the behaviour of other animals, just to give an example, the human being has not only been able to trace an ethographic profile, but has also understood the specificity of the human perspective. Doubt, which has always played a major role in the cognitive processes, has been fed by anthropo-decentring processes that have been possible thanks to the epiphanic encounter with animal otherness and to the shift of intersection brought about by the advent of new technologies. We are led to believe that a new technology is only an enhancement of the intersection: sometimes it is, but most of the time it looks more like a virus than a probiotic. New technology tends to reorganize our image of the world, contributing to fuel in us that sense of partiality or, if you like, that awareness of our "domain of validity", which is the basis of counterintuition and research, transforming our heritage into a sort of "epistemological obstacle", to use Bachelard's words (2002).

But this does not mean that knowledge is a departure from our phylogenetic nature or that our legacy is simply an obstacle to the most daring scientific hypotheses, such as Einstein's physics or Darwin's evolutionary theory. It is thanks to our nature that we have been able to identify, from the heritage perspective, a domain of validity and not an absolute, however paradoxical such a statement may seem. Where, then, does doubt come from? Perhaps from the fact that we have taken leave of our bodiliness? From our nostalgia for our supposed ancestral sidereal dwelling? From our ecstatic access? I do not think so. Doubt arises from the experience of going beyond, from the aesthetic dimension of the Aristotelian *thaumàzein*, where the human being feels overwhelmed, thrilled and amazed in front of their own smallness. We could say that doubt is akin to the sense of the sublime, with the languor that leads us to leave our native country to venture into unknown lands. Doubt arises from dealing with an overpowering nature that fascinates and frightens us, with the experience of the gap between our resources and the desiring openness. It is a feeling that is strongly rooted in the body and cannot be disembodied in any way. More than a cause for reflection, an epoché of the already given, a fantasy about the possible, or a hallucination… doubt is a thrill.

As such, it is an oscillation that shakes the entire foundation of the body. Long before being rationalized into what we call critical thinking, the intersection has undergone a shift, an excentralization with respect to the given orbit; it has undergone the "*clinamen*" with respect to the gravitational pull so dear to Epicurean thought. It is neither a coincidence nor a necessity that dictates its coordinates, but an encounter that takes place on a precise threshold. Doubt arises from an epiphany, from the appearance of the non-already-given possible, which only on the edge of the connection can have its own predicative value. The body is a relational structure, an entity that already foresees in itself the heteronomic deviation operated from the outside. Knowledge is therefore precisely this act of forcing the meshes of the already given through the relational emergence, a deviation from the predetermined trajectory, which is therefore neither accidental nor necessary. One can say, then, that doubt is the manifestation of a condition of uncertainty and the testimony of a decentralization that has taken place: therefore, it is not detachment from the body, but a state of the body itself, so that the dubitative act always has its deepest reasons in the somatic dimensions of the emotional and motivational involvement.

It is therefore necessary to consider the process of decentralization, that is, to investigate the somatic movements that remove the individual from the gravitation of the past, and take them away from already travelled roads. In my opinion, we cannot ignore the motivational principle that, in human beings, is expressed in curiosity, in the interest to probe and test, in the imitative propensity, or in the need for social affirmation. No research is ever an end in itself, just as no one ventures into unknown territories without emotional support. Intelligence is born from the global involvement of the body, albeit through different levels of intersection with external reality. And it is precisely at the relational threshold with the world, and not in reflexive detachment, that decentralisation occurs: this eccentric and deviated trajectory is made possible not in a solipsistic and epurative closure, but in the participation in the symphony of the world. Therefore, there can be no decentralization in self-reference: only a hybridation, i.e. a decentralization from the outside, can lead this centrifugation to move away from the heritage. Doubt, therefore, is nourished not by a sublimation of the human spirit from bodiliness, but by giving strong voice to the somatization of otherness, in that process of projection into the other that allows us, for example, to fly with the wings of a bird and to project ourselves into the motion of the planets through a telescope. Doubt arises when we suddenly try to change the fixed point of our epistemology and realize that reality consents.

I believe that all this is part of human nature: we are naturally inclined to question our perspective, because, once again, we are naturally inclined to build hybridizations with other beings. Knowledge passes through hallucinations that often derive from psychotropic effects, related to substances that we introduce with food or that are transmitted to us by states of metabolic alteration attributable to diseases or infections. Phylogeny has given us a brain that has incredible functional and structural plasticity, a redundancy that allows it to introject everything around it. I am astonished when I still hear people say that the human being is an incomplete, partial entity, endowed with primitivisms, because saying such things means fully ignoring the anatomical-physiological findings of the human being. This brings Darwinism back to the fairy tale of Epimetheus and Prometheus, so as to continue to live within the comfortable humanistic mansion. And it is not just about a structural evaluation, so to speak, of the human being—involving, for example, the complexity of the brain, the particular specialization of the eye and of the oculo-manual coordination, the adaptive structure of the pelvis or of the occipital hole, the differentiation between posterior and anterior train (also the result of a specialization), or the neoteny and the loss of the mating season. Rather, the point is to acknowledge a precise ethological heritage due to which we are a species that can be well represented in terms of prevailing motivations, communicative and interactive patterns, parental and social structures in various rituals, and feeding methods modified by the agricultural revolution, but still well traceable in some practices.

Just to give an example, what has always amazed me is the total absence of serious studies on the motivational structures of the human being—obviously I mean the intrinsic propensities, like the predatory instinct in a cat—except for some praiseworthy exceptions which, all in all, seem not to fully grasp the specific character of the human being. First of all, the motivations are mistaken for the objectives, that is, for the external causes capable of arousing a motivational behaviour or of acting as targets in the expression of the motivation. It would be as if an ethologist believed that cats are motivated by balls instead of their propensity to chase everything that is in motion. In order to list the motivations of a species, i.e. the copulative tendencies which characterize it, one must not focus on the target or the accidental elicitor, but on the type of expressive orientation—one could say the transitive-verb underlying a given behaviour. In order to teach a dog something, for example, it is common practice among dog educators or trainers to use motivational involvement, perhaps within a playful framework but still in a motivational key, starting from good knowledge of these aspects about dogs. If I work with an animal like a cat, I will try to act on the predatory instinct, whereas with a dog I'll focus on the collaborative or competitive one (as well as the predatory one), and with a rabbit I will certainly not choose the predatory one. Why, then, in psychology and pedagogy, is there no strict and detailed analysis of the motivations prevailing in the human being? The answer I give myself is that we do not even contemplate that human beings might have intrinsic motivations, especially if we remain attached to the humanistic framework of Prometheanism.

Working with school children (Marchesini 2016), I realized how important motivational involvement is: teaching should rely on these intrinsic motivations if it cares for children's interests and acknowledges their active role in knowledge. In the 1990s I noticed that kids had no problem learning the name and corresponding images of thousands of Pokemons, yet they struggled to learn the far simpler things they were told in class. When I discussed the issue with the teachers, I was told: "it's because they have fun with Pokemon". Of course, I thought, but this is no answer, because I could have said: "Yes, that much is obvious, but *why* do they have fun with that?". The basic point is that in Pokemon-didactics—let me call it that—children are completely involved because they are doing things that are deeply rooted in the motivational prevalences of the human being. Children have fun because there is coherence between their inclinations and the activities that lead them to acquire information—which, alas, in this case is related to Pokemon.

If we consider human beings in their most frequent activities, if we do not dwell on the target of the given action but on the verb-copula of the action itself, we will realize that the acts of gathering, putting together and collecting represent a motivational coordinate that underlies many human occupations, and is the prevailing conjugation of the intersection between individual and context. A child will pick daisies in a meadow, shells on the beach, stones in a riverbed, mushrooms in a forest, berries or fruit from a plant, and so forth. We can put in brackets the what and the where, as well as the way of performing the action; what remains is one same verb: to collect. Even as adults, this motivational coordinate is an integral part of our hobbies and our work; it is a source of both commitment and fun for us, since the two things are not mutually exclusive. We collect spontaneously—almost by internal motion—withstanding efforts, nuisances and burdens, because desire prevails over the expressive cost, and in so doing we learn. And we do so by putting the whole body at the service of knowledge: from the necessary perceptual specialisations to the emotional markings, which are the foundation of the memories that will define a given experiential horizon and a relative biographical characterization.

So, I decided to try to list ten motivational coordinates that predominate in human beings, adopting a properly ethological approach. These are the motivations that I have found to be prevalent in children:

1. *sillegic*: a tendency to collect, gather, catalogue, put together by similarity, compare, pay attention to differences;
2. *epimeletic*: a tendency to take care of and help others, encourage growth, care for and protect, clean and groom, reassure, adopt, put in order;
3. *mimetic*: a tendency to imitate, re-produce, represent, get infected, be fascinated by diversity, transform each phenomenon into an inspiration;
4. *social*: a tendency to form a group, conform to common styles, desire the recognition of others, seek consensus and accreditation, participate in shared tendencies;
5. *exploratory*: a tendency to go beyond appearances, to look in depth, to evaluate the opportunities of a context, to widen the space of knowledge, to look for the requirements of a problem;
6. *introjective*: a tendency to incorporate external elements in order to achieve one's objective, to use objects as tools, to use external phenomena in a predictive sense;
7. *collaborative*: a tendency to build team dynamics in order to achieve results, to create systems of belonging, to think in a cooperative and reciprocal way;
8. *perlustrative*: a tendency to discover new territories, to move nomadically in order to seek new opportunities, to venture outside known places;
9. *possessive*: a tendency to defend one's resources, to seize new ones, to put in place mechanisms of subtraction from others, such as hiding or threatening;
10. *competitive*: a tendency to implement competitive behaviours, to compete with others, to try to win or surpass others or to excel, to emulate the best performances.

When I speak of somatic involvement, I mean to underline that the body as a whole participates in the process of projection-hybridisation with the external world which is the basis of decentralization, doubt and problem creation, i.e. the systemic state that pushes the individual to *intus-legere*. If, for example, we take into consideration the epimeletic projection, which is the basis of all care behaviours—such as dedication, diligence, completeness, order, empathy, organization—we immediately realize that cognition is not only the cold processing of incoming data, but it is endocrine, affective, tactile, parental (just to make some examples). The much-reproached aggressiveness is actually a fundamental resource in the stubbornness, competition, and fighting that play a large part in all research activity and which feed on adrenal surrenal and gonadic flows, but are also affected by the immune and metabolic system. Indeed, it is not possible to understand the intellectual work of research by separating it from the desire for social affirmation, and this in turn is affected by sexual and affective coordinates. Experience is not introduced through a data deposit, but on the basis of matrices of differential growth in the body, so that even the senses are not already-given channels but, like the foliage of a tree, respond to the coordinates of light that determined their morphogenesis.

6.4 Cognitive Plurality in the Animal World

If intelligence is the manifestation of being-a-body, it naturally follows that there is cognitive plurality among different species. Of course, plurality does not mean discontinuity, if one understands Darwinian evolutionism, because every species shares some characteristics with other species by homology (common heredity) and by analogy (adapted convergence). In other words, it is necessary to abandon the concept of "cognitive universal", so dear to the promoters of the computational vision. Talking about animal intelligence also means abandoning other interpretative keys of animal behaviour, including: (1) the analytical view of behavioural motives, which is found in the concepts of instinct and conditioning; (2) the deterministic view of these motives, inherent in the idea that given a drive there is an instinct or given a stimulus there is a response; (3) the timelessness of the animal's positioning, i.e. the negation of its here-and-now and its *telos*; (4) the exclusively sensory vision of its interface with the world.

The 20th century was a time of counter-reformation with respect to the Darwinian revolution, bringing Cartesian mechanism back into fashion. The legacy of the great English naturalist, in fact, urged the overcoming of the "human vs non-human" dichotomy, leading instead to a comparative view based on similarity-continuity in difference. The goal was to refute the absurd dichotomic disjunction between the human and other living beings, and to sanction the groundlessness of such categorical separation. However, the affirmation of interpretive mechanism carried out a real humanistic restoration, which aimed to undermine—but I would rather say to annihilate—the philosophical assumptions and ontological considerations put forward by Darwin. The aim was to confine his work within well-circumscribed fields so as to prevent it from overflowing and upsetting the anthropocentric system on which the humanistic tradition was and is founded. First, we tried to subtract the human being from the coordinates of phylogeny, with a very successful operation carried out by Arnold Gehlen through the idea of the human being's biological poverty or incompleteness, thus relegating the Darwinian explanation within a very narrow field. Contiguity with other animals was then projected into a past from which man allegedly emancipated himself thanks to his exonerative power, and was therefore linked to concepts such as those of ancestrality or partiality.

The subsequent move was to use the Cartesian automaton to explain three things. First (i) the interface of non-human animals: this was reduced to Jakob von Uexküll's *Umwelt*, that is, the total immersion and clouding of the non-human in the elicitative universe, as opposed to man's sole access to full perspective, in Heideggerian terms. Second (ii) the innate, which was explained through the determinism of instincts, as prefigured by Niko Tinbergen's psychoenergetics: in this view, the non-human is denied foreshadowing-reflection-decision so as to reduce its expression to consumption mechanisms. Finally (iii) learning was explained through the associationistic concepts of stimulus-reflection structured by the idea that the non-human does not build knowledge but develops conditioning mechanisms, as claimed by Skinner's behaviourist school. Based on these three coordinates, the anthropocentric restoration closed the gap opened by Darwinian thought, shifting the dichotomy onto other differences.

These three operations have held up for almost a century, and even today we insist on explaining the non-human in terms of closed interface, deterministic imperative of instincts, and cogent conditioning. And these three interpretative coordinates are the best that Descartes could have hoped for to support his idea that the animal is a machine. In fact, the puppet-animal that comes out of these postulates is not open to external reality, lacks an internal world, is devoid of a behavioural system, is driven by switches and, with the ultimate paradox, functions like a cybernetic machine, i.e. through feedback mechanisms. Which means that, in this paradigmatic logic, any computer machine has a much more complex functional-explanatory matrix than an animal organism, which is reduced to the status of a behavioural thermostat. In fact, both psychoenergetics and behaviourism explain the animal through switches, based on a 1:1 structure-function ratio, by which the structure translates into its function in an algorithmic way. The drive released by the energy inherent in the consuming act and the reflex triggered by the stimulus of the associative explanation imply that other species lack the presuppositions of the intellectual faculty, which relegates them to the status of a trigger-based mechanism.

My approach, instead, attributes a series of abilities to the animal:

1. the ability to construct a prefigured objective, emerging from the dialectic between one's motivation and the range of possibilities offered by the external conditions;
2. the ownership of a positional state that is internal with respect to the external context and is systemic, therefore not based on disjoint elements;
3. the continuous relationship with everything that surrounds it, adapting its responses not to the modulation of already-given performativities, but to the realization of new endowments;
4. presence, i.e. being able to be present in the moment not by being completely immersed in the here-and-now but presenting a diachronic structure;
5. the ability to conceive problems or go beyond appearances to bring out new opportunities;
6. the capacity for problematization or evaluative immersion in a problem, which requires an evaluation of the gap between the current condition and the pursued one, but also full understanding of the structural requirements of the problem;
7. the ability to solve or reduce the scope of a problem through creativity and flexibility.

If we address the expressive and evolutionary question within a problematic model—the animal as a creator of problems and not as a reflex-driven consumer—we give back to the non-human its full power of expression. This power, in fact, arises from a subjective experience in the here-and-now, based on a properly biographical heritage and projected into an inherent *telos* revolving around individuality, i.e. the diachronic singularity of the heterospecific. To acknowledge "animal intelligence" means to modify the explanatory coordinates of behaviour, interface and learning in non-human species, and not to simply place the mentalistic explanation side by side with the current paradigms. With respect to this, one can say that considering the heterospecific as a subject endowed with a mind means several things. It means (i) having a systemic approach to animal ontic, i.e., considering the different behavioural components—whether or not they are phylogenetic—as parts of a system that evolves as a whole and creates a dialogical environment between the components. It also means (ii) having a monistic-emergential approach that considers intelligence as an expression of the body as a whole. Likewise, it involves (iii) considering heritage components as multi-functional endowments: maps and not threads that give the subject freedom both in the present experience and in its projection into the future, but also in its way of recapitulating or giving meaning to the past. Attributing a mind to animals means (iv) considering the mind as an internal world that has a precise positioning in time, and therefore a here-and-now, and as a project that has to do with a biographical structure.[7] Also, this involves (v) evaluating the internal world as an interface that makes the subject partial to the world, but not closed-off in a predetermined way within an *Umwelt* established by phylogeny. Rather, the interface is a mobile threshold that changes continuously on the basis of the subject's experience and state in the present (proximal threshold of experience) in a logic of non-fixity of, and power over, the interface.

In the light of these considerations, it becomes evident that the modification of the explanatory paradigm in the cognitive sense does not only concern the problem of consciousness, which is important but not decisive and does not disambiguate the reflection on intellectual activity as a whole. Indeed, consciousness is one of the *many* levels of intersection between the body and the world, probably specialized in focusing attention on particular states of the body (suffering, well-being, functions, pain), on somatic markers (emotions, motivations, arousal), on representational or reflexive structures, on external events obtained through sensory windows (sensations), and on the overall biographical state or self-consciousness. This function or

---

[7] Here biographical structure or being the "subject-of-a-life" does not mean the banal unravelling of the ontogenetic process, but ownership of one' s present (how to live one's here-and-now), projection into the future (what objectives to prefigure and pursue), and the memory of one's past. The expression "subject-of-a-life" was used by Tom Regan. According to him, «individuals are subjects-of-a-life if they have beliefs and desires; perception, memory, and a sense of the future, including their own future; an emotional life together with feelings of pleasure and pain; preference- and welfare-interests; the ability to initiate action in pursuit of their desires and goals; a psychophysical identity over time; and an individual welfare in the sense that their experiential life fares well or ill for them, logically independently of their utility for others and logically independently of their being the object of anyone else's interests». Regan (1983: 243).

state of focus allows the subject, in certain situations, to engage as many cognitive resources as possible within a specific cognitive function, wherever the novelty of the situation-problem requires it. In fact, the best cognitive performance, when one requires access to the entire database of the brain and has to perform serial operations in parallel or in a flow diagram, finds an impediment in the apparatus of consciousness, precisely because of its processual module based on perimetration and sequentiality. If we want to have an insight or remember a word we are struggling to find, the best way is to sleep, walk, or think about something else, which means to *exclude* consciousness from the problem. It is not correct, therefore, to circumscribe the act of *intus-legere* within the domain of consciousness.

The problem of knowledge raises some important questions: (i) what kind of intelligence is present in different species?; (ii) is it possible to measure different intelligences or should one stick to comparisons, as in anatomy? and if so, what kind of comparisons can be made?; (iii) is it possible for a human being to know "what it is like to be a bat" or to empathize with a different intelligence? To the first question I'd answer that each species necessarily has its own intelligence, because it is called to dwell in its own problematic spaces. To the second question I reply right away that it is not possible to make a sort of hierarchical assessment of the intellectual performances of different species, but it is possible to compare them with respect to individual intellectual functions, using the operators of homology-analogy already used to evaluate other anatomical-functional structures. As for the third question, I think that the far-too-rhetorical doubt proposed by Thomas Nagel is misleading. Animals have species-specific peculiarities, but they are not aliens that come from a universe completely different from our own. The diversity of each species is also a function of similarity and, once again, Darwin can help us here, with his book *The Expression of Emotions in Man and Animals*. If it were impossible to understand the *intus-legere* of an animal, not only would it be impossible for us to relate to any other species than the human one, but ethologists would have to find another job, as the analysis of animal welfare would be vain talk.

The basic assumption of cognitive ethology is that mental performance is an adaptive function just like any other organic performance, and is therefore calibrated to the specific needs encountered along a species' phylogeny. As such it is based on the concept of fitness, i.e. of reproductive advantage related to the lifestyle and environment of a species. From this point of view it does not make sense to speak of mind and cognitive processes in the singular, using the human being as a measure of the variability of cognition in the world: intelligence, as a complex set of trends and cognitive properties, cannot be subjected to a comparison referred to a model, but must be considered in a plural way. Intelligence is a biological function that—like sensoriality, the anatomy of the limbs, or digestion—presents itself in the animal universe through a multiplicity of vocations and attitudes, most of the time so different that they cannot be superimposed on each other. However, these various attitudes also bear a similarity that descends from three things. First, (1) from the universality of the physical characteristics of the context, which is why a dog playing frisbee must know how to calculate its trajectory exactly as a baseball player does with a ball. Secondly, (2) from the phylogenetic closeness between two species, i.e. from the rate of trans-specific homology, so that the parental and social attitudes of chimpanzees and bonobos have characteristics in common with human beings for the simple fact that such predicates descend from a common ancestor. Finally, the similarity comes (3) from sharing the same fitness selectors, i.e. from the level of analogies or adaptive convergence, since all arboreal animals must know how to perform a detour, if they do not want to perish when a given branch no longer has any use.

Studying the multi-shapedness of living beings, even in terms of cognitive variability (similarity-difference), is good training to understand that diversity is not inferiority and that it is always misleading to face multiplicity with an obsession with hierarchies. Cognitive activities are therefore not excellences that bring other animals closer to the human being, but functions that are strictly related to the adaptive setbacks, be they simple or complex, and which for this reason tend to give each species a particular idea of the world and of its own being in the world. Therefore it is necessary to refer to them in the plural, but also to admit that they are a further biological tool to realize the diversity of phylogenetic heritage, just as the sensorial multiplicity endows each species with a peculiar way of immersing itself in reality. However, despite the apparent simplicity of this statement, admitting that cognitive performativity has plural expressions means suspending two of the most important clichés on the mind, i.e.: (1) that man represents the measure with which to judge other animals; (2) that man sums up all the cognitive faculties present in the world of non-human animals.

When analyzing the different morphology of the limbs, the specialization of the fur, the various eating peculiarities, the different kinds of access to the world operated by the senses, etc. we have no prejudices or difficulties in applying Darwin's dictate of comparison by homology (common descent) and analogy (adaptive convergence). Yet this is not the case for cognition, which appears as the only performance that i: (i) denied to animals altogether, by those who consider them driven by automatisms and wish to maintain a clear ontological separation between human beings and other species. Alternatively, cognition (ii) framed in terms of graduality, so that between the human being and all other species there would be quantitative differences, with man being the peak. Finally, cognition is sometimes only (iii) attributed to other species by projection through trivial anthropomorphisms. I personally disagree with all three of these options, because I believe that each species has its own intelligence, comparable but not measurable in hierarchical terms with that of others. If we accept that human cognitivity is a point of view, adaptively calibrated in terms of peculiar selective coordinates, then we can understand that the universe of the living is made up of a multiplicity of species-specific intellectual qualities.

Cognitive plurality, in other words, means that the paths and specializations taken by different animals in "knowing the world" have followed various adaptive logics through several paths, made of divergences and convergences, so that the intellectual perspective is a further instrument called to actualize a particular configuration of reality. This view admits that an animal's performativity in its interaction with reality fully responds to the risks and opportunities that the particular adaptive situation presents, in relation to the singular evolutionary trajectory of its species. Therefore, there is a cognitive plurality because the challenges faced by different species are plural. The intelligence of a species is therefore not a function that brings it closer to the species *Homo sapiens*, but a characterization of its intersection with reality calibrated to its specific needs, just like all other functional devices. The mind is called to reorganize reality and bends it according to peculiar configuration coordinates. Cognitive performativity therefore increases perspective partiality: the mind reinforces singularity as the unrepeatability and irreversibility of the individual. Cognitive plurality means, therefore, that two species facing the same situation will problematize it in a different way, leading to experiences, ideas, solutions and lessons that cannot be superimposed.

We must therefore conclude that both the denial of animal cognitivity and the attribution to other species of a quasi-human cognitivity (the animal as *minus habens*) are the result of anthropocentrism, a paradigm that does not leave room for an objective and selective evaluation of the intellectual function. When we talk about cognitive performativity we mean precise ways of "reading reality" or different "perspective configurations" able to give ad hoc adaptive rewards, which are not valid in an absolute sense but only in relation to the lifestyle and context of the species. Therefore, every cognitive performance, being a vocation, always carries strengths and weaknesses with it: in fact, if we transport the cognitivity of a species outside its context/lifestyle, i.e. outside its cognitive demand, most of the time a strength becomes a weakness. For example, a dog's cognitive ability to focus on its objective and pursue it regardless of context variables sometimes makes its behaviour appear stubbornly stupid. If we take one species as the cognitive model of another, we will inevitably condemn the latter to insufficiency. There are no quantitative differences when it comes to adaptive correlation, but there are differences in terms of specialization and functional constraints.

Intus-legere is never unrelated to a certain way of being, that is, to the cognitive system characterising a particular species and denoting a peculiar form of intelligence. The ability of dogs to interpret relationships and to operate like sophisticated politicians within the dynamics of a team makes them capable of consultation and collaboration: this makes them appear gifted with great intelligence in our eyes, and we are bewitched by how they place themselves at the service of the group. Cats, on the other hand, being operational soloists, present a resolvent intelligence, different from the social one typical of dogs: they are better at solving puzzles, but less inclined to learn collaborative or obedient practices. Therefore, analysing dogs and cats through the anthropocentric lens, we will alternately reward the former or the latter But this evaluative metric is simply wrong: if we do not use it for the sensory apparatus we should refrain from doing so for the processing apparatus. The different ways of reading reality—experiencing, learning, resolving, evaluating, judging, etc.—can be considered nothing more than different ways of "asking questions to the world". In the same ion, the subject's propositional attitudes towards their here-and-now—how to fear, believe, hope—come from compositions of endowments that are strongly correlated to their species, for example, type of emotional structure, motivational coordinates, the orientative, evaluative and operative representations involved, etc.

A species' intellectual peculiarity lies precisely in the fact that, according to the specific cognitive endowments, the mind will ask specific questions to the world and will make use of equally adaptive biases in cataloguing the problematization patterns; consequently, it will construct propositional attitudes that are affected by its phylo/onto-genetic history. A prey animal, such as a horse or a rabbit, is no less intelligent than a predator, such as a dog or a cat: it simply has a different processing ability. Both the questions asked and the propositional attitudes used are different: there is no doubt that a prey species has been adaptively rewarded in the proposition "fear that", i.e. a state of mind of distrust, rather than one of "dare to", which is indispensable in a predator. The different cognitive activities refer to: (1) questions endowed with a specificity of their own, therefore very often not superposable between one species and another; (2) ways of undertaking actions that relate to the style of a particular species in relation to other variables, such as social character, bodily characteristics, biorhythm, etc.

Each species has its own "order of questions", i.e. it is urged to respond to different challenges because these are given by the context and lifestyle, i.e. they are elements of the adaptive profile. If therefore the functional universal (to experience food) is common to all species, the order of the question (the challenges encountered when trying to get food) is different depending on whether the animal is folivorous, omnivorous, carnivorous, nocturnal or diurnal, whether it lives in a forest or in a savannah and so on. Each species faces different problems also in relation to its type of sociality and level of intraspecies competition, but also to its position inside the trophic chain of an ecosystem, depending on the type of competitors, predators, parasites, environmental risks, the size of its territory, seasonal variability, fluctuations in the environment, migratory or sedentary tendencies, reproductive modalities, etc., (just to give some examples). Therefore, there are different solicitations not only in the parameters of cognition but also in the metacognitive exercise, so that some species are perlustrative, others are more skilled in distinctive abilities, and others rely more on their map memory than on the kind relative to social rank. The lifestyle and the environment of a given species, in their multiform characteristics, therefore identify very different "setback horizons" that raise different orders of questions—i.e. different problems to which each species must respond. Predators, for instance, are much more inclined towards resolvent and prefigurative structures, while herbivores are specialized in kinesthetic and orientative cognitivity, and omnivores acquire articulated and flexible distinctive and categorical repertoires. Social animals develop complex relational structures with strong empathic and communicative vocations, while solitary animals are more inclined to the areas of solution and reflection.

It is not only the orders of questions that differentiate the cognitive structures of different species, but also the specific "modes of undertaking" of the cognitive performance required. In fact, a relational question can be realized through an articulated communicative production or by facilitating the empathic dispositions and therefore the circumstantial structures. In the same way, a question can be solved by expressing the heuristics or by working ad hoc on the problem (insight). In the end, what follows are different cognitive performativity structures—specific collections of cognitive activities, i.e. vocations and cognitive attitudes that differ from species to species, which together identify a specific way of knowing the world. Some species tend to develop complex performances of practognosis or manipulation of the world, especially in the case of primates; others are able to build communication networks so superindividual as to give rise to collective intelligence or cognitive superorganisms.

The undertaking of the inventive capacity also presents marked differences in different species, as demonstrated by research on the intelligence of crows compared to that of primates, also in relation to problem-solving activities that require the manipulation of tools. It is not possible to pose a problem which presents particular operational challenges to two different species—perhaps favouring one for operational consistency compared to the other—and then to conclude that the former is more intelligent than the latter. Even when we offer operational tools to an animal or prepare a particular setting for it, we should always keep in mind the animal's modality of undertaking, which does not only concern the conformation of its body, but also the way in which that species faces a given problem. To put a chimpanzee in front of a mirror means to put it in a good condition to achieve perceptive intersection. The same cannot be said for a dog, who investigates reality through olfactory monitoring. To conclude that the chimpanzee has self-awareness because it passes the mirror test and the dog does not means failing to account for this important aspect.

Both the different orders of questions and the different modes of undertaking give rise to only partially overlapping knowledge structures, which can be defined as "species intelligence"; these are characterized by the prevalence of some cognitive activities both in terms of presence and in terms of relevance, i.e. in terms of effectiveness and expressive efficiency. Animals that live in herds and that entrust their survival to the concertative context—e.g. baboons—have developed relational cognitive activities and therefore present a cognitive identity based on the decoding of the interactions between active referents (social intelligence). Conversely, solitary animals or lone predators—like many felines—have developed problem-solving-type cognitive activities and therefore have a cognitive identity based on the manipulation of passive referents (enigmatic intelligence). From the different presence and specialization of various cognitive activities we can deduce the type of overall intelligence of the species, in a plural evaluation of the intellectual character where it makes no sense to speak in quantitative terms. The "plural intelligences" model I proposed (2008)—reinforcing Howard Gardner's concept of multiple intelligence by applying it to phylogeny—starts from a very performative view of the cognitive function; cognitivity is not a luxury, but a function like any other, which rewards the individuals of a species and as such must be consistent with the other biological characteristics of the species under examination.

There is no doubt that it is impossible to talk about risks and opportunities for the subject in a non-specific way, i.e. without referring to the species dimension. An individual's objectives are closely related to its species-specific motivational system, so that a predator might aim at reaching a moving target while a collector is probably more interested in collecting objects to carry to its den. If it is true that natural selection specializes different animals by giving them a body structure capable of adapting to particular ecological niches, it is also true that it predisposes them to specific behavioural styles, and an important role in the definition of a style is precisely a species' type of intersection. This is given by several factors. There are (i) perceptual specificities, i.e. the way in which that animal views the world through its type of sensory channels and the use it makes of them. Then there are (ii) motivational coordinates, i.e. the interest and sensitivity towards particular entities-events that occur in a given context. There is (iii) the emotional structure, i.e. the type of participation that the subject presents in the face of the events. Finally, one must consider (iv) the elaborative and representational endowments, i.e. the orientation maps and the appraisal resources, and (v) the cognitive meta-components, especially attention, curiosity, and memory in the process of intersection with the context.

Adopting a neurobiological conception according to which the neural set is mediated by specific neuromodulators, one can see the orientation in two ways. First, as (1) a propensity for particular activities (collecting, preying, building a nest, looking after the offspring, etc.) that are expressed in a preferential way and with a strong intrinsic component. Second, as (2) an electivity towards particular targets, able to evoke the motivational expression through specific conformations. The orientation defines, therefore, the type of involvement that the subject feels in its intersection with external reality, specifying elective sensibilities, degrees of satisfaction, orientations with respect to the world, expressive tendencies: in other words, the objectives of every species. An opportunity, therefore, is never objective, but always referable to the specified objectives and, more generally, to the species-specific orientation. In the same way, risks have a very different shape for a chick, a cat or a seal, so there is no such thing as an overlap in the analysis of challenges. We may also find opposite situations where, for one species, a situation counts as a risk, whilst for another species the same situation indicates an opportunity. This is not as far-fetched a possibility as one might think.

As for the type of challenge, the analysis of the structural requirements of a problem is inevitably species-specific because it implies an assessment of the divergence between the present and a potential situation. Animals used to searching underground (e.g. a dog or a pig) will not hesitate to consider the problem of the object hidden under a cup, unlike other animals that get food by collecting or simply chasing. The nature of the obstacle to be removed should therefore be framed through a series of parameters. For example: (i) the target could be under a lid or inside an object, whence a totally different perception of the obstacle. Alternatively, (ii) the target needs to be reached through a deviation or a detour, in this case the obstacle is perceptible differently by arboreal species compared to non-arboreal ones or, in any case, to species that live in particular contexts, like polar bears. In a different scenario, (iii) the obstacle could be a constraint on the target or a constraint on the subject's body: in this case there is a profound difference between species inclined to practice, species that use tools and species that reach the target in a direct way. Finally, one could make it so that (iv) the obstacle imposes a given practice to be overcome: this would present different levels of coherence in different species.

The concept of gap-distance also varies from one species to another: for example, animals that move differently in space have a different conception of reaching their target. If for an arboreal species reaching a hanging object defines a gap of a vertical order, for a species which has another life dimension the gap may not even exist or may be thought of according to other geometries. This becomes even more striking if the setback is a risk related to an opportunity, since in this situation it is more likely that high arousal will lead the animal to express innate defaults of a generic order rather than context-referred responses. In this case, some species present innate behaviours which, in our eyes, appear more reflective than others, but only because of our anthropocentric evaluation: a frightened animal emits an alarm sound which seems more intelligent to us than another one which simply runs away, but both could be acting by impulse.

What about the level of intersection of the stimulus? Every ethologist knows that what is relevant for one species in terms of urgency or non-derogation can be irrelevant and therefore not even activate the arousal system in another. If a species is not used to looking directly at its counterpart we should not be surprised if it fails the mirror test, but this does not mean that it does not have self-awareness. Likewise, if for a species olfactory orientation is more important than the visual one, for these animals a mirror will be irrelevant. How would we behave in the face of a test of olfactory specularity? Surely the experimenters would conclude that we have no self-awareness. What I want to underline is that the theory of the stimulus-response association rests on an erroneous assumption, i.e. that there is an objective stimulus condition capable of triggering the responsive system. In reality, different conditions are never objective setbacks: it is the species dimension that defines the presence of setbacks. For a prey, the activation of a setback is more significant in the panorama of risks, while for a predator it is the opposite: for this reason, to us humans, predators seem more intelligent than prey, but this view has no foundation. If then a setback condition determines high arousal, it is easy for the subject to lean towards irreflexive behaviours, and once again the level of intersection, responsible for the emotional activation, varies from one species to another.

One aspect that is closely related to the setback problem is the way in which a species normally achieves its objectives. The species that face problems through practices have totally different practognostic styles—think of the manipulation of primates with respect to the use of the mouth in canids or of the hooked grip of felids, as well as the refined use of the nose in suids or of the beak in corvids and in psittacidae. The use of instruments also differs from one species to the next, as demonstrated by the use of sticks in anthropomorphs compared to other techniques found in shrikes, otters, etc. The strategy used by various species therefore does not respond to a universal canon, and the learning processes follow paths of approximations that cannot be superimposed. The setback-response link uses a system of coping techniques that are strongly defined by the species dimension, and learning cannot be dissociated from the initial response style. By relegating the innate and the species dimension to the sphere of prefigured and therefore unalterable behaviour, we have created an unacceptable division between learning (seen primarily in the light of a simple associative grammar) and species-specificity. In reality, a learning process is nothing more than the evolution of the innate modules, following species-specific interpretative (appraisal) and operational (coping) styles. Therefore, the species dimension cannot simply be considered as the binding perimeter within which it is possible to acquire an individual identity, nor can it be reduced to an apprenticeship which, through social relations, gives the individual standardised competences: the species dimension is part of all learning processes because it provides the individual with the evolutionary matrix that produces the different experiences and learnings of ontogeny.

6.5 Dialogical Epistemology and Layers of Reality

Science rightly speaks of a world that contradicts our common sense and our natural interpretations: the decentralization of the Earth inverts our relationship with the stars, objects lose their appearance of solidity to break up into force fields, the living are no longer brought back to a repetitive genealogy but to a mutant creation of new forms, the dimensional ellipses envisioned by mathematics annihilate the ternary fundament of experience, and time itself seems to rest on a spatial sheet instead of being the simple flow of before and after. How did this come about, and why? How did this counter-intuitive reality emerge, and what does it mean? Are there real predicates and apparent predicates as Galileo claimed, or are there primary and secondary qualities as Lock supposed? It is fundamental to know the human epistemic apparatus and the basic coordinates that have decided its functions in order to investigate the concept of stability of knowledge, which today seems to be anything but stable.

Every living being inhabits its own world, made up of specific risks and equally peculiar opportunities, which is why its way of questioning reality has conformed to specific needs and not to an objective or neutral assessment. Life is essentially a partial perspective: being a predator is different from being a collector or a grazer, just as being prey requires special measures of interface with reality. And these are but a few examples of the complexity of the intersections of the living within the flow of life. The emergence of a living form is always a definition of a specific immersion in the context-world, both in a perceptual and in an operational sense. Uexküll defined this immersion as *Umwelt*, but to all intents and purposes it is better defined as a layer of reality, i.e. a specific way of interrogating reality and weaving it according to a specific organizational matrix. Every living being must therefore rest its presence on a seabed of stability as a background of risks and opportunities, i.e. those elements that have consistency and relevance with respect to the replicating potential of the subject.

Every event, in order to be understood, must therefore be transformed into its kinetic, metamorphic, and epiphanic dimension—that is, in terms of appearance and presence in a well-determined temporal range—but the evaluation of movement requires a still background, that is, it relies on a presupposition of stability. As much as we are aware of the Copernican revolution, we continue to say that the sun rises at a given time and sets at another, just as despite Darwinian nominalism we still try to understand the essence of "cats" by differentiating it from other species. Stability is what is intuitively given to us by our apparatuses to understand the world, and these apparatuses, obviously, are not only sensorial but also cognitive and metabolic. The canvas of stability is essential to be able to question reality, i.e. to bring out what interests us, otherwise everything would vanish in a continuum of energy fields that would be very unhelpful in terms of bringing out risks and opportunities. The question that might come up is whether we therefore *construct* what we call external reality or whether there is some effective *correspondence* between what appears to us and what there is. Epistemic foundations are undoubtedly prerequisites to grasp an event, representing its cliché. Should we therefore be satisfied with an instrumental vision of epistemology?

My answer may seem ambivalent, and perhaps it is: I think that the world as it appears to us is the result of organisational arbitrariness, but at the same time it is not mere illusion or invention; it is the result of the particular organisation that *we* have imposed on the virtuality of reality. When I speak of virtuality, I do not mean an empty and amorphous entity which can be subject to any imposition of form, but rather a content which can be organized into several schemes, but which also presents a degree of resistance, and therefore cannot take all conceivable forms. In this sense, I speak of species-specific image apparatuses as dialogical structures and not as functions capable of inventing completely relative worlds.

In this sense, dialoguing means asking specific questions that involve correlated answers on the part of reality—answers that therefore are not universal. The world thus appears to us according to *our* specific questioning and presents us with a partial view of itself. Indeed, the very act of asking questions implies some actions of partialization such as: (i) overshadowing some findings; (ii) giving relevance to others; (iii) conjugating them in a certain way. Our questions can be regarded as organisational information packages that impose a specific configuration on reality. Having said this, it is also evident that questioning does not happen in dissociation from the questioned, as if the dialogue were taking place on a threshold of mutual impermeability. To use a metaphor, one could say that a question is more like a recipe, where reality provides the ingredients and the image we get from it is the cake. *Intus-legere* is therefore a dialogue that harmonizes with the answers that the questioner receives from reality, and it is the dialogue itself that brings out the layer of reality where each organism is positioned.

The world appears to us as a collection of predicates that we acquire through our sensory organs: the texture of a piece of furniture, the taste of a strawberry, the colour of a flower, the smell of freshly cut grass. So it seems, but we know that the predicate perceived does not belong exclusively to the object—for example, not everyone agrees on the predicates of a particular body/event, and different species always detect different predicates in the same phenomenon. On the other hand, the predicate is not even a total invention of the organs that interface us with reality: indeed, one can easily rely on the semiology of predicates to answer one's needs. Moreover, we experience synchronic or diachronic relations between predicates, and we can infer norms that, even if not necessarily stable, give us a good predictive probability: often the acquired correlations offer us the possibility of favouring the occurrence of a phenomenon as well as knowing how to anticipate it. On the other hand, we know just as well that often our image apparatuses lead us into error and illusion, when we move away from the phylogenetic operative field that marks our fitness.

Now, if predicates neither fully belong to the world nor are not fully the product of our sensory organs, one can say, with a good approximation, that they are the emergent outcome of the interaction between our interface peculiarities—how we face the world—and some overall qualities of reality. This emergence, our way of "seeing" reality, is therefore not an expression of the world itself, but of our particular dialogue with the world. And why do we dialogue with the world? For a very simple reason: we need answers to live, we need order to stabilize our improbability. Any living being is a thermodynamically unstable system—a dissipative structure, as Prigogine put it (2015)—which owes its temporary maintenance in the complexity of the non-equilibrium to its openness, to its interface functions. The comings and goings of a plasmalemma, of an intestine, of a luxuriant arboreal foliage or of a blood vessel are in fact an emergence of the world, whose dialogue is capable of delimiting an environment, a domain of metabolic validity, through interchange thresholds. We must therefore stabilize our presence on a layer of reality, as we need this no more and no less than a plant needs to take root.

A living being is not autarkic: it must dialogue if it wants to preserve the internal order that characterizes it, and information is the most precious resource for it to trade. On the other hand, information is not *about* the world, it is not an objective, manifest or cryptic reality that can be accessed; information is found in dialogue *with* the world, at the point where a particular domain intersects a field of possibilities. The answers do not lie outside our exposure to the world, which would be tantamount to saying that they are of the world, nor do they emerge from imaginative or projective constructions, which would be like saying that they are simply ours: the answers are the result of precise questions, that is, of ways to reassemble the world around a flow of key words. Dialogue therefore has its own ecosystem structure and it is there that the living dwells and finds its true nourishment: information. I would like to make it clear, however, that this information does not ultimately or even roughly define the consistency of reality, but rather reveals the organisational possibilities available to the dialoguer. The dialogue is therefore always creative, not because it invents reality but because, through the possibilities that reality offers, it is able to bring out an unprecedented form in reality, just as a sculptor carves an unprecedented statue out of marble.

A question does not show what *is*, but what *is possible*; it does not show a questioning face that is objectively foreign to me, but reflects my own way of sketching a profile within the list of possible ones. Therefore, information is not something detached from the operational content of the interrogating entity; in other words, it always requires a scale or gradient of information that brings out the correlated clues and causes required of the particular operational space of the epistemic entity. This space indicates the fitness field of the questioner, whose success does not depend on a generic or absolute presentation of reality, but on a precise catalogue of risks and opportunities to reveal. In other words, the questioner lives on a precise layer of reality, given by a specific organisation of reality where the organisation itself produces binding effects and emergences, i.e. qualities that cannot be subsumed by other organisational levels. Human common sense is therefore a way of inhabiting reality; it is neither a projection nor an invention, but it is not even the only possible reality.

Charles Darwin has shown us that reproductive success is the key to understanding the form, the *hic et nunc* of *natura naturata*, because only those who have successfully replicated themselves are now here to show off their characteristics. But reproductive success, capable of giving rise to perfect coherence between form and function, between organism and environment, between ecological niche and behavioural style, between anatomy and physiology—a coherence that is above all congruity and adherence to a style, an admirable balance between the demands of effectiveness and efficiency—has precise questions for the world. Any living being needs the acquired information to maintain its internal order, its negentropy: and that is its life, i.e. its expression of its here-and-now, its replication and evolution. Can the energy a living being feeds on be reduced to an epistemic function? The questions asked by a living being concern the events that intersect the living being's specific order and therefore require a correlated epistemic foundation, that is, a peculiar dialogue with the world. We can therefore consider the epistemic apparatus of a species as the layer of stability that allows it to grasp only the events that are important to it from the point of view of its fitness.

It is dialogue that defines the epistemic apparatus, so we should not be surprised at its partiality—the given dialogue reflects the subjectivity of the question, the specific choice of the event to be recorded, the type of information that is required—but this does not justify falling into epistemic relativism or into the total disarticulation of the functional result from the texture of the world. For example, we are led to view an entity in motion against its background, so we say e.g. that it is the lion that approaches and not the savannah that recedes. However, this preconceived interpretation is not true in absolute terms—hence the illusion that the moon, and not the clouds, is what moves in the sky, as reported by Eibl-Eibesfeldt (2017). Rather, this way of seeing the world makes sense in the context in which *we* live and, moreover, it has a significant adaptive meaning. The epistemic apparatus calibrated by natural selection does not respond to a need for absolute knowledge of reality but for "species-specific action towards reality on the basis of strongly characterized needs". The epistemic apparatus, therefore, does not reflect reality, but the needs and the setbacks related to the reproductive success of the subject.

Nevertheless, the epistemic apparatus has been calibrated on the resistances of reality. The wings of a bird speak of ascending currents, the fins and silhouette of a fish reflect the viscosity of fluids, and gills testify to the presence of oxygen dissolved in water: likewise, epistemic apparatuses are the result of the peculiar adaptive needs of a species in its interface with external reality. Epistemic apparatuses are not, therefore, simple means to access reality that, like glass, owe their functionality to transparency, to the ability to reflect reality in the most objective way possible, but they are instruments that *act on* the world, just as a wing has a particular function related to the element (air) with which it deals. And yet, though confronted with a particular predicate of resistance—not all the resistances of reality have acted as selectors—the epistemic apparatus reflects setbacks caused by processes of negation or constraints caused by reality.

Epistemic apparatuses therefore respond to two selective pressures. The first lies in (i) the particular needs for information related to that living being, within a well-defined "living environment" (living in the trees of a rainforest poses different information needs compared to living in the open space of a savannah) and a "lifestyle" (being a predator attentive to motion is different from being a gatherer, which comes with a need for gathering-related information). The second is (ii) the counterpart with which the subject must deal, i.e. the world as made up of discrete and solid entities and not as constituted by force fields. As such the world is characterized by two things. On the one hand, there are (1) "common norms"—such as the law of gravitation, the surface-to-volume relationship, the common motion of parts—due to which it is probable that several species will deal with it using the same epistemic strategies. On the other hand, there are (2) by "specific layers", because each species needs different backgrounds against which to move its actors: hence the "plurality of dialogic levels" that we observe in different animal species.

Therefore it is necessary to recognize not the fallacy of the senses but the meaning of the phylogenetic dialogue. The processes of cutting out, connecting, attributing, ordering, asking—which together constitute the manifestation of the phenomenon, its emergence from the virtual domain—must be traced back to the adaptive ontic that is based on partiality in terms of entitlement over reality. Both sensory access, i.e. phenomenal virtuality, and the actual intersection with the event and the actualization of the phenomenon, can rely on a contextual cushion to the extent that they partialize reality. If we think in creationist terms we inevitably tend to separate between the "how of knowledge" and the "what of knowledge", falling into the inevitable dichotomy between inductive regimes and deductive regimes, with the aporetic drift that comes with it. In my opinion, this dualism results from an anthropocentric view of the epistemic praxis and from a failure to embrace the evolutionary view of the cognitive process. The interrogation is never positioned on a polarity but is always the product of the dialogic interaction between the two poles: dialogue gives rise to a mirroring, so we must recognize the recursion of the process where the what determines the how and vice versa.

If, in fact, we recognize not only the partiality of the intersection with reality, which derives from phylogenetic specialization, but also the introjection of the very predicates responsible for this partiality—just as the gills reflect the water environment of fish—we will realize that the "how of knowledge" is the result of a prior introjection of the "what of knowledge", even if the latter is not achieved through an individual evolutionary process. The image apparatuses of every species partialize reality, organizing it according to a given matrix of use of the world, but they themselves are forged *by the world* and not by a capricious demiurge—let alone by a demon who has made fun of the living by building an illusory reality around it. Knowledge is always a way to build partiality within the totality of reality, so it is not possible to achieve absolute knowledge of reality, but always only of a single level of it. Still blinded by the scientific revolution of the 17th century, we are led to believe that science offers us an objective mirror image of the world when, on the contrary, it does nothing but establish domains of validity referring to the particular technologies of interrogation of the world developed by technopoiesis.

Knowledge therefore implies an epistemic apparatus that is "capable of" or "called to" making the life context of a given entity partial or usable. Two things are clear in this respect. First, (1) it is impossible to intersect the world without questions and theories; second (2) any intersection, in its emergence, is in turn a source of questions and theories of a different order. The how and the what of knowledge are only apparently two separate realities because each individual, called to the epistemic act, already in its first intersections with the world is provided with an apparatus of philological heritage. This "how" results from a previous selective exposure to the context, i.e. to the "what" of knowledge, in its predicates of resistance and influence. The way (how) of knowledge has therefore already introjected certain predicates of the object (what), namely those involved in the adaptive process, so that the how reflects the what. However, it does not reflect it in its being as such, but in the predicates which are influential from an adaptive point of view, thereby inevitably constructing a phenomenal partiality. The how is therefore an a priori what, redefined and partialized on the basis of the adaptive meaning of some of its predicates, just as breathing only captures the oxygen part of the air.

Therefore, the first organizational dimension of the epistemic apparatus refers to phylogeny and can be traced back to a process of partial or distorted configuration of knowledge. The how of knowledge is thus not an a priori transcending the object of knowledge, but is the result of the selective dialectic between the object itself and the ontic style of the knowing subject. And this process refines the ability to partialize the thing on the basis of relevant predicates, introjecting them like a key into the corresponding lock. The epistemic apparatus is therefore the result of a dialectic function that intersects the objectives of the taxon in phylogeny and the resistances-influences that are virtually present in the context and can be identified through precise predicates. The epistemic apparatus is ultimately a resolvent organ, called to solve the problematic contents encountered by the subjects of a given species.

6.6 Layers of Reality as Aspects of Virtuality

Dialogue is therefore a way of organizing reality, and epistemology is a way of positioning oneself on a layer of reality in order to draw useful information from it. This means that it is not an approximation but an emergence: it is not an arbitrary or paradigmatic construction, but a way of eliminating noise and bringing out certain contents. Mental categories and sensory organs are wired together to organize a layer of reality that is consistent from an adaptive point of view, as it can support that species in its daily need to maintain a highly improbable internal organization. A layer of reality is therefore a single aspect of reality, which is not invented, not illusory, not commensurable. Rather, it is the answer to a precise question, an answer that necessarily has to cut corners, fade out inappropriate details, arbitrarily put together different points into prearranged Gestalts, bring out correlations, hypothesize causal relations by cutting off third variables, and operate background oscillations. In short, the answer can never be exhaustive; on the contrary, the very nature of the answer is elective.

The order or, if you like, the predictability of a description-explication depends on the scale of observation, that is, on the layer on which our epistemic apparatuses rest. To quote Zbilut and Giuliani: "The universe can therefore be seen as a picture with different levels of apparent determinism and noise depending on the scale of observation" (Giuliani 2009: 72). This scale-dependence essentially tells us that if we modify our access, for example through a technological apparatus, by positioning ourselves on a different layer of reality, we will be faced with a different organization of reality and this will require that we modify our descriptive-explanatory structure.

Organizing reality does not mean inventing it but extracting scale emergences within "a field of defined or resistant possibilities". In this sense, I find Maurizio Ferraris's concept of "the resistance of reality" (Ferraris 2002: 139–165) particularly appropriate. This notion, moreover, helps us understand an important epistemological principle, that is, that there are many ways to read reality, but only one way in which reality falsifies your assertions. This also means that an instrument is not an amplification of the phylogenetic endowment, but a transformation of the same, capable of modifying the access scale and therefore the gradient of determinism and noise that we find in it, i.e. the type of information that we can obtain. It also means three things: (1) even if reality exists, there is no privileged layer of reality nor one layer capable of subsuming others; (2) it is not possible to access more than one level of reality at the same time; (3) human beings cannot reduce reality to a formal scheme. In this sense, I agree with Hilary Putnam's "common sense realism" in saying that what exists is independent of its knowability (…) and that there can be many correct descriptions of reality (Putnam 1992).

On the other hand, if reality can be organized into several binding-emergent structures, each one endowed with particular resistances, it is evidently impossible to reach a single and objective knowledge of reality—one can only broaden the information space. Basically, it is not possible to say that a given description of reality is true in the absolute sense—and this is part of a long epistemological tradition—but only that one can broaden one's field of information about reality by multiplying the number of questions, that is, by accessing several levels of reality, where each tells something about the entity-event. I believe that every layer of reality defines domains of validity, so we are not faced with approximations but with a plurality of views on the world. Every questioning progression, whether it has a theoretical basis or results from the advent of a new technology—I don't see much difference between the two—increases our ability to access new levels of reality and therefore offers us new opportunities to broaden the horizon of our knowledge.

So let's go deeper into this concept-project of actualisation of reality. As I said, I call these different organizations of reality, each incomplete from an explanatory point of view if analyzed on the level of components, (Deacon 2001) "layers of reality". Each layer of reality can only be explained as a centrifugation from the same, which implies the explicative self-insufficiency of all epistemic layers. The organisation behind any reality level depends on the potential of reality and on the type of question that is being asked. In this sense, if we wanted to find an absolute conception of reality, we could define it not as a specific state but rather as a space of virtuality, one within which the different levels of reality are all admissible. Every layer of reality brings out phenomenal predicates that, despite being emergent entities, are not to be considered projective.

Epistemology. therefore, is not a way of "inventing a context" but a particular intersection with reality on the basis of a temporary dwelling in reality. In order to act on a particular context, it is necessary to define correlation gradients and therefore to emphasize some aspects and to encrypt others. Assuming a positioning scale, the questioner brings out an order that is not illusory or invented, but that nevertheless depends on the relational space in which the observation-dwelling is positioned. This hunger for well-defined answers therefore leads to the emergence of a dialogue or information exchange, electing a specific interface level that actualizes one of the many possibilities of reality, coagulating one of its possible organizations within the field of virtuality. The subject is situated in its species and individual dimension precisely because of this positioning of the interface, which allows the subject to detect only those events that have some relevance to its problem.

On the other hand, it is wrong to believe that only the experiential level or the direct observation level—common sense or phylogenetic epistemology—are entitled to define reality, as is often stated by those who profess common sense realism. A scientific theory is as entitled as direct experience to describe reality: it has the same ability to bring out predicates that tell us something more about a given field of virtuality of reality, and the same ability to deal with the fields of resistance of the external world. Moving away from phylogenetic epistemology, new phenomenal predicates and new resistances of reality will emerge through anthropodecentrative mechanisms—the counterintuitive character of science refers precisely to this centrifugal process. Every time the questioner hybridizes their logical endowments with a tool, every time they gain particular knowledge, every time they open up to new existential dimensions through animal epiphany, they modify their dialogical structure with reality. Decentralizing themselves from common sense, the questioner receives a different organization or scale gradient, giving rise not to a construct, but to a new level of reality.

The concept of "layer of reality" recalls a passage from Aristotle referring to Speusippus' theories: "Speusippus starts from the One and adopts principles for each of his levels of reality." The concept of multiplicity of principles implies a positional choice on the part of the questioner. Following this indication, in a dialogic conception of the descriptive-explanatory experience, the phenomenal predicates depend not only on the intersection between the properties of the observer and those of the observed, but also on the layer of reality on which the observer stands. One can summarize what has been said so far in three points: (1) not only does the observed adhere to reality, but it was built on the basis of reality itself; (2) a level of reality depends on both the field of virtuality of reality and on the characteristics of the observer; (3) a level of reality also depends on the type of interrogation that the observer poses at a given time to the field of virtuality of reality.

We access and recognize a status of reality through the partiality of its appearance, its giving itself to us in a circumstantial way. Therefore, we intersect reality's predicates of resistance to, or influence on, our specific way of being. But even within this limited access, what we define as a phenomenon or phenomenal predicate is ultimately an aspect of reality, a possible expression of it. The truth/falsehood of any statement depends on the questioning structure, in other words on how I intersect reality or on what level of reality I ask the question, that is, on "what interests me" and "how I ask it". If the question for some reason does not intersect some specific resistances, these do not emerge: our epistemic apparatus is not structured to bring out the Copernican universe because the type of interrogation that could make it emerge is not pertinent to the "epistemic positioning" of our species. It is necessary to hybridize this epistemic apparatus to bring out new interrogative structures, which means initiating new dialogues through the aid of otherness or by *becoming* otherness, and this happens whenever the human being constructs anthropodecentrative mechanisms.

On the other hand, there is no doubt that in order to question reality it is essential to position oneself on a precise level, to stabilize certain parameters of noise and determinism, and to use a scale gradient that brings out specific findings. The anthropodecentrative mechanism—for example, observing the sky through a telescope—does not allow a structure of reality to emerge, but rather brings out a new organization of it, showing us new resistances and thus widening our observation of reality. Therefore, theories are not approximations, but specific questions that allow new resistances to emerge, providing more information about reality and allowing us to formulate new questions. The predictability of a theory thus has to do with the congruence between the questioning structure and the type of phenomenon to be investigated, so that the level of predictability can be referred to the space of probability in which we are operating. Each interrogative structure necessarily intersects some resistances and not others: this depends on the type of scale positioning adopted by the questioning, that is, on the ability to bring out and order only some events.

Common sense is therefore nothing more than *one* of the many possible views of reality, capable of providing our phylogenetic trajectory with the support level on which to formulate the questions about reality that are useful to us. This means that common sense is not "reality" but a presentation of it, that is, a layer of reality—one that has intersected our species' questions. *A fortiori* it is not an illusory projection or construction, let alone an approximation. One could say that the level of reality occupied by common sense experience represents the small observatory on reality that has been indispensable for us to solve the problem of our species-specific fitness. Therefore, phenomenal predicates certainly do belong to the observed object, i.e. they are not constructions, since they have acted on the epistemic apparatus to calibrate its "experiential how"; however, their phenomenal peculiarity emerges from the way in which *we* question reality, i.e. from the horizon and the depth of field of our observation.

6.7 In Conclusion

The proposal that I have put forward in this essay can be summarised in a few basic points, which I list below.

i. Reality does not have an ultimate-defined configuration, but is characterized by a "virtual field", which is finite but indefinite in terms of possibilities, and therefore presents very precise limitations, i.e. resistance points, but allows for many possible configurations.
ii. Each act of knowledge is a precise organization of this virtuality, giving life to a "layer of reality", much like carving a statue starting from the virtuality of marble—it is a process of actualization.
iii. Each level of reality is the result of a specific dialogue between the subject and reality, i.e., it is the emergent outcome of a "dialogical mode of intersection", i.e. a given question posed to the world.
iv. Every living being, on the basis of its own somatic and ecological condition, is not confronted with the field of virtuality of reality, but with an "epistemic dimension", i.e. a more restricted virtual field, so that every epistemological perspective is necessarily constrained to the specific dialogic process of that body with its world-context.
v. Every epistemic dimension necessarily determines the emergence of a species-specific layer of reality, because no species can prescind from the phylogenetic and somatic heritage that defines its intersection mode—whence "intellectual pluralism".
vi. Given this restriction due to phylogeny, every individual is able to go beyond the already-given, namely to *intus-legere* or go beyond appearances, building a "singular intersection" with reality.
vii. Every time an entity organizes its own intersection, through processes of experiential singularity, it actualizes a level of reality that did not exist prior to the epistemic act. It does so through a "predicative emergence" that gives rise to reality predicates of which the subject is an active protagonist, just like a sculptor.
viii. All that the individual encounters as an immutable law is a "resistance of reality": not what is given to it as a possibility, but what is denied to it, that is, what and where it is unable to actualize.
ix. Knowledge is not detached from the other dimensions of subjectivity, but is affected by the ownership, desire and sentience that characterize animality, so that there can be no knowledge without the full involvement of the somatic multilayer.
x. Cognitive processes can progress not only by virtue of the creative and singular action of *intus-legere* or of the multiplicity of points of view, able to bring out new resistances of reality, but also by widening the epistemic dimension through two main avenues, namely: (1) modifying the dialogic level of

intersection, for example through new technologies, and (2) using counter-intuition, for example by questioning some assumptions taken for granted only because they are the result of a given phylogenetic heritage.

References

Bachelard, Gaston. 2002. *The Formation of the Scientific Mind*. Manchester: Clinamen Press.
Bostrom, Nick. 2014. *Superintelligence: Paths, Dangers, Strategies*. Oxford: Oxford University Press.
Bozzi, Paolo. 1998. *Fisica ingenua*. Milano, Garzanti: Studi di psicologia della percezione.
Darwin, Charles. 1859. *The Origin of Species*. London: John Murray.
Darwin, Charles. 1872. *The Expression of Emotions in Man and Animals*. London: John Murray.
Deacon, Terrence William. (2001). *Incomplete Nature: How Mind Emerged from Matter*. New York: Norton.
Eibl-Eibesfeldt, Irenäus. 2017. *Human Ethology*. New York: Routledge.
Ferraris, M. (2002). Esistere è resistere. In *Bentornata realtà. Il nuovo realismo in discussione*, ed. M. De Caro and M. Ferraris, 139–165. Torino, Einaudi.
Giuliani, Alessandro, and Joseph Zbilut. 2009. *L'ordine della complessità*. Jaca Book: Milano.
Hobbes, Thomas. 1909. *Leviathan* (1651). Oxford: Clarendon Press, ch.V.
Marchesini, Roberto. 2008. *Intelligenze plurime*. Bologna, Alberto Perdisa Editore: Manuale di scienze cognitive animali.
Marchesini, Roberto. 2016. *Il bambino e l'animale. Fondamenti per una pedagogia zooatropologica*. Rome, Anicia.
Mayr, Ernst. 1997. *Evolution and the Diversity of Life: Selected Essay*. Cambridge: Harvard University Press.
Nagel, T. 1974. What Is It Like to Be a Bat? In *Philosophical Review, LXXIII* (4): 435–450.
Nietzsche, Friedrich. 1917. *Thus Spoke Zarathustra*. New York: Modern Library.
Prigogine, Ilya, and Dilip Kondepudi. 2015. *Modern Thermodynamics: From Heat Engines to Dissipative Structures*. Chichester: Wiley.
Putnam, Hilary. 1992. *Renewing Philosophy*. Cambridge: Harvard University Press.
Rosenblum, Lawrence D. 2010. *See What I'm Saying: The Extraordinary Powers of Our Five Senses*. New York: Norton.
Kuhn, Thomas. 1962. *The Structure of Scientific Revolutions*. Chicago: Chicago University Press.
Regan, Tom. 1983. *The Case for Animal Rights*. Oakland: University of California Press.
Bauman, Zygmunt. 2000. *Liquid Modernity*. Cambridge: Polity Press.

Part III
Ethical Dilemmas Beyond the Borders of Species

Chapter 7
Contributions of Ethology to the Birth of a Post-Anthropocentric Ethics

Marco Celentano

Abstract From the time of Aristotle until the first half of the twentieth century, with rare exceptions, the Western tradition has excluded non-human animals from the community of beings worthy of an ethical consideration, assigning them the role of "animated tools" of the human will (Aristotle, *Politics*: 1254 b 10). Only from the 1960s did an anti-speciesist, post-anthropocentric, bio-centric ethics begin to spread. This chapter documents the role played in promoting this ethical revolution by ethological research, the important discoveries it has led to and the changes in perspective they generated. What these pages are trying to focus on is, therefore, the link between the scientific revolutions introduced by ethology (daughters of the Darwinian revolutions), with its various stages and articulations, and an ethical revolution: that of contemporary anti-speciesism.

7.1 Introduction

The first section frames the singular relationship that the dominant currents of modern Western thought established with "the animal", declined in the singular and intended as any non-human animal, starting with the century of scientific revolution. Since the Cartesian turning point, animals began to be considered a non-sentient, and even unthinking natural "machines", breaking with the Aristotelian tradition which, while reducing them to the role of human instruments, at least recognized in them a "sensitive life".

Despite the transformism that began to spread in the eighteenth century, the Darwinian revolution that marked the nineteenth century, and the birth of ethology in the early twentieth century, this reductionist approach characterized the prevailing orientations, both in the philosophical and the scientific sphere from the early mid-seventeenth to the first half of the twentieth century. The section concludes by showing how, in the 1960s, new teaching techniques experimented on apes, while remaining theoretically, methodologically and ethically in the wake of Cartesianism, led to an anthropocentric self-refutation, thanks to their surprising results.

The second section focuses on an epochal turning point, even more radical and significant from an ethical and theoretical point of view: the one also introduced in

R. Marchesini and M. Celentano, *Critical Ethology and Post-Anthropocentric Ethics*, Numanities - Arts and Humanities in Progress 16,
https://doi.org/10.1007/978-3-030-74203-4_7

the '60s by field studies of apes and the beginning of a hard fight for their protection and the defence of their natural environment. A turning point and a struggle that found their emblems in the faces of the "three angels of Leakey", the three young primatologists sent by the anthropologist Louis Leakey to carry out the first field research on the life of the three great apes, our "sister" species (chimpanzees, gorillas, and orangutans): Jane Goodall, Dian Fossey and Birutè Galdikas.

The third section focuses on the rediscovery of the pathic, affective and cognitive dimension of animal existence, that is, of its rejoicing and suffering, of the affections and forms of thoughts, introduced by the pioneers of cognitive ethology, and on the important impulse it gave to the birth of anti-speciesist ethics. Particular attention is paid to the essay *Have Animals an experience?*, published by Konrad Lorenz in 1963 (Lorenz 1963), which, by discussing the question of "pleasure" and "pain" in non-human animals, anticipated issues that would come to the fore in the scientific and ethical debate only in the following decade, with the birth of cognitive ethology, as promoted by Donald Griffin (1976), and the publication of the first contemporary anti-speciesist manifesto, namely, the book *Animal Liberation* (Singer 1975) by Peter Singer.

7.2 Modernity and "The Animal"

The teaching of Aristotle, according to which the "sensitive life is common to the horse and the ox and to every animal" (Aristotle, *Nicomachean Ethics*, I, 6, 1098a), reaches modernity modified and expanded by important thinkers, such as Bernardino Telesio, Giordano Bruno and Tommaso Campanella, for whom the ability to feel sensations is, indeed, widespread in every kind of material entity.

The century of the scientific revolution marks, however, with respect to this tradition, a drastic caesura. The paragraphs IV and V of the fifth part of the *Discours de la method* by René Descartes served as a symbolic watershed introducing the idea that non-human animals are a sort of natural "automatons" (Descartes 1637, V, 4) and do not actually experience real "feeling", "passion" or "thought" (Descartes 1637, V, 5). Then, 41 years later, the scientist, philosopher and theologian Nicolas Malebranche, heir to the Cartesian tradition, will formalize this approach by stating that "animals do not feel" because they are machines "devoid of soul and completely incapable of perception" (Malebranche 1678). This theoretical obliteration of the sensorial, emotional and volitional dimension of animal life, and in particular of animal suffering, was to be met, in the following centuries, with a great success in the philosophical and scientific field, also for its ethical and practical implications. In fact, it favoured a removal of the spontaneous empathy that the suffering, and more generally the expressiveness, of other animals can raise in human beings (it is known the anecdote according to which Malebranche, while conversing with a friend, kicked a pregnant whining dog justifying his act by such a conviction). Precisely for this reason, it offered legitimacy both to the traditional forms of exploitation and to the use of non-human animals and to a research method that, starting from that time and for more than three centuries to come, would not have stopped marking a dizzying increase: experimentation on living animals. Nicolas Fontaine, after observing some experiments carried out at the Jansenist seminary in Port-Royal, wrote in his memorial, published posthumously 99 years after the Cartesian *Discours*: "They dealt blows to the dogs with perfect indifference, and they made fun of those who pitied the creatures believing that they felt pain. They said that animals were like watches [...] but that the body could not feel anything. They nailed the four legs of the poor animals to boards, in order to vivisect them and see the blood circulation, which was an important topic of conversation" (Fontaine 1736, II: 52–53).[1]

Regarding the specific problem of animal sensitivity and intelligence, the era of the scientific revolution inaugurated a kind of approach that, in the philosophical, anthropological and psychological fields, would have survived, despite Darwin and the birth of ethology, at least until the first half of the twentieth century. It consists in measuring the cognitive and communicative abilities of non-human animals on the basis of a single parameter, explicitly anthropocentric and logocentric: their ability or inability to articulate discourses, their aptitude or ineptitude to understand and reproduce the sounds and rules of human languages. For Descartes, the hypothesis that even the most "idiotic and stupid" men surpass in intellect the most intelligent animals appears confirmed, a priori, by the fact that the first ones "know how to combine together different words and compose a speech to be understood" (Descartes 1637, V, 5), while, in his opinion, magpies and parrots would speak but not "think what they say" (ibid.). This conviction leads the father of rationalism to reject the ancient idea, again defended by Michel de Montaigne, "that the beasts speak to each other, but we do not understand their language: since, if this were true, they, having several organs which correspond to ours, could be understood equally well by us as by their fellows" (ibid.). It may be surprising that, at the time, not even the philosopher of "methodical doubt" was touched by the doubt that, in a scientific investigation into animal communication skills, the burden of deciphering and learning the language of the other should fall on the scientists and not on the animals. But it is even more surprising to notice that, for about three centuries, the hypothesis that the anthropomorphic apes, and other non-human animals, are capable of thinking and using symbolic languages were to be discarded starting from approaches, methods and arguments similar to the Cartesian ones.

In the first half of the twentieth century, several failed attempts to teach some chimpanzees to reproduce human language through the phonic medium followed one another. American psychologist Robert Mearns Yerkes was to be, in the first decades of the century, one of the pioneers in this field (Yerkes 1916, 1943). In the following years the studies of Kellog and Kellog (1933), Hayes (1951), Hayes and Hayes (1952) were to confirm the failure of this kind of attempt.

A different kind of experiment was instead to begin, in 1967. While remaining in the Cartesian perspective of measuring the intelligence of non-human animals assuming by as its parameter their ability to learn human languages, it was to lead, thanks to the new techniques adopted, to surprising discoveries and to a progressive

[1] The volume was published posthumously, Fontaine died in 1709.

dismantling of the Cartesian assumptions. To obtain this result, it was enough to stop asking of chimpanzees a performance that is impossible for them, like the vocal articulation of words and phrases drawn from human languages, and let them learn the sign language commonly used by deaf-mute people in the USA: the American Sign Language (ASL).[2]

Allen and Beatrice Gardner, two psychologists of the University of Nevada, launched these experiments by starting to teach Washoe, a young female chimpanzee, to learn this language. Washoe, thanks to the extraordinary efforts of Roger Fouts, to whom she was later entrusted, learned to properly use about 250 different signs and to compose with them meaningful sentences, as well as teaching the same language to other chimpanzees, with no human incentives and mediation (Fouts 1997: chap. X). Many similar experiments, supported by different techniques, were successfully performed from the early 1970s onwards, with chimpanzees, bonobos, gorillas and orangutans.[3] Even the self-recognition tests in the mirror (mark test or mirror self-

[2]As it is known, already many human populations that have not used writing, and among these different tribes of North American Indians, invented and adopted manual languages that functioned as a sort of universal language through which different tribal members or linguistic groups could understand each other.

[3]I will limit myself here to mention some of the most well-known experiments in teaching human languages to apes produced since the 1970s:

- The case of the chimpanzee Nim Chimpsky, instructed, in the 1970s by Herbert S. Terrace of the Columbia University, who learned, over the course of 44 months, to communicate using the sign language and combining 125 different signs.
- The new interactive communication technique introduced in 1972 by Sue Savage-Rumbaugh and Duane Rumbaugh of the Yerkes Primate Center of Atlanta (Georgia), based on the use of lexigrams, usable through a portable keyboard with buttons marked with different geometric symbols, each of which reproduces the sound of a given word. The chimpanzees Lana, Austin and Sherman were among the first apes to be trained with this technique, then also applied by Rumbaugh to bonobos (Savage-Rumbaugh and Lewin 1994).
- The studies conducted by David Premack with the chimpanzees Sarah, Peony, Elizabeth and Nim instructed the use of plastic signs (Premack 1986) and the paper he published together with G. Woodruff in 1978 entitled *Does the chimpanzee have a theory of mind?*. In it the authors introduced, the hypothesis that very intelligent animals such as chimpanzees possess "a theory of mind", here understood as the ability to attribute, to other individuals, mental states, expectations and desires and to use these hypotheses or information to adapt their behaviour to the other, basing on their own experience. The application of this concept of the theory of mind to the study of animal behaviour has proved to be of great use when, from the 1980s onwards, ethological research began to document, in a conspicuous manner, phenomena of concealment and deception in apes and in social birds like crows.
- The work led by Gary L. Shapiro between 1973 and 1975 with the orangutan Aazk using training techniques similar to those of David Premack.
- The performances of Koko, a gorilla trained from a young age to use the American Sign Language (ASL) by Francine Patterson, current director of the Gorilla Foundation, located in California: after a few months, Koko began to invent, for the objects that were unknown to her, new names composed of two words, such as "match bottle" to name a lighter, "baby elephant" to indicate a wooden Pinocchio, "hat eyes" to characterize a mask. The young ape has also developed the ability to lie through this form of language. Patterson reports, among others, the case in which

recognition test: MSR),[4] introduced in the '60s by Gordon Gallup (1970), while arousing a long and heated debate concerning the interpretation of the performances observed, confirmed ability to recognize their own mirror image in chimpanzees, bonobos, orangutans and gorillas.[5] The logocentric, anthropocentric and at the same time mechanistic and reductionist approach introduced back-then by Cartesianism will finally reach its auto-confutation after over three centuries. In the new millennium, the ability to recognize one's own image reflected in a mirror has been attested, through the Mark test, also in Asian elephants (Plotnik et al. 2006), rhesus macaques (Rajala et al. 2010), horses (Baragli et al. 2017), bottle-nosed dolphins (Reiss and Marino 2001), orcas (Delfour and Marten 2001), in birds such as magpies (Prior et al. 2008) and in cartilaginous fish like the mantas (Ari and D'Agostino 2016), in a tropical cleaner fish like the *Labroides dimidiatus* (Masanori et al. 2019) and even in social insects such as ants (Cammaerts and Cammaerts 2015).

7.3 The Discovery of "Anthropoid" and "Anthropomorphous" Cultures and the Fight for Their Survival

The studies based on the teaching of human made languages to non-human animals and on the mirror test, although leading to innovative and surprising results, still remained linked in various aspects to the Cartesian tradition; in fact, they appeared:

- marked by the anthropocentric criterion of considering our ways of expressing intelligence as the only possible ones, or at least superior to all others;
- oriented to a priori assume the hypothesis that the average linguistic skills of an adult man, belonging to a civilization that has codified his own communication system in written rules, constitute the only channel, or the only form, through which thinking and reasoning beings can manifest and articulate themselves;
- set up, theoretically and methodologically, starting from the conviction that removing an animal from its environment and imprisoning it in a laboratory, to

Koko, caught gnawing a pencil by an instructor, responded to his reproach by making the sign corresponding to the word "lips" and simulating putting on lipstick (Patterson 1978, 1987).

- The results obtained by the orangutan Chantek who, trained by the American anthropologist and primatologist Lynn Miles (Miles 1994), assimilated the understanding of spoken English and the active and passive use of ASL, coming to understand and use about 150 signs to communicate desires, coin new terms, ask questions and lie.

[4]In its typical form, the mark test consists in marking the face or another part of the body of an animal with a spot of colour which it can see and touch while it is sedated and, when it wakes up, verifying if the animal locates, examines and tries to remove the spot by looking in the mirror.

[5]Unlike other apes, gorillas, in which individual recognition is based more on olfactory factors than on visual factors, show greater difficulties and often fail in the mark test. However, there are studies that attest that some individuals belonging to this species have passed it (Posada and Colell 2007; Allen and Schwartz 2008).

impose on it experiments conceived with parameters and motivations completely unrelated to its own nature, is a way of making science more rigorous than the observation of its behaviour in a natural environment.[6]

If today most scholars have dismissed the dogma that, from Descartes to behaviourism, forced scientist to see in animal intelligence a mere mechanicalness, and if the behavioural sciences were at least partially detached from the lazy arrogance and the reiterated violence implicit in the traditional methods of studying behaviour in the laboratory, this was due to a synergy between the type of research we have mentioned and a very different way of studying animal behaviour: field research combined with the commitment to the protection of wild animals and of the environments in which they live.

A first important step in this direction had already derived from the birth of classical ethology, whose method was based on the idea that to understand the valences and functions of animal activities, it is necessary to observe them in their natural environment. Finally, animals were no longer uprooted from their environment and forced to live in laboratory enclosures. Instead, it was the scientist who went to the places where his study "objects" lived, to dive or climb with them, to spend days, months and years observing them, learning their languages and discovering their meanings and functions in the species-specific ecological and social context. The achieved results opened up an entire world to scientific understanding, that of animal intelligence, emotionality, sociality and communication, for centuries invisible to most Western scientists and writers, too concentrated on the effort to reduce every element of nature to a human instrument. From the '50s, ethological research would have crossed the borders of Europe, beginning to spread throughout the world.

These developments also led to an increasingly interweaving of the anthropological and primatological studies and to the possibility of for the first time launching projects for a long-term observation of apes in their natural environment. The main promoter of the latter was Louis Leakey, the most authoritative paleoanthropologist of the time. The trio of young primatologists to whom the scholar entrusted the arduous task would have contributed to irreversibly modifying the perception that the human world had of apes and, more generally, of non-human animals.

In 1960, Jane Goodall began to study chimpanzees at the Gombe Stream Chimpanzee Reserve, in the western Tanzania, a place that, thanks to her efforts, would have become since 1968 a protected area and no longer a safari destination.

Having already mentioned, in the third chapter of the volume, the scientific merits of Goodall, I will here briefly dwell on the ethical commitment in the anti-speciesist, environmentalist and humanitarian field that this indomitable woman has demonstrated for over half a century now.

---

[6]These critical findings do not apply to the path taken by Roger Fouts who after being compelled for a long time to study chimpanzees in unsuitable conditions for them, fought hard over fifteen years, finally succeeding in 1993 to fund the Chimpanzee and Human Communication Institutes, in whose natural reserve Washoe has been able to live happily, together with the other chimpanzees of the group that Fouts has cared for, for the last fourteen years of its existence.

Working in the field for decades, Goodall soon became aware of the problem of the gradual destruction of the natural environments and of the risk of extinction to which apes, other mammals, many other animal and plant species, and even several human cultures, were and still are exposed, in Africa as elsewhere. This awareness led her to become, from the 1970s, one of the most well-known activists for the defence of the natural environments, of the organisms that inhabit them and of the human populations in need, damaged by a development model, like the one still dominant today, which is blind to the environmental, biological and social damage which it itself produces.

In 1977, Jane founded the Jane Goodall Institute (http://www.janegoodall.org/): a structure that, through a wide range of activities and projects, aims to practice, spread and teach respect for and the protection of the environment, living organisms, human populations and cultures exposed to the "collateral damage" of the current development model. In fact, if the primary mission of the institute is the conservation and/or re-insertion of chimpanzees in their natural environment, and the improvement of their living conditions in the cases of individuals or groups that cannot be legally freed from captivity, the "holistic" approach promoted by the JGI implies that this objective is always pursued in close intertwining with projects of development of the human communities present in the places where the institute operates, of the education of young people to environmental sustainability and of sustenance and start-up education and work of the most disadvantaged. In short, the commitment to get to know and protect chimpanzees and their natural environment is understood by the founder and members of the JGI as an important segment of a broader commitment aimed at spreading the awareness of the interdependent relationships between man and the rest of nature, at introducing young people to the discovery and respect of animal minds and cultures, and at safeguarding the entire community of the living, men included, by the devastating effects of anthropic impact and capitalist development.

I will limit myself to mention five of the international projects on which the Institute, that now has offices in 26 countries, works:

- the Jane Goodall's Institute Center for Primate Studies, mission of which is the non-invasive study of the behaviour and cultural traditions of primates in their natural context of life, based in the Gombe National Park, Tanzania;
- the Tchimpounga Chimpanzee Rehabilitation Center, also known as the largest wildlife oasis founded in Africa to fight against illegal trafficking in chimpanzees, which is located in Congo and has another recovery center dislocated in South Africa;
- the ChimpanZoo International Research Project, aimed at studying chimpanzees in captivity to improve their living conditions.
- the TACARE project, funded by the European Community, which provides support to the populations of 30 African villages through reforestation programs, environmental education, water purification and health care;
- the "humanitarian and environmental" program Roots & Shoots, now active in more than 80 countries, goal of which is to educate young people about respecting

the environment, knowing and understanding other cultures, about the importance of an individual commitment to rediscover the affective and social bond between man and other animals and the fight against exploitation and social inequalities.

Today, around 290 great apes, including chimpanzees and gorillas, are housed in natural "sanctuaries" managed by the JGI and about 130 African human communities are assisted by institute volunteers with education and training programs for environmental protection activities, or start-up to the eco-compatible management of agricultural works. About 5800 school projects related to Roots & Shoots are currently underway in the most diverse areas of the world.

Jane Goodall, who was named Messenger of Peace for the United Nations and awarded the UNESCO Gold Medal, today, at the age of 85, still travels for most of the year, to promote such projects and find the funding needed to support them.

Dian Fossey, the first contemporary "martyr" in the fight against the exploitation and killing of the wild animals, was also a primatologist and ethologist. In the early 1960s, Fossey started, with the gorillas of the Rwandan forests doing work similar to that of Jane Goodall with chimpanzees in Tanzania. In 1967 she founded the Karisoke Research Center, an institute that still exists today, based in the Volcanoes National Park of Rwanda. Its mission is to protect and study mountain gorillas, an endangered species. Dian's commitment in the fight against the illegal suppression and trade of gorillas led, in 1985, to his barbaric killing by poachers, probably covered by the complicity of politicians or businessmen linked to the illegal trade in gorillas. The film *Gorillas in the Mist*, released in 1998 and based on the book by Fossey of the same title (Fossey 1983) and some other volumes (Gordon 1993; Mowat 1987; Nienaber 2006), made his story known to the world, even though his murder remains unpunished until now. The Dian Fossey Gorilla Fund International, an institute that aims at raising funds to safeguard gorillas and their natural environment, was founded after his death.

Birutè Galdikas, the third woman sent by Leakey to discover the world of our sister species, dedicated herself to the protection and study of the orangutans in nature by going, in 1971, to Borneo, to the almost uncontaminated territory of the Tanjung Puting Reserve, working in the field for more than thirty years.

Galdikas founded the Orangutan Foundation International, an institute aimed at finding funds and volunteers to support a center for the recovery and reintegration in nature of captive-living orangutans and for the protection of the orangutans of Borneo and the rainforest where they live.

Still, this ethologist-zoologist-activist, who is currently a professor at the Simon Fraser University in Burnaby, British Columbia, and a Professor Extraordinaire at the Universitas Nasional in Jakarta, Indonesia, continues his anti-speciesist and ecological efforts.

Thanks to the courage, passion and talent of these researchers, to the enormous discomforts and dangers they had to face, man began to discover and protect the cultures and the expressiveness of our sister species, today more and more exposed to the risk of a complete destruction, due to deforestation, climate change and poaching.

The commitment of these women was at the origin of all the main projects of protection, rehabilitation and re-insertion in nature of the primates born from the '60s onwards and, more generally, it contributed to promoting a new sensitivity to the environmental problems and to the protection of animal and plant species at risk, and biodiversity.

Another ethologist-primatologist who has dedicated his life to the study and protection of the great apes, and should be mentioned here, is Roger Fouts. I have already discussed in the first section of the chapter some of the results that Fouts achieved in his work with Washoe, the first ape trained in the use of the American Sign Language. I would now like to highlight some of the ethical implications of his work and the important impact it had on global public opinion. In the preface to the book in which Fouts recounts his experience (Fouts 1997), Jane Goodall focuses on the "indissoluble bond" between him and Washoe, which lasted until the latter's death in 2007, and on the fact that Fouts, in order to protect the young chimpanzee, sacrificed "his career prospects". After a battle that lasted fifteen years, Fouts finally succeeded in saving Washoe and other chimpanzees from living out their lives the experimental laboratories and transferring them to a place suitable for their natural characteristics in which they could spend the rest of their existence; it was the nature reserve annexed to the Chimpanzee and Human Communication Institute of Central Washington University, founded in 1993 by him and his wife Deborah Fouts. As Goodall writes it was one of the "most extraordinary events of our time", because Fouts was a pioneer in the fight against another crucial aspect of the human exploitation of non-human animals and animal testing and has become one of the best-known animal rights advocates in the world, focusing on the legal rights of great apes and organizing, along with Jane Goodall, a worldwide campaign for their protection which has had a wide echo.

Finally, I would like to mention Ruth Harrison, a pioneer in the fight for the rights of farm animals, who, although not an ethologist, has become a point of reference both for anti-speciesist movements and for those involved in the applied ethology (van de Weerd and Sandilands 2008). In 1964, 11 years before Peter Singer's book *Animal Liberation* (Singer 1975) that would become the first manifesto of contemporary anti-speciesistism, Harrison was among the first to denounce, in the book *Animal Machines* (Harrison 1964), the atrocious living conditions of animals locked up in intensive farms and to launch the fight, still in progress, for their disposal. Three years later, in 1967, Harrison founded the Farm Animal Care Trust (FACT) and began to work as a consultant for many committees, institutes and associations dedicated to combating the exploitation of meat animals. Her pioneering commitment helped spread the message all over the world. Her complaint shocked global public opinion by stimulating, among other things, also the drafting and approval of the *European Convention for the Protection of Animals Kept for Farming Purposes*, a document drafted in 1976 that gave basic principles for the keeping, care and housing of farm animals, especially in intensive breeding systems, and inspired the current national and international legislations on the subject.

In short, as I have tried to demonstrate in this section, all the main battles promoted the anti-speciesist movements and ethics born in the 1970s (from the protection of the biodiversity and endangered species to the fight against animal experimentation and the daily harassment of billions of farm animals) have taken a fundamental impulse from that journey to the discovery of animal minds, societies and languages which began with the birth of ethology. In particular, most ethologists, from the 1960s onwards, have conceived the study of animal ethology and the commitment for the protection of animal life and well-being as an inseparable unicum. More generally ethologists have had, and still have today, the merit of collecting, sometimes in extremely difficult conditions, an enormous amount of empirical knowledge related to the behaviour and psyche of animals, which new technologies such as sound and film recordings have made possible starting from the twentieth century onwards. This way, they made accessible to the scientific community and to the general public the data that show, with immediate evidence, the cognitive resources, the social complexity, the individual and cultural differences, the ability to suffer and rejoice for causes not only physical but also social and emotional, that are widespread in the world of non-human animals.

Anyone who observed, albeit only in a documentary, a couple of grebes performing their wedding dance, or a chimpanzee mother help her cub learn to open coconuts, will have no doubts about the priceless value of these contributions. Anyone who watched the movie of the liberation of the chimpanzee Wounda which, removed from human exploitation, rehabilitated by the volunteers of the Jane Goodall Institute, and finally brought back to its natural environment and liberated, first makes a small run, then stops, turns back, hugs Jane Goodall closely for several seconds and then finally goes away, will have an immediate confirmation of the ethical and theoretical importance of this kind of documentation.

Anyone who watched the shots of the female gorilla Koko expressing the pain for the death of her kitten friend first through the American Sign Language, and later, alone in her shelter, with a heartbreaking crying, immediately understands that between this kind of animals and human beings there are profound similarities that also invest the plane of social and emotional ties.

7.4 From the Evidence of Animal Pain to the Discovery of the Animal "You": A Journey That Has Just Begun

As it is well known, promoters of the anti-speciesist movement, which began to spread in the 1970s, first and foremost focused their attention on the matter of animal pain, insisting that contemporary society could no longer go on denying its existence. That is, to behave as if the suffering systematically inflicted by men on the animals kept in intensive farms, in the many places used for their commercial use and in the scientific laboratories, did not exist. It was no coincidence that these developments towards meta-specific and anti-speciesist ethics were emerging precisely in the years in which a new area of ethological research such as cognitive ethology placed at the center of its research programs the question of animal "sentience", "consciousness" and thinking. Rather, it was a historical conjuncture, a synergy between a scientific orientation that affirmed the need to recognize to at least the so-called "higher animals" a volitional capacity and a cognitive life and an ethical movement that demanded respect for their ability to suffer or rejoice. Even in this case, ethological research offered ethical reflection with the materials on which to reason in the first place, as well as the empirical evidence of what it was postulating. Ethical and scientific reasons were undoubtedly intertwined in the reflection of the scholar who was the main promoter of cognitive ethology: Donald Griffin (1976). His radical critique of the model of Cartesian ancestry which equated "the animal" in general to a natural automaton, offered those who rebelled against the exploitation of animals, their killing for commercial purposes and the destruction of their natural environments, the first scientific support on which to base their claims.

However, one of the pioneers of this turning towards a post-mechanistic ethology, willing to attribute to non-human animals an "experience", a qualitative reception of their physical states and an ability for thought, had been, even before Griffin, Konrad Lorenz, founding father of comparative ethology together with Nikolaas Tinbergen. Lorenz came only at a late age to the environmental commitment of which his "spiritual testament in defense of man and nature", collected in the volume of talks with Kurt Mündl *Rettet die Hoffnung (Save hope)*, bears witness (Mündl 1988). Furthermore, the father of classical ethology was not a vegetarian and his contribution to the birth and development of anti-speciesist ethics was, therefore, only indirect. It derived primarily from its activity as a popularizer and from the capacity that all those who knew him, or read his books, found in him: his ability to understand and describe animals "from within", as Frans de Waal observed (de Waal 2001). That is: grasping and explaining their internal "motivations" (a term that in ethology has both physiological and psychic values) that lead them to behave in one way or another. Secondly, Lorenz contributed to the diffusion of a new sensibility and a greater attention towards non-human animals, as early as a dozen years before the official birth of cognitive ethology, he introduced and anticipated many of the themes that would have been at the centre of research programs in this area of study. His essay *Haben Tiere ein subjektives Erleben? (Do animals have a subjective experience?)*, initially written for a radio conference (Lorenz 1963), inaugurated the reflection on a subject that was at the time considered taboo in the field of behavioural sciences: animals' experience, "subjectivity" and thought.

After illustrating the limits of both the separatist and anthropocentric approaches and of each model that limits itself to anthropomorphically homologating the feeling and thinking of other animals to those of the humans, Lorenz affirmed that the anatomical and neurophysiological comparison represented, in his opinion, the most promising way to obtain, at least indirectly, clues to the existence, or non-existence, in non-human animals of the capacity to have "qualitative" experiences and on the possible degrees of similarity between their feelings and ours. At the same time, the ethologist surprised his listeners and readers by confessing that *for him*, the decisive proof of the existence of an internal experience in the higher animals did not come from anatomical, physiological, or morphological findings, but from an intuitive experience. He had noticed "evidence" for an animal "you" the first few times he had found himself playing or interacting in non-constrictive conditions with intelligent creatures like mammals and birds, spontaneously and incontestably. As the scholar said: "The knowledge of the subjective experience of my fellows and the belief that even a superior animal, like a dog, has its own experience have grown in me at the same time" (Lorenz 1963. It. transl.: 63).

But, ultimately, what response did Lorenz offer to the question posed in the title of the essay? According to his approach, do all non-human animals have an experience? Do they experience forms of pleasure and pain, pleasant and unpleasant sensations, motions of attraction or repulsion? Do they have emotions like fear and curiosity, mental states such as expectations or desires, feelings of disappointment, frustration, discomfort or well-being, bonds of affection like family ties or friendship? Do animals remember, imagine, dream, think?

The ethologist's answer, despite the title of his essay seemsg to rule out a positive or negative response concerning the entire animal kingdom, was not generic. It sounded indeed like: not all!

Lorenz specified that we will never be able to say anything certain about the experience of another living being, because even the sensations of another *human* being remain, for each of us, strictly speaking, unattainable. But, in his opinion, comparative anatomy and physiology enable us to identify at least some organs, and a level of systemic complexity, in the presence of which we can consider extremely probable this type of experience. A condition of the possibility of having feelings and perceptions, and experiencing emotional and mental states, is, in fact, according to the hypothesis that the ethologist had already proposed in his Russian Manuscript in the '40s, "a relatively very high organization of the central nervous system and sensory organs" (Lorenz 1992: 201). A level of organization that, in his opinion, we can find only in "a small part of the living beings", that is, in the so-called higher animals, endowed with a central nervous system (CNS).

We understand "at a glance that a dog is sad, even if we do not know how to explain its motivation", Lorenz wrote, we recognize as such the "crying" of a mammalian puppy, or the chirp of an abandoned chick. Regardless of their species, these animal calls, like children's voices, can empathically involve us and induce us to behaviour of parental care. In other words, towards this kind of animals, we tend to behave, spontaneously, as we do with human beings: we interpret their experiences, behaviour and languages, imagining that they are at least in part like ours; we actually establish emotional and social relationships with them through these models. However, the ethologist pointed out that this "analogical conclusion" (inferring from the similarity of the organs a similarity in the basic sensory and emotional experiences) becomes more unlikely when the differences between our sensory and cognitive apparatus and those of the organisms we observe increase: "The more the structure of their perceptive organs and nervous system are different from that of mine, the more their functions will be different. How the experience associated with these functions may present itself is basically impossible for me to know. The «evidence of the you» leads me to attribute to my dog an experience somewhat similar to mine, but the more we enter in the lower steps of the organic realm, the less it is possible to apply these analogical conclusion" (ibid.: 64).

In fact, just by comparing our perceptive and cerebral apparatuses with those of other mammals, we already discover both surprising similarities and significant differences. Realizing that a dog is feeling affections and emotions and establishing with it an empathic and affective bond are experiences that spontaneously mature in children and adult human beings, if they have not undergone the effect of intraspecific traumatizing relationships. Men and dogs can socialize and understand each other in many areas. And yet, we will never experience the olfactory world and the complex scented maps, for us ineffable, that 'our' dog discovers at every corner of the street; all the more so, the ways in which the external environment may be perceived or mentally represented by organisms phylogenetically more distant from us escape us completely or almost completely.

The problem is then made more complex by the fact that the similarities between species, as Lorenz liked to remember, can be the result of homologies, or of the derivation from common ancestors, but they can also derive from a convergent evolution, that is, from a mutually independent adaptation to similar functions. Convergent evolution can, in turn, give rise to very similar performances in organisms which are phylogenetically very distant and morphologically very different (for example the flight in insects and birds) and therefore it can produce similar results with apparatuses, organs and neurophysiological activities that are enormously different. The eye of a bee, anatomically very different from ours because, like that of any other arthropod, it is composed of many small photosensitive units, allows a tri-chromatic vision analogous to ours, but more extended because it also perceives the ultraviolet light. On the contrary the eyes of many mammals, incomparably closer from the phylogenetic point of view and more similar to ours in many aspects of anatomical organization, being adapted to other environments and functions, only allow dichromatic vision.

What does Lorenz mean when he says that the superior animals have their own *Erleben*? What meaning is attributed here to the word "experience"?

The ethologist does not avoid this question and, referring to Wilhelm Busch, suggests that "the primary form of every experience" is the "capacity to feel pleasure and pain" (Lorenz 1963. It. transl.: 88, 89). Lorenz here referred to a characteristic of the animal organisms that the later cognitive ethologists would have called "sentience" (Griffin 1976, 1984): the ability to perceive the changes in the physical-chemical states and energy gradients of one's own body and their relationship with some environmental factors and that to respond to them with behaviours of distancing or approaching, avoiding or searching for certain external stimuli. The basic form of the experience is, in short, according to Lorenz, the alternation between the "message of pain", or of the unpleasant feeling, which intimates "do not do it again", and the voice of pleasure that suggests "do it again" (Lorenz 1963. It. Transl.: 63) or, more simply: "stay in this situation". Basing on this dialectic, well before the appearance of man, animals began to build, through the different ways and degrees of individual and social learning, a baggage of acquired "knowledge" that was going to be added to their hereditary endowments, making their behaviour more flexible and modifiable. Through this basic form of experiencing, animals capable of associative learning begin, in fact, to experiment resources and explore their environment, memorizing what is learnt not in a conceptual form, but in terms of implicit memory and behavioural attitudes. Precisely in this sense, according to Lorenz, it is correct to attribute experience "even to the higher animals": there are, without doubt, pre-human and pre-conceptual forms of experience of pleasure and pain, sensations such as nervous excitement and relaxation, emotions such as fear or attraction, emotional and social needs such as receiving parental care and being accepted by the reference group, and forms of thinking older than rationality, which are widely widespread among the "higher" animals. Therefore, according to him, both anthropomorphic homologation of perceptive and cognitive forms of other species to human ones and the anthropocentric supposition that can be no analogy between our basic sensations, perceptions and emotions and those of another mammals, or birds are completely misleading. In other words, the concept of experience must be understood, for Lorenz, in a way that is at the same time as less anthropomorphic and less separatist as possible. Thus, in a nutshell, according to the approach proposed by Lorenz in 1963, the similarities between their neurophysiological organization and ours, widely attested by comparative anatomy and neurophysiology, force us to recognize *at least* in the animals that have a Central Nervous System the ability to feel pleasure and pain, attraction and repulsion. The anatomical, physiological, and experimental evidence in favour of this hypothesis accumulated by neurophysiological research in the following years means that, today, it is widely shared in the scientific community.[7]

However, the thesis that the possession of a centralized nervous system is a *conditio sine qua non* of the possibility to experience suffering and its opposite, sketched by Lorenz in that essay, seems to reveal itself, in the light of the research carried out in the last decades, if not fully overcome, at least too restrictive and prudential. Scholars who have given important contributions to the development of cognitive ethology, such as Donald Griffin or James and Carol Gould, anti-speciesist intellectuals such as Peter Singer, Italian ethologists such as Danilo Mainardi and Giorgio Celli, have argued that this approach arbitrarily traces a border, based more on our inability to recognize experiences and thoughts in organisms that are very different from us than on the actual lack of empirical or experimental evidence of their existence. Already at the time of Lorenz, neurophysiological research had shown that animals traditionally classified as "inferior", like social insects, are in

[7] This position was re-launched just a year later in the book *Animal Machine* by Ruth Harrison (1964), which contained a passionate denunciation of the living conditions of animals kept on intensive farms and which was to give rise the current trend of contemporary animalist ethics. The attribution of the experience of pain to mammals, birds, reptiles, amphibians and fish, animals endowed with a real Central Nervous System, is today almost unanimous among scholars of biological formation and new studies continue to accumulate evidence in this sense. For example, confirming and completing research produced in the previous decade by other researchers, biologist Victoria Braithwaite of the University of Pennsylvania has collected a neurophysiological, ethological, experimental and comparative documentation, which can be considered exhaustive in showing that fish experience pain (Braithwaite 2010).

fact capable of creating cognitive maps and, therefore, forms of representation of their environments that are not only perceptive, but also mental and mnemonic, used to communicate information to other con-specifics and to locate nutritional sources. Over the past thirty years, based on a growing corpus of studies, the belief that "even mollusks like octopus and cuttlefish, or insects like bees and ants, possess a mind […] and have developed sophisticated communication systems and thinking skills" (Mainardi 2010) has been spreading among cognitive ethologists. This conviction first of all concerns the very large group of animals like arthropods (insects, arachnids, crustaceans), annelids and molluscs which, even if they do not possess a Centralized Nervous System, are equipped with a metameric system, formed by chains of ganglia that allow highly complex performance. Less investigated, with respect to the topic of cognitive maps, is the more subtle question posed by Lorenz, concerning the possible existence, in these organisms, of forms of experience of pleasure and pain. But new elements have also emerged in this field in recent years. For example, it had already been pointed out, years ago, that the nervous systems of some insects produce analgesic opioids when tissue damage is taking place, similarly to higher animals. More recently, in the drosophila fruit fly that was the protagonist of the major biological discoveries of the twentieth century a receptor, known as TRPA1, has been identified that plays an important role in decoding painful stimuli. It seems to be present also in all other animals and to originate, almost identical to its current form in an organism that lived about half a billion years ago and from which all existing invertebrates and vertebrates derive (Kang et al. 2010).

In short, with our current knowledge, we can with Cartesian certainty no longer exclude that our neurophysiological and cultural limitations prevent us from seeing in creatures so different from us as wasps or octopus traces of experience, sensation, emotion or thought, which instead exist. This could depend on the fact that their expressiveness is too different from ours, that they lack, for example, bodily warmth, voices, mobility of the gaze, facial mimicry and expressiveness, having in their place hormonal, chemical, electrical messages and other forms of communication that we do not grasp. The cognitive barrier could also depend on emotional barriers. Not only with insects, arachnids, crustaceans, worms and mollusks, but also with animals that are more complex and closer to us in the basic neurophysiological organization, such as fish, amphibians and reptiles, it may seem impossible for us to develop anything more than a simple habit of mutual presence, or a mere ability to interact. It seems, in other words, to be missing the possibility of establishing a bond of affection that in mutually non-stressful conditions, we can instead easily establish with animals belonging such as birds and mammals.

It is, in short, really difficult to understand in which cases, in spite of what Descartes thought, it is the limits of our cognitive and communicative channels and our prejudices and presumptions preventing us from recognizing that in animal life other forms of experience and representation of reality may occur, and conversely, in which cases, as seems to happen for in the vegetable realm, life takes place without mediation of sensations and experiences, as pleasant or unpleasant as they may be. As we can see, these are questions in which philosophy has always been committed, in which the effort to understand the resources and limits of our mind is intertwined, for its own internal need, with questions aimed at investigating the characteristics of "other minds" that differ from ours without having to be considered "inferior" to it.

Developments in neurophysiological, ethological and psychological studies seem today to substantially confirm the critical value of the approach that Lorenz took in 1963, recognizing in many non-human animals a sensitive and affective life and an ability to think (Allen and Bekoff 1997), but at the same time they suggest that it would be a mistake to suppose that animal thought can manifest itself only in the anthropomorphic (and culturally Eurocentric) form of the conscious elaboration of projects and their subsequent execution. One of the hypotheses that Lorenz, following a tradition which went from Helmholtz to his friend von Holst, proposed in this essay has been in the meantime widely corroborated: the animal and human brain continually perform a series of extremely complex calculations, on which all phenomena of perceptual constancy are based without these entering the threshold of awareness. In addition to the phenomenon of "unconscious deductions" (as Helmholtz called them) or "ratiomorphic" processes (as Lorenz, following Egon Brunswick, preferred to say) more recent research has also confirmed the hypothesis launched by von Holst and Mittelstaedt and resumed from the ethologist in *Haben Tiere ein subjectives Erleben*?, that our brain elaborates, from time to time, a sort of planning of the motor activities that the body is about to accomplish, building an anticipatory scheme of action, without us being aware of it. Accurate studies like those presented by Christopher Frith in the book *The Cognitive Neuropsychology of Schizophrenia* (Frith 1992) confirm, in keeping with what von Holst, Mittelstaedt and Lorenz maintained, that the production of this "efferent copy" of an incipient action "does not happen on a conscious level" (ibid.). And yet, according to Frith, this unconscious process is indispensable for a thinking subject to recognize themselves as an executing agent of their own behaviour and precisely the jamming of this process would be one of the causes of the inability to think of themselves as subjects responsible for their own actions, often manifested by schizophrenic patients.

As we have seen, even today, the ideas that then Lorenz put into play, and the knowledge he was spreading are intertwined with the questions that empirical and experimental research raises for a theoretical reflection and conceptual elaboration that identifies the points where the intersection and critical exchange between scholars of different backgrounds and the removal of disciplinary boundaries are required by the characteristics of their own objects of study. That is, from the internal complexity and the unexpected reciprocal relations that these "objects" reveal. To conclude, the possibility of a further deepening of our knowledge of animal minds, cultures and societies today more than ever depends on two aspects:

- a concrete commitment aimed at safeguarding their survival and independence and, therefore, the natural environments that host them;
- the development of a philosophical and ethical reflection oriented towards an eco-centric vision, which considers the community of the living as a whole and which does not separate itself with a disdainful gesture from the empirical and natural sciences, nor does it submit to them, but challenges them and allows itself to be challenged by them in a practice of knowing capable of adapting itself to its

object, to the places where it exists, lives or manifests itself and does not pretend that the object came to our laboratory or computer, to our desk or just in front of our chair by itself. All the more if the "object" in question is another perhaps sentient living being, perhaps like us, moved by attractions and fears; perhaps thinking, certainly less harmful and dangerous for all the living species, than Man and his contemporary society.

7.5 Cognitive Ethology and Anti-Speciesist Ethics: Two Revolutions in One

As its title makes clear, this chapter is aims at documenting the role ethological research has played in promoting *the birth* of contemporary post-anthropocentric and anti-speciesist ethics. Its purpose is to show the links between the *scientific revolution* introduced by ethology and the *ethical revolution* which began with current anti-speciesist movements. On the contrary, an illustration of the enormous and multiform developments of the latter that have taken place in the last decades, precisely because of their vastness and internal complexity, remains outside the intent of the chapter.

Although, the brief notes that follow cannot of course overcome this structural limitation, they should offer a small glimpse on the intertwining of cognitive ethology, ethological philosophy and anti-speciesit ethics since the 1970s, which results obviously incomplete and oriented to report only some exemplary cases.

But, above all, I would like to attract the reader's attention on an aspect that has perhaps not been emphasized enough in the previous pages: if ethology has offered the nascent anti-anthropocentric ethics a good part of the empirical and scientific documentation on which they base their claims, pioneers of anti-speciesism like William Russel and Rex Burch (the first to introduce the well-known criterion of the three Rs),[8] Ruth Harrison, Jane Goodall, and later Peter Singer in turn played a very important role in promoting *an emancipation of the ethological thought*. They in fact promoted, its connection to the great *emancipationist* tradition that historically emerged from the long wave of Enlightenment, and becoming at the same time an extension, and a problematization, of this tradition. In this respect, the philosophical and ethical revolution introduced by contemporary anti-speciesism and the scientific revolution introduced by cognitive ethology, *both still in progress*, embody two faces

[8]The Three Rs principle, formulated for the first time in the volume *The Principles of Humane Experimental Technique* (Russel and Burch 1959), aims at providing guidelines for the least invasive and least extensive use of animals in scientific research that technologies make possible. The Three Rs indicate, respectively, the concepts of:

Replacement:	methods that avoid or replace the use of animals in research.
Reduction:	use of methods that enable researchers to obtain comparable levels of information from fewer animals, or more information from the same number of animals.
Refinement:	use of methods that alleviate or minimize pain, suffering and distress in the animals and improve their living conditions.

of the same epochal revolution that has profoundly, and positively, changed the way in which human beings relate to other animal species.

From the second half of the seventies, the anti-speciesist-environmentalist-humanitarian commitment of Jane Goodall, the spread of Singer's libertarian ethics and the descent into the field of numerous other scientists and activists gave a more general contribution to the inclusion of ethological knowledge in the field of ethical and philosophical reflection. They brought out, in fact, in both its problematic nature and indisputable significance, the connection between *speciesism* and phenomena such *as racism, sexism, homophobia and discrimination against the "disinherited" of the earth.* This in turn linked the question of human oppression against the animals, and the need of its overcoming, to *all the great ethical and social struggles for emancipation that have crossed modern and contemporary history.*

In other words, the scientific revolution launched by cognitive ethology and the ethical revolution triggered by anti-speciesism have been intertwined since the beginning. Peter Singer, the philosopher who perhaps more contributed to spreading the anti-speciesist issues internationally, used in his first book, *Animal Liberation* (1975), much data and several references taken by some of the major ethologists of the time: Konrad Lorenz, Niko Tinbergen, Irenäus Eibl Eibesfeldt, Jane Goodall and George Schaller. The following year, he published the volume *Animal Rights and Human Obligations* (Regan and Singer 1976) with his colleague Tom Regan. In that same year, Donald Griffin, with the book *The Question of Animal Awareness* (Griffin 1976), began to revolutionize the science of animal behaviour by proposing a new research program focused on the study of different forms of "sentience" (the ability to feel and memorize sensations) and intelligence, widespread in the animal world.

As Colin Allen and Marc Bekoff observe, these two events were meant to interact with one another since the beginning: "In *Animal Liberation*, Peter Singer sought to revolutionize societal treatment of nonhuman animals by arguing that animal agriculture and animal experimentation cause conscious pain and suffering that is real and morally significant. With *The Question of Animal Awareness*, Donald Griffin sought to revolutionize the science of animal behaviour by insisting that questions about animal consciousness should be placed firmly in the foreground of a new program of research he labelled «cognitive ethology» . Both proposals immediately evoked a range of reactions, from enthusiasm to virulent attacks. In the ensuing three decades, little consensus has been reached about either. Although Singer's arguments about ethical treatment of animals preceded Griffin's arguments about scientific understanding of animals, it is obvious why ethicists concerned with the former should be interested in the latter. Singer himself based the case for animal liberation on scientific evidence of behavioural and neurological homologies between humans and other animals. Cognitive ethology, by rejecting behaviourist strictures against attributing subjective states of awareness to nonhuman animals, offered the prospect of increased scientific support for the claim that animals are conscious in the ways that matter ethically" (Allen and Bekoff 2007: 2).

Soon after, these developments would, in turn affect the birth of new sectors of the ethological research. In Chapter 2 of the collective book *Animals and us. 50 years and more of applied ethology* (Newberry and Sandilands 2016), Ruth Newberry and

Victoria Sandilands observe that Harrison, Russell and Burch and Singer's works contributed to create the scientific and cultural milieu in which the first pioneeristic forms of applied ethology oriented to increase an "animal welfare" developed.[9]

In 1993, Singer himself, together with the philosopher Paola Cavalieri launched the *Great Ape Project* (GAP): an international organization of primatologists, anthropologists and ethicists who elaborated a *United Nations Declaration of the Rights of Great Apes* (chimpanzees, bonobos, gorillas and orangutans) aimed at conferring them with basic legal rights.

Subsequently, the Australian philosopher continued to deal with antispecies issues but also extended his commitment to questions like environmental emergency, the growing imbalance between rich and poor, anti-Semitism, exploitation of the less industrialized countries and the damage and social imbalances caused by capitalist globalization. Books as *The Expanding Circle* (Singer 1981), *Ethics into action* (Singer 1998), *One World: The Ethics of Globalisation* (Singer 2002), *Pushing Time Away: My Grandfather and the Tragedy of Jewish Vienna* (Singer 2003), *The Life You Can Save: Acting Now to End World Poverty* (Singer 2009), *The Most Good You Can Do* (Singer 2015), *Ethics in the Real World* (Singer 2016), *Why Vegan?: Eating Ethically* (Singer 2020) attest to the continuity and breadth of this commitment.

Meantime, the utilitarian approach to anti-anthropocentric ethics that he introduced was reworked and re-proposed by various scholars and thinkers. Among these, American philosopher James Rachels (1942–2003) made important contributions to the development of an anti-speciesist ethical orientation. Discussing the issues of ethical vegetarianism and animal rights, he adopted an approach according to which human choices and actions are to be evaluated on the basis of their effects on both human and non-human beings. Furthermore, in many of his writings, from the short essay *Do Animals Have a Right to Liberty?* (1976) to the volume *Created from Animals* (1990), he fully grasped the historical and conceptual link between the Darwinian revolution, which also runs through the entire history of ethology, and the development of a new ethical sensitivity towards non-human animals.

According to Rachels, animals—like some human beings with disabilities that render them unable to understand and respect certain ethical rules—cannot be subjected to moral duties and constraints, but despite this they should be recognized as bearers of moral *rights*: "like the retarded person, they lack characteristics

[9] An accurate account of the positive but at the same time always problematizing influence exerted in the following two decades by this intertwining of animal welfare ethology and "animal rights" commitment was offered in 2009 from the issue 118 (3) of the scientific journal *Applied Animal Behaviour Science*, entirely dedicated to the topic. The issue illustrates *inter alia* the main changes in the national and international legislation that these ethical battles had inspired. In one of the papers, entitled *Ethology applied to animal ethics*, the biologist and professor of Animal Welfare Hanno Würbel emphasized what he considered the most important aspect of this turn: "According to modern animal welfare legislation, animals should be protected from suffering and lasting harm not for the benefit of us humans as in earlier anthropocentric conceptions, but in their own interest. [...] Moreover, from an ethological perspective, protecting animals in their own interest represents true altruism which places considerable ethical demand on us" (Würbel 2009: 118). In 2016 the volume *Animals and Us. 50 Years and More of Applied Ethology* offered further updates and reflections on the subject.

necessary for having obligations; but they may still be proper beneficiaries of our obligations. The fact that they cannot reciprocate, then, does not affect our basic obligations to them" (Rachels 1976: 223).

Since the 1990s and even more so in the new millennium, the intertwining between cognitive ethology, philosophy of ethology and active commitment in the fields of wildlife protection, "animal rights" and ecological commitment has become increasingly thicker.

In the first two decades of this new millennium, ethologist Jonathan Balcombe was one of the scholars who gave important contributions to the study of animal sentience offering extensive documentation on the ability to experience pleasure and pain in vertebrate and invertebrate animals. From 2015 to 2019 Director and Associate Editor of the "Interdisciplinary Journal of animal feeling" *Animal Sentience*, he dedicated to these studies books as *The Use of Animals in Higher Education: Problems, Alternatives, and Recommendations* (Balcombe 2000), *Pleasurable Kingdom: Animals and the Nature of Feeling Good* (Balcombe 2006), *Second Nature: The Inner Lives of Animals* (Balcombe 2010), *The Exultant Ark: A Pictorial Tour of Animal Pleasure* (Balcombe 2011), *What A Fish Knows: The Inner Lives of Our Underwater Cousins* (Balcombe 2016), *Super Fly: The Unexpected Lives of the World's Most Successful Insects* (Balcombe 2020).

Among the scientists who have made important contributions to the developments of both cognitive ethology and anti-speciesist ethics in the last twenty years, it is therefore impossible not to mention ethologist and activist Marc Bekoff. Since the nineties he has been offering significant empirical and theoretical contributions to both our understanding of animal experiences and minds and their protection and defense. In 1997 his collaboration with the philosopher Colin Allen led to the joint editing of the volume *Species of Mind: The Philosophy and Biology of Cognitive Ethology* (Allen and Bekoff 1997). In 1998 he was editor of *Encyclopedia of Animal Rights and Animal Welfare* (Bekoff 1998) and co-editor with John Byers of the book *Animal Play: Evolutionary, Comparative and Ecological Perspectives* (Bekoff and Byers 1998). In 2000 he co-founded the association "Ethologists for the Ethical Treatment of Animals" with Jane Goodall, the purpose of which is "to develop and maintain the highest of ethical standards in comparative ethological research that is conducted in the field and in laboratory" (Bekoff and Goodall 2000: 277). A few years later they published the book *The Ten Trusts: What We Must Do to Care for the Animals We Love* (Goodall and Bekoff 2002).[10] In 2007 his collaboration with Allen gave rise to a new fruit: the still much quoted paper *Animal minds, cognitive ethology, and ethics* (Allen and Bekoff 2007). His commitment to both the areas of scientific research and ethical reflection continued uninterruptedly in the following years. Recent testimonies of it are the volumes *The Animals' Agenda: Freedom, Compassion, and Coexistence in the Human Age* (Bekoff and Pierce 2017) and *Canine Confidential:*

[10]Even in a recent text, co-edited with Dale Peterson, *The Jane Effect: Celebrating Jane Goodall* (Peterson and Bekoff 2016), Bekoff returned to highlight the enormous ethical impact that Goodall's research and commitment have had.

Why Dogs Do What They Do (Bekoff 2018), as well as his participation to the collective paper *Recognizing animal personhood in compassionate conservation* (Wallach et al. 2020).

The meaning and the ethical relevance of the debate arisen in those decades from the intertwining of ethology, philosophy and ethics was well understood and exposed in the collective book *Animal minds and animal ethics*, edited by Klaus Petrus and Markus Wild (Petrus and Wild 2013). In the introduction, titled *Big issues in animal philosophy*, the authors wrote:

"What may, in very general terms, be called 'the animal issue' has drawn wide academic and public attention in the past thirty years. The issues at stake are our (Western) perception of animals, our interaction and involvement with animals, the differences between ourselves and other animals, our moral obligations towards animals and the practical consequences that a moral standing of animals would have. After the turn of the 21st century, animal ethics are very much on the mind of philosophers, ethicists, professionals who use animals, politicians, lawmakers, pet-owners, and the public. A related phenomenon is the explosion of research into the cognitive abilities of animals, as seen in the inspiring work being done on the science of animal cognition and behaviour. This development has not remained without a direct influence on philosophy, especially regarding not only the philosophy of mind but also the moral philosophy. Clearly, the animal issue has engaged philosophers in two related but distinct ways. On the one hand, there has been a growing interest in the question of animal minds. Can we attribute mental states to non-human animals? If so, what kinds of mental states? What does the mental life of a non-human animal look like? On the other hand, there has been a growing interest in the question of animal ethics. Do we have direct moral obligations towards animals? Do animals have rights? Should states enact strong legal policies with regards to animals? Philosophers working on questions of animal ethics usually draw on research into animal cognition and subscribe to strong positions regarding animal minds" (Petrus and Wild 2013: 7; 8).

An overall reflection on this debate, on the studies that supported it and on the great questions it has raised is finally proposed in the very recent Judith Benz-Schwarzburg volume *Cognitive Kin, Moral Strangers? Linking Animal Cognition, Animal Ethics & Animal Welfare* (Benz-Schwarzburg 2019). In the preface, Benz-Schwarzburg presents her work with these words:

"This book is an interdisciplinary study at the intersection of the humanities and natural sciences. It deals primarily with questions of animal ethics, animal welfare, and cognitive ethology, but it also includes, for example, insights from evolutionary anthropology, cognitive archaeology, comparative psychology, developmental psychology, theoretical philosophy, linguistics, and veterinary medicine. Based on the results of empirical research in the area of mental ability and performance, the book is intended to explore something that has been under investigation since the introduction of Darwin's theory of evolution: namely, the possibility that the difference between people and animals is only gradual. This entails an examination of crucial aspects of the evolutionary relationship between humans and animals and their practical importance for human-animal relations today" (Benz-Schwarzburg 2019: IX).

7.6 Conclusion

Naturally, as clarified at the beginning of this section, in the decades that separate our time from that of the birth of cognitive ethology and anti-speciesist ethics, many ethologists, philosophers and groups of activists have contributed, together with those mentioned, to enrich these international debates and social movements. For philosophical contributions, just see the works of scholars such as Gerald J. Massey (who coined the term "philosophical ethology". See: Massey 1999), Elliot Sober, Mary Midgley, Christine Korsgaard, Luisella Battaglia, Roberto Marchesini, Massimo Filippi, Martha Nussbaum and Bernard Rollin.[11] Unfortunately, it would have been impossible to follow the threads of all these developments in this work, given their richness and complexity. Nonetheless, with this book the authors hope to have been able to give to the ongoing debate related to this contemporary ethical-scientific revolution, fuelled by a historical feedback between cognitive and cultural ethology, evolutionary and developmental studies, ethical reflection and philosophical ethology, a small contribution.

To conclude: in this era of global ecological crisis, the possibility of a further deepening of our knowledge on animal behaviour, languages, minds, cultures and societies that is the object of ethology is inextricably linked to a concrete commitment aimed at safeguarding their survival and independence and the natural environments that host them, just as it is linked to the struggle for the abolition of intensive farming—which is still one of the main causes of pollution—and of other forms of exploitation suffered by animals today.

The situation makes the development of a philosophical and ethical reflection oriented towards an eco-centric vision that considers the community of the living as a whole and does not separate human communities from the environments in which they live, necessary and urgent.

Ethology has certainly offered all disciplines that deal with living organisms, their environments and their societies, a lesson in humility that even a part of the philosophical community has begun to assimilate: when the "objects" of our research are other, definitely sentient living beings moved, like us, by attractions and fears, perhaps thinking and certainly less harmful and dangerous than man and his contemporary society, *it is necessary to develop a practice of knowing capable of adapting itself to the object* and to the places where it lives, capable of respecting its autonomy and protect its survival, and to not, on the contrary, authoritatively and lazily expect

[11]For a first approach see: Massey (1999), Sober (2000), Midgley (1978, 1983, 1984, 2010), Korsgaard (2006, 2018), Battaglia (2009, 2011, 2016), Marchesini (2008, 2011, 2018), Filippi (2011, 2016), Nussbaum and Sunstein (2004), Rollin (1981, 1995, 2006). In turn, I have dealt with these issues in some essays. See: Celentano (2000, 2011, 2013, 2017a, b, c).

the object to appear in our laboratory, on our computer, our desk, or just in front of our comfortable chair, to be examined.

References

Allen, Colin, and Marc Bekoff. 1997. *Species of Mind*. Cambridge, MA: MIT Press.
Allen, Colin, and Marc Bekoff. 2007. Animal Minds, Cognitive Ethology, and Ethics. *The Journal of Ethics* 11 (3): 299–317.
Allen, Melinda, and Bebbet Schwartz. 2008. Mirror Self-Recognition in a Gorilla (Gorilla Gorilla Gorilla). *Journal of Integrated Biosciences* 5 (1): 19–24.
Ari, Csilla, and Dominique D'Agostino. 2016. Contingency Checking and Self-Directed Behaviors in Giant Manta Rays: Do Elasmobranchs Have Self-Awareness? *Journal of Ethology* 34 (2): 167–174.
Aristotle. 2012. *Nicomachean Ethics*. Chicago: University of Chicago Press.
Balcombe, Jonathan. 2000. *The Use of Animals in Higher Education: Problems, Alternatives, and Recommendations*. Washington: Humane Society Press.
Balcombe, Jonathan. 2006. *Pleasurable Kingdom: Animals and the Nature of Feeling Good*. London, New York: Macmillan.
Balcombe, Jonathan. 2010. *Second Nature: The Inner Lives of Animals*. London, New York: Macmillan.
Balcombe, Jonathan. 2011. *The Exultant Ark: A Pictorial Tour of Animal Pleasure*. Chicago: University of California Press.
Balcombe, Jonathan. 2016. *What a Fish Knows: The Inner Lives of Our Underwater Cousins*. New York: Scientific American/Farrar, Straus and Giroux.
Balcombe, Jonathan. 2020. *Super Fly: The Unexpected Lives of the World's Most Successful Insects*. London: Penguin Books.
Baragli, Paolo, Elisa Demuru, Chiara Scopa, and Elisabetta Palagi. 2017. Are Horses Capable of Mirror Self-Recognition? A Pilot Study. *Plos One* 12 (5): e0176717.
Battaglia, Luisella. 2009. *Bioetica senza dogmi*. Soveria Mannelli: Rubbettino.
Battaglia, Luisella. 2011. *Un'etica per il mondo vivente*. Roma: Carocci.
Battaglia, Luisella. 2016. *Uomo, natura, animali. Per una bioetica della complessità*. Campospinoso (PV): Altravista.
Bekoff, Marc. 1998. *Encyclopedia of Animal Rights and Animal Welfare*. Westport: Greenwood Press.
Bekoff, Marc. 2018. *Canine Confidential: Why Dogs Do What They Do*. Chicago: University of Chicago Press.
Bekoff, Marc, and John Byers (eds.). 1998. *Animal Play: Evolutionary, Comparative and Ecological Perspectives*. Cambridge: Cambridge University Press.
Bekoff, Marc, and Jane Goodall. 2000. Announcement. *Journal of Applied Animal Welfare Science* 3 (3): 277.
Bekoff, Marc and Pierce Jessica. 2017. *The Animals' Agenda: Freedom, Compassion, and Coexistence in the Human Age*.
Benz-Schwarzburg, Judith. 2019. *Cognitive Kin, Moral Strangers? Linking Animal Cognition, Animal Ethics & Animal Welfare*. Leiden: Brill.
Braithwaite, Victoria. 2010. *Do Fish Feel Pain?* Oxford, New York: Oxford University Press.
Cammaerts, Marie-Claire, and Roger Cammaerts. 2015. Are Ants (Hymenoptera, Formicidae) Capable of Self Recognition? *Journal of Science* 5 (7): 521–532.
Celentano, Marco. 2000. *Etologia della conoscenza*. Napoli: Città del Sole.
Celentano, Marco. 2011. *Konrad Lorenz e l'etologia contemporanea*. Milano: Franco Angeli.

Celentano, Marco. 2013. Etologia e antropologia. Utilità e limiti del principio di continuità. In *Una prospettiva evolutiva sulle emozioni. Da Charles Darwin alle neuroscienze*, a cura di. Fantini, Bernardino, Destro Bisol, Giovann, Rufo, Fabrizio, 47–66. Pisa: Edizioni ETS.

Celentano, Marco. 2017a. From Konrad Lorenz's "Phylogenetic Apriorism" to the Birth of Evolutionary Epistemology. In *Readings in Numanities*, ed. Oana Andreica and Alin Olteanu. New York, Berlin: Springer.

Celentano, Marco. 2017b. Chi mangiamo a cena stasera? Scelte alimentari e forme rimosse della violenza. In *La grammatica della violenza*, a cura di. A. Sannella, M. Latini, and A.M. Morelli. Mimesis, Sesto San Giovanni (MI).

Celentano, Marco. 2017c. Both for Ethics and Health: Non-Animal Technologies: An Achievable Goal. *Journal of Theoretical and Applied Vascular Research* 2.

Delfour, Fabienne, and Kenneth Marten. 2001. Mirror Image Processing in Three Marine Mammal Species: Killer Whales (Orcinus orca), False Killer Whales (Pseudorca crassidens) and California Sea Lions (*Zalophus californianus*). *Behavioural Processes* 53 (3): 181–190.

Descartes, René. 1637. *Discours de la method*. Leide: Ian Maire.

de Waal, Frans. 2001. *The Ape and the Sushi Master*. New York: Basic Books.

de Waal, Frans. 2006. *Primates and Philosophers: How Morality Evolved*. Princeton, NJ: Princeton University.

Filippi, Massimo. 2011. *I margini dei diritti animali*. Aprilia (Lt): Ortica Editrice.

Filippi, Massimo. 2016. *L'invenzione della specie*. Verona: Ombre corte.

Fontaine, Nicolas. 1736. *Memoires pour servir à l'histoire de Port-Royal*, II. Cologne: Michel Tronchay.

Fossey, Dian. 1983. *Gorillas in the Mist*. London: Hodder & Stoughton.

Fouts, Roger. 1997. *Next of Kin*. Scranton: William Morrow & Company.

Frith, Christopher. 1992. *The Cognitive Neuropsychology of Schizofrenia*. Hove: Lawrence Erlbaum.

Gallup, Gordon Jr. 1970. Chimpanzees: Self Recognition. *Science* 167 (3914): 86–88.

Goodall, Jane, and Marc Bekoff. 2002. *The Ten Trusts: What We Must Do to Care for the Animals We Love*. San Francisco: Harper San Francisco.

Gordon, Nicholas. 1993. *Murders in the Mist: Who Killed Dian Fossey?* London: Hodder & Stoughton.

Griffin, Donald. 1976. *The Question of Animal Awareness*. New York: Rockefeller University Press.

Griffin, Donald. 1984. *Animal Thinking*. Cambridge, MA, London: Harvard University Press.

Harrison, Ruth. 1964. *Animal Machine*. London: Vincent Stuart.

Hayes, Catherine. 1951. *The Apes in Our House*. New York: Harper and Row.

Hayes, Keith, and Catherine Hayes. 1952. Imitation in a Home-Raised Chimpanzee. *Journal of Comparative & Physiological Psychology* 45: 450–459.

Kang, Kwonyoon, Stefan Pulver, Panzano Vince, Erich Chang, Leslie Claire Griffith, Douglas Theobald, and Paul Garrity. 2010. Analysis of Drosophila TRPA1 Reveals an Ancient Origin for Human Chemical Nocioception. *Nature* 464: 597–600.

Kellog, Winthrop Niles, and Luella Agger Kellog. 1933. *The Ape and the Child*. New York: Mc-Graw Hill.

Kohda, Masanori, Takashi Hotta, Tomohiro Takeyama, Satoshi Awata, Hirokazu Tanaka, Jun-ya Asai, and Alex Jordan. 2019. If a Fish Can Pass the Mark Test, What are the Implications for Consciousness and Self-Awareness Testing in Animals? *Plos Biolog* 17 (2): e3000021.

Korsgaard, Christine. 2006. *Morality and the Distinctiveness of Human Action*. In Frans de Waal. Quot.: 98–119.

Korsgaard, Christine. 2018. *Fellow Creatures: Our Obligations to Other Animals*. Oxford: Oxford University Press.

Lorenz, Konrad. 1963. Haben Tiere ein subjektives Erleben? In *Jahrbuch Technischen Hochschule München*: 57–68, and in Id. 1965. Über tierisches und menschliches Verhalten. Muenchen-Zurich: Piper. Band II: 359–374. Ital. transl. 2011. *Gli animali hanno un'esperienza?* Fondi: Eupolis.

Lorenz, Konrad. 1992. *DieNaturwissenschaftvomMenschen: Eine Einfiirungindievergleichende Verhaltensforschung. Das «RussischeManuskript» (1994–1948).* Ital. transl. *La scienza naturale dell'uomo.* 1993. Milano: Mondadori.
Mainardi, Danilo. 2010. *Quello che c'è nella mente degli animaliC Corriere Della Sera*, August 8.
Marchesini, Roberto. 2008. *Intelligenze plurime: Manuale di scienze cognitive animali.* Bologna: Peridsa.
Marchesini, Roberto. 2011. *Modelli cognitivi e comportamento animale.* Isernia: Eva.
Marchesini, Roberto. 2018. *Etologia cognitiva: Alla ricerca della mente animale.* Bologna: Apeiron.
Malebranche, Nicolas. 1678. *Recherche de la vérité.* Paris: Chez André Pralard.
Massey, Gerald J. (ed.). 1999. *Philosophical Topics.* Fayetteville: University of Arkansas Press.
Midgley, Mary. 1978. *Beast and Man: The Roots of Human Nature.* London: Routledge.
Midgley, Mary. 1983. *Animals and Why They Matter: A Journey Around the Species Barrier.* Athens: University of Georgia Press.
Midgley, Mary. 1984. Biological and Cultural Evolution. *Institute for Cultural Research Monograph Series* 20.
Midgley, Mary. 2010. *The Solitary Self: Darwin and the Selfish Gene.* Durham: Acumen.
Miles, Lyn. 1994. Me Chantek: The Development of Self-Awareness in a Signing Orangutan. In *Self-Awareness in Monkeys and Apes: Developmental Perspectives*, ed. S. Parker, Sue, Robert Mitchell, and Maria Boccia. Cambridge: Cambridge University Press.
Mowat, Farley. 1987. *Woman in the Mists: The Story of Dian Fossey and the Mountain Gorillas of Africa.* New York: Warner Books.
Mündl, Kurt. 1988. *Rettet die Hoffnung.* Engl. transl. 1990. *On Life and Living: Konrad Lorenz in Conversation with Kurt Mündl.* New York: St. Martin's Press.
Newberry, Ruth, and Vittoria Sandilands. 2016. Pioneers of Applied Ethology. In *Animals and us: 50 Years and More of Applied Ethology*, ed. Jennifer Brown, Yolande Seddon, and Michael Appleby. Wageningen: Academic Publishers.
Nienaber, Georgianne. 2006. *Gorilla Dreams: The Legacy of Dian Fossey.* New York-Lincoln-Shanghay: iUniverse Inc.
Nussbaum, Martha, and Cass Sunstein. 2004. *Animal Rights: Current Debates and New Directions.* Oxford: Oxford University Press.
Peterson, Dale and Bekoff Marc. 2016. *The Jane Effect: Celebrating Jane Goodall.* San Antonio: Trinity University Press.
Patterson, Francin. 1978. The Gestures of a Gorilla: Language Acquisition in Another Pongid. *Brain and Language* 5 (1): 72–97.
Patterson, Francin. 1987. *Koko's Kitten.* New York: Scholastic.
Petrus, Klaus, and Markus Wild (eds.). 2013. *Animal Minds and Animal Ethics.* London: Transcritpt.
Plotnik, Joshua, Frans de Waal, and Diana Reiss. 2006. Self-Recognition in an Asian Elephant. *Proceedings of the National Academy of Sciences* 103 (45): 17053–17057.
Posada, Sandra, and Montserrat Colell. 2007. Another Gorilla (Gorilla Gorilla Gorilla) Recognizes Himself in a Mirror. *American Journal of Primatology* 69 (5): 576–583.
Premack, David. 1986. *Gavagai! Or the Future History of the Animal Language Controversy.* Cambridge: MIT Press.
Premack, David, and Guy Woodruff. 1978. Does the Chimpanzee Have a Theory of Mind? *Behavioral and Brain Sciences* 4: 515–526.
Prior, Helmut, Ariane Schwarz, and Onur Güntürkün. 2008. Mirror Induced Behavior in the Magpie (Pica Pica): Evidence of Self-Recognition. *PLoS Biology* 6 (8): e202.
Rajala, Abigail, Katharine Reininger, Kimberly Lancaster, and Luis Populin. 2010. Rhesus Monkeys (Macaca Mulatta) Do Recognize Themselves in the Mirror: Implications for the Evolution of Self-Recognition. *PLoS ONE* 5 (9): e12865.
Rachels, James. 1976. Do Animals Have a Right to Liberty? In *Animal Rights and Human Obligations*, ed. Regan Tom and Singer Peter. Englewood Cliffs: Prentice-Hall.
Rachels, James. 1990. *The Moral Implications of Darwinism.* New York: Oxford University Press.

Regan, Tom, and Peter Singer. 1976. *Animal Rights and Human Obligations*. Englewood Cliffs, NJ: Prentice Hall.

Reiss, Diana, and Lori Marino. 2001. Mirror Self-Recognition in the Bottlenose Dolphin: A Case of Cognitive Convergence. *Proceedings of the National Academy of Sciences* 98 (10): 5937–5942.

Rollin, Bernard. 1981. *Animal Rights & Human Morality*. Buffalo, NY: Prometheus Books.

Rollin, Bernard. 1995. *Farm Animal Welfare: School, Bioethical, and Research Issues*. Iowa city: Iowa State Press.

Rollin, Bernard. 2006. *Science and Ethics*. Cambridge: Cambridge University Press.

Russel, William, and Rex Burch. 1959. *The PrinciplesofHumane Experimental Technique*. London: Methuen.

Savage-Rumbaugh, Sue, and Roger Lewin. 1994. *Kanzi: The Ape at the Brink of the Human Mind*. New York: Wiley.

Singer, Peter. 1975. *Animal Liberation*. New York: New York Review/Random House.

Singer, Peter. 1981. *The Expanding Circle: Ethics and Sociobiology*. Oxford: Oxford University Press.

Singer, Peter. 1998. *Ethics into Action*. Lanham: Rowman and Littlefield.

Singer, Peter. 2002. *One World: The Ethics of Globalisation*. New Haven: Yale University Press.

Singer, Peter. 2003. *Pushing Time Away: My Grandfather and the Tragedy of Jewish Vienna*. New York: Ecco Press.

Singer, Peter. 2009. *The Life You Can Save: Acting Now to End World Poverty*. New York: Random House.

Singer, Peter. 2015. *The Most Good You Can Do*. London: Yale University Press.

Singer, Peter. 2016. *Ethics in the Real World*. Princeton: Princeton University Press.

Singer, Peter. 2020. *Why Vegan? Eating Ethically*. New York: Liveright.

Sober, Elliot. 2000. *Philosophy of Biology* (2 ed.). Boulder: Westview Press.

van de Weerd, Heleen and Sandi-lands VCictoria. 2008. Bringing the Issue of Animal Welfare to the Public: A Biography of Ruth Harrison (1920–2000). *Applied Animal Behaviour Science* 113 (4): 404–410.

Wallach, Arian, Chelsea Batavia, Marc Bekoff, et al. 2020. Recognizing Animal Personhood in Compassionate Conservation. *Conservation Biology* 34 (5): 1097–1106.

Würbel, Hanno. 2009. Ethology applied to animal ethics. *Applied Animal Behaviour Science* 118 (3): 118–127.

Yerkes, Robert. 1916. *The Mental Life of Monkeys and Apes*. Cambridge: H. Holt.

Yerkes, Robert. 1943. *Chimpanzees*. New Haven: Yale University Press.

Chapter 8
A Re-evaluation of Animal Interests Starting from a Critique of Maslow's Pyramid

Roberto Marchesini

Abstract In this essay, I wish to demonstrate that, in order to understand the interest of the subject as an intentional entity, it is not enough to refer to sentience, to welfare parameters or to Maslow's Pyramid. Similarly, it is not enough to refer to compassion or sympathy, interpreted in the etymological sense of "being in the same dispositional feeling". In order to preserve the interests of non-human animals as subjects, it is indispensable to: (1) accept their existentiality and avoid the mechanism of heritage; (2) carry out an empathic approach, that is, the ability to reproduce a condition or an inclination different from our own, which requires a suspension of anthropomorphism. It is possible to know the intentional dimension of non-human animals only if we apply Darwin's criterion of adaptative resemblance and distinction. But in order to do that, it is necessary to strengthen both the scientific knowledge and the ethical reflection, because neither of the two approaches is able to make the interest of animal subjectivity emerge.

8.1 Premise

Being an animal means having two inherent qualities: (i) feeling, that is, the ability to live events in an emotional way, somatizing them, i.e. transforming them into a state of the body and advocating them for oneself; and (ii) desiring, that is, being projected towards the world, through interest and curiosity, i.e. the desire to act-operate on the world in a proactive way, through intrinsic motivations. The sentient and desiring condition of the animal is at the centre of the inalienable character of its subjectivity for which events involve it in the first person. Emotions and motivations endow the animal-subject with "interests that concern it", which is not the case with the innumerable mass of objects, even if they are performative (e.g., a machine).

Owning a "for oneself" means assuming a condition of self-ownership, and this is a predicate that we find immediately when we interact with an animal: in fact, an animal is never passive, but rather surrenders or gives itself. The "for-oneself" establishes, therefore, some inherent characters that beg to be fulfilled, for which the subject appears to us as a patient that is exigent with respect to what happens or can happen to it. On the other hand, this very core of inertia makes us understand the

R. Marchesini and M. Celentano, *Critical Ethology and Post-Anthropocentric Ethics*, Numanities - Arts and Humanities in Progress 16,
https://doi.org/10.1007/978-3-030-74203-4_8

subjective expression (that is, the condition of partiality and infidelity, of adherence to a here-and-now), which we normally translate with phrases such as "it wants/doesn't want to" or "it's in a good/bad mood"—states that clearly do not apply to the universe of things or machines. This core of subjectivity is therefore expressed through the dispositions of feeling and desiring, that is to say of intrinsic pre-reflexive entities that we call emotions and motivations.

Hence the need to consider this sentient and desiring nucleus as the engine of the reactive-proactive action of the animal, the flywheel of its "inevitable protagonism" on the stage of existence. For this reason, the emotions and motivations that bring out animal expression must also be considered the "motives of subjectivity". Owning an internal here-and-now, however counterintuitive it may seem, means not being completely at the mercy of the external here-and-now. So far, it's all quite clear: the same event can be experienced in totally different ways, but above all it may be more than what the animal is experiencing in the here-and-now: in other words, an animal can be above its own momentary condition. This means that an animal is never repetitive, but singular in its expressions, so that every piece of knowledge always gives rise to a unique and unrepeatable behaviour. Both innate and learned endowments therefore are tools, and not automatisms, that animal subjectivity uses to reach its projections of feeling and desire. So, subjectivity also manifests itself through the ownership of its endowments.

An animal does not wait for a stimulus or an event to ignite its movement, because the peripatetic character is its inner quality: it moves, it is not moved, and it is in motion even when it is stationary. The motivational character of animal subjectivity constantly leads it to go into the world to carry out activities that are part of its nature. Each animal brings itself into the world through coordinates of actions, i.e. verbal propositions—such as chasing, collecting, owning, etc.—which represent acts: the act of chasing, collecting, owning. In other words, the interaction that the subject has with the world is never generic, but always takes place in the name of precise acts that are part of the animal's species-specific characteristics. The acts foreseen by the identity of a species represent a sort of set menu, which contemplates the type of "expressive consumption" that that individual wants to achieve.

Motivations are internal entities, what is outside is only the target towards which the motivation projects the individual and the space-time where the action is carried out. Placed in a meadow, a child will pick daisies while a cat will chase butterflies: the practicability of a context is assigned first of all by the motivational prevalences in place in a given species. This allows us to understand that experience is something very different from the passive exposure to stimuli: it is rather a way to explore the objects present in the external context—obviously those that can be perceived—to achieve some very precise results. The motivations are like "copulae"—conjugative elements—that lead us to particular targets through very specific acts: chasing, in fact, is very different from collecting, and not only from a postural or kinetic point of view, but also as a disposition towards the world.

The desiring condition—as projection towards something and proaction in an activity, not yet formalized in a real desire—is therefore to be considered as the first *movens* of animality, a need to act that is independent of the specific objective and the result of its action, which are simply contingent: a cat does not want the string, but wants to project itself towards the string's movement and to act by chasing and grasping it. This is an expressive need that arises from a character of exuberance that does not allow the animal to remain imprisoned in the here-and-now, but always makes it tend towards a desirable something, engaging in some activity so as to bring out its intrinsic industriousness. Not understanding this, thinking that the only needs of an animal are those related to comfort or welfare, means locking up the soul in a golden prison.

8.2 The Desire of Self-Expression

The expressive appetence that arises from the motivational condition is undoubtedly a driving force directed towards the world, a languor of active life that seeks consumption through: (i) orientation, (ii) expression, (iii) conjugation. It is obvious that bringing oneself into the world implies the formation of objectives as a consequence. But the cat's string, as an objective, is the consequence of the cat's desire to chase, not the cause of that desire: we can therefore speak of "expressive exuberance", not of lack. It is important to understand this. It is not through comfort or replacement (I will be the one who cares about you and gives you what you want) that the languor can be extinguished, because it is not a "lack of something" but an "excess of expressive resources" that require to be spent. For this reason, we can say that the objective emerges, rather than being given objectively, whenever a motivational appetence is brought into the world and searches the context in search of an opportunity that can make that expression possible. Being desiring also makes the subject susceptible to the appeal exercised by the entities that make up the external context, provided that they have some consistency with the motivation aroused. Ultimately, the objective is nothing more than the outcome of the encounter between an attractive entity and a present entity that provides an expressive opportunity.

Proactivity indicates the tendency of an animal to actively interact with the outside world, taken by a state of uneasiness, restlessness, projection, as if it had to bring out an internal energy exuberance. While in reactivity the subject is invested with a stimulating element to which it responds, in the proactive case we see a real search process, like a languor that leads it to seek specific targets so as to express certain behaviours. However one may define them in exemplifying or metaphorical terms, motivations are not to be considered as the lack of something—as if the subject acted because it lacked something—but as expressive desires that overflow into the individual and set them in motion. This can be clearly seen in play, an activity governed by motivational appetence, where it is not the lack of something that counts, but the possibility of expressing one's proactive coordinates. It is therefore necessary to speak of expressive needs, which arise from innate predispositions to perform certain actions under the pressure of precise endocrine and neuromodulation structures.

In order to understand the phasic and homeodynamic system of the motivational state, i.e. its being correlated, i.e. sensitive to different influences, but at the same time autonomous with respect to physiological needs on the one hand and representational needs on the other, let's try to understand the links that bind it to these two factors. With respect to the relationship between motivation and need, we can note two aspects: (1) the link of remote causation, because motivation is a dispositional character that has been affirmed in phylogeny within the population of a species, since it is able to give fitness to its carriers, in the sense of favouring the fulfilment of a particular physiological need; (2) dependence on internal factors of somatopsychic influence, since a given motivation is physiologically related to a given need, so that any dystonia of the need produces variations in physiological gradients that can make the subject more susceptible to motivational calls. Therefore, there is no doubt that although it has its own phasic cyclicity, held within the relationship between appetence and satisfaction, in any case the physiological need retains its elicitative influence.

As for the relationship between motivation and "desire in the strict sense", that is to say, desire accompanied in a modal sense by the specifications of what-where-how-when, whatever the level of intentionality evolved in representational terms, we can say that the definition of a target, of a context, of an operative module, conjugated in planned actions and in development times, can also be a factor of elicitation, acting in the opposite sense to the canon described so far. In other words, it is true that the propensity to collect can stimulate these activities in a spontaneous way, bringing out a post-experiential modal based on the potential action offered by the context. However, in the same way, a person who has developed the passion of mushroom picking can be induced to start an activity of searching-harvesting mushrooms only because it is part of a context that presumably offers the same potential actions. In this case, it is a thought that leads the subject to activate their motivational dispositions in activities related to previous experiences.

Ultimately we can say that despite being different—(i) need, understood as a physiological necessity, (ii) desire, understood as a projection towards something very precise or towards an equally well-defined activity, (iii) motivation, understood as a desiring projection, definable in an act but not specified in the contents—these three factors of the individual's "operational opening" on the world are undeniably connected. However, if it is true that need, motivation and desire—this is the most correct sequence to define these factors—are undeniably connected and exert reciprocal influences on each other, it is equally true that they respond to mechanisms of homeostasis that cannot be superimposed. This is easy to see in the relationship between food needs and the motivations related to them in the "motivational system of foraging", the one, to be clear, that has established itself within a species because it has helped find food resources. In the predatory expression of cats the relative autonomy between motivation and physiological need is very evident: a cat continues to chase, and to look for opportunities to exercise this projection-proaction, even after having satisfied its food needs.

We can say, in the final analysis, that the two homeostatic mechanisms of hunger and the related motivational expression (chasing), although connected, respond each to their own or partially independent homeostatic systems, so for a cat it is not enough to eat to turn off the desire to chase. Most probably, if the cat did not have food at its disposal and had to find it using the activities for which it is motivated, well, certainly the need for food and the need for expression would find simultaneous satisfaction. On the other hand, even if that were the case, it is not certain that there would be no exorbitant motivational resources left to spend on playing a game or simply on a pastime. This is a situation that can easily be seen in wild animals. Of course, if the need is extinguished when the body has re-established certain physiological gradients, as far as the motivational aspect is concerned, the appetence decays thanks to the fact that the subject has put in place those particular behaviours up to reaching a state of psychological satiety called satisfaction. We must therefore admit that there are such things as expressive needs.

The same can be said of the relationship between motivation and desire. If it is true that a desire that is well represented on its target contents recalls some motivations, however, even in this case, we can notice that it is not the "desired", i.e. the desired object or result, that turns off the motivational projection, but the activity that allows one to reach it. This is the reason why, if the target is too easily reached or if the expressive appetence exceeds the action carried out, the object is not able to turn off the appetence and bring the individual into a condition of satisfaction. If too easily caught, a mouse does not extinguish the cat's predatory desire, so it seems to us that the kitten indulges in the art of sadism when, on the contrary, it does nothing but try to extinguish its expressive desire. In other words, it is not objects that give motivational satiety, but the actions we perform to reach them. The desiring condition can, therefore, be defined as a proactive projection regulated not by the result of the action, but by the effort put into the action itself. Often, indeed, we notice that easy gratifications, or those not achieved through action, stimulate the longing languor instead of soothing it, in a vicious mechanism that makes the subject more and more dependent on repeated doses of gratification.

This leads to some important considerations about the expressive needs that dwell in the human soul and that are often bypassed and neglected, because of the mistaken idea that reaching a goal can extinguish our desiring restlessness and that comfort is the best we can expect from life. A comfortable bed, an abundance of food, the feeling of security, the ability to protect ourselves from the elements, in other words all the factors of welfare are certainly important, but they cannot compensate for the frustration of an inactive life. We appreciate them, that is, if we have earned them with the sweat of an active life. Just as rest has true value and refreshes us after we have grown tired, but otherwise becomes boredom, in the same way a world full of objects within reach that denies us the projection, the languor, the effort of reaching soon becomes a golden prison. Motivations, therefore, are undoubtedly affected by, but also prescind from, both the fulfilment of needs and the achievement of results, determining expressive appetences that can only be understood by admitting the expressive needs of specific species.

Each species is therefore equipped with a well-circumscribed menu of these basic verbs that I have called motivations, which support the different propositional expressions that an animal belonging to a species can produce through multiple compositions and modal declinations. The motivational menu, together with other parameters, gives ethological recognisability to a species, making it identifiable not only from a morphological point of view, but also in terms of this basic proactive tension. On the other hand, being the bearer of a certain menu of proactive acts addressed to the outside world, not only offers a further element of ethographic recognisability, but also establishes the projective and proactive appetences that the subject manifests, defining a system that bases its homeostasis on activity, that is, on being able to give voice to those particular expressions of operativeness on the context. This aspect is generally ignored because of the mistaken view that the desiring condition is not a projection but a lack, so it is believed that the target, in itself, is able to turn off the desire and provide satisfaction.

Instead, I would like to underline the difference between the concept of projection and that of lack. Projection, in fact, indicates the subject's tension towards something, a tension that implies an action on the context and a proactive exuberance to support the effort put into the action. On the contrary, lack indicates an element of which the system is deficient and the addition of which would bring the system into a condition of balance. In a condition of absence I have to add something to turn off the languor, while in exuberance it is necessary to open up the system to allow for expression to calm it. If, in fact, we consider the desiring condition only in terms of lack, we miss the importance of this tension, the soothing meaning of action as such, the gratification that comes from having discharged proactive exuberance. In other words, the target or result is undoubtedly gratifying, but it is not at all sufficient to calm the desiring languor; on the contrary, gratification (the achievement of the target), if lacking in satisfaction (expressive satiety), does nothing but increase the desiring condition with all that follows. It is a gross error to believe that our projections towards the world are governed by a principle of absorption, a mistake fraught with consequences when we look at the state of pleasure that an individual derives from their life.

Let's consider this aspect: the more I gratify myself through cheap targets-results, which do not require effort along the motivational coordinates, the more the desiring languor will increase instead of fading out. In other words, a sort of vicious circle is created that breaks the homeostatic mechanism, so that the more the subject receives, the more they feel the languor. If pleasure is limited to gratification without ever achieving satisfaction, which vice versa depends on action, the individual will soon suffer from a sort of addiction to the target, that is to say, it will increasingly depend on receiving gratifications in this sense and will increasingly need larger doses of that particular rewarding entity, be it an object or a result. The target, in fact, will have acted on the subject as a gratifying and implementing factor, without ever satiating its projective condition, i.e. its link with action and only secondarily with the implied goal. This is what we are unfortunately witnessing in our society, where people are increasingly psychologically bulimic in relation to resources to consume, even if they seem to be disinterested because they are completely dependent on gratifying sources and totally out of touch with any possibility of true satisfaction.

8.3 Reviewing the Concept of the Hierarchy of Needs

In 1954, American psychologist Abraham Maslow conceived his principle of "hierarchy of needs" based on the fact that there is a sort of progression in the satisfaction of individual needs. In this hierarchy of needs, each level of the lower order requires to be satisfied so as to access the higher level: first comes the need for food and only after is there the need for security. In this view it is therefore necessary to speak of a sort of path from the bottom to the top, where the first level is reserved for basic physiological needs, while the peak is represented by self-realization, passing through intermediate stages such as security, belonging, or the esteem profused by one's social microenvironment. The image that Maslow offers us is thus a pyramid, where the different stages must find progressive response. And at the bottom of the pyramid there are the physical needs that we could define as essential, such as eating, drinking, resting, reaching physiological homeostasis and correct homeothermia. In other words, according to this view, if the body is compromised in its general functions and therefore does not find satisfaction in its needs, all other needs lose importance.

The second step identified by Maslow is the sense of security, that is, a predominantly emotional condition that involves feeling safe in the usual context and therefore free from fear, conflict, precariousness, anxiety, anguish, anger, or panic. We can see the principle of security as freedom from fear caused by entities that put our lives at risk, from anger that forces us to fight against the entities that disturb us, from sadness so as not to feel the burden of abandonment that also in this case creates a sense of precariousness and insecurity. Now, there is no doubt that the need for safety is important, but are we really sure that it only comes after the physiological needs have been met? An animal that is afraid usually stops feeding and when a negative emotional condition bothers it, it can even let itself die. In human beings, too, we see similar situations, especially in cases of depression, grief, panic attacks and borderline cases. In my opinion, rather than a hierarchy between the needs of first and second order, there is a correlation, which causes the lack of one to affect the other, creating varying priorities depending on the actual lack in place at any given moment.

The third step consists of what Maslow indicates as belonging, that is to say, being inserted into a microcosm of social acceptance, that is, of bonds of friendship, family affection, sexual intimacy. This step could be interpreted, in a more all-inclusive sense, as feeling part of a group and not abandoned to oneself or in any case not marginalized. The last two steps—that is to say: the fourth relative to esteem and the fifth to self-realization—are easily attributable to the social relationship extended beyond the affective and affiliative microcosm and, above all, to the last step, i.e. the idea-project that the subject builds of themselves, as well as affirmative feedback from others. When people speak of realization, I immediately think of the desiring condition, which Maslow puts at the apex of his pyramid, as a less urgent need that only emerges if all the others have been satisfied. There are some aspects whose validity undoubtedly goes beyond the human being and that can be applied in the investigation of basic needs related to animal welfare. It is no coincidence that when we talk about animal welfare we refer to the five freedoms listed in the Brambell Report (Brambell 1965), which basically refers to the first two steps of Maslow's pyramid. But also in this case, the observations we can make deny the idea of a hierarchy and a linear progression of needs.

There is, in fact, much that does not add up in Maslow's vision. Indeed, in nature there is no hierarchical staticity of needs, but there are different priorities depending on the specific situation in which the animal is found. For example, in the mating season, social affirmation, which Maslow places at the top of the pyramid, comes well before safety and often also before the fulfilment of basic physiological needs. The individual is not a static entity that is always in the same condition, but a fluctuating entity that, case by case, dictates its changing priorities. What Maslow places at the top and which, with a good approximation, represents the free expression of individual characteristics, has a strong influence on the animal's physiological characteristics. For example, a dog that cannot express its motivations and therefore fulfil its natural talents enters a state of restlessness that undermines its sense of security and alters its mechanisms of physiological homeostasis. Maslow's pyramid is incorrect, in that it considers each level as unidirectional, as if every step simply rested on the previous one, whereas, on the contrary, the different needs of an animal have bidirectional or, better, multidirectional relationships between them. In other words, the different needs seek a reciprocal equilibrium, so that, just to give an example, the relationship between physiological needs and self-expression must find a specific point of balance. For an individual to experience wellbeing, very often they should lower their level of comfort and increase their level of expression: one can never improve one's condition by increasing comfort in spite of expression.

By criticising the hierarchy of needs and Maslow's pyramid, I wanted to underline how incorrect it is to think that needs proceed in one direction only. I also wished to point out that in reality they interact with each other in a multidirectional flow, so that the lack of satisfaction of one area affects all the others. Speaking of motivations, this aspect becomes particularly evident. An individual who is kept in a condition of non-exhaustion of their motivational tendencies will inevitably manifest compensatory behaviours in this field. However, their discomfort will also spread on their emotional aspects, on their sense of security, on their ability to adapt to the environment, on their socio-relational propensities, sometimes leading to alterations of a physiological order. The expressive need and the consequent discomfort deriving from a lack of expression have become even more striking today, in an age that tends to increase the fruition power of the subject through the practice of consumerism and comfort, while eliminating the expressive possibilities of the subject, who as a result is in a constant state of restlessness.

The expressive need can be framed as the need to bring out or give fulfilment to an interaction with the external world that comes from the inside and where the external reality offers only the opportunity or the pretext for implementation. It is therefore an expressive need that, if not fulfilled, places the subject in a state of restlessness. This is an exuberance that, if not satisfied, puts the subject in a condition of non-equilibrium due to inactivity, as the individual is always oriented and projected towards the outside. The motivations therefore indicate this "movement towards the world", which we could represent as a verbal structure that puts the Self in contact with everything that surrounds it. We should not, however, think of the expressive need as the need to perform an action that is already predetermined in all its components, because a motivation always produces an open behaviour which is structured in the expression itself. The motivational orientation needs to be specified, that is, it is not a closed-off action: on the contrary, it introduces a creative event, that is it puts the subject in a condition of experiential experimentation.

So, back to our exemplary cat. Being endowed with a predatory motivation, i.e. the verb *to chase*, means: (1) being conjugated to a large number of targets and therefore experiencing different entities in terms of the characteristics they present as opportunities and challenges; (2) trying out a large number of motor choreographies and kinaesthetic expressions, which allow the subject to experience not only the bottom-substrate but also their own body; (3) knowing different environments, knowing how to distinguish the different areas of agency and the constraints that they present, what they are like and what can be found in them; (4) having an idea of the durations, synchronizing in actions, assuming reaction times, openings and closures. The expressive need therefore affects all the others, in a process of progression that is far from unidirectional.

Therefore, each need is interrelated to the others and there is no hierarchy of needs because the most important need, at a certain moment, is the one that finds the least possibility of being satisfied and becomes the most urgent according to the particular physiological and psychological condition of the subject. This means that it is not possible to understand the interests of a species by pretending to draw up a table of "priority and propaedeutic" needs through a universal canon, because it is a question of recognizing different interests among the various species, which vary between individuals and according to the particular moment. The expression of one's subjectivity is certainly the pivotal point of any discussion of animal ethics, which can only be done by abandoning anthropocentric universalism. The latter, even when it seeks to include other species in the list of moral patients, always risks extending rights to the different, so to speak, while failing to recognize different rights.

References

Brambell, Roger F.W. 1965. *Report of the Technical Committee to Enquire into the Welfare of Animals Kept Under Intensive Livestock Husbandry System.*

Broom, Donald M., and Ken G. Johnson. 1993. *Stress and Animal Welfare.* Netherlands: Springer.

Keverne, Barry, Felicia A. Huppert, and Nick Baylis. 2005. *The Science of Well-Being.* Oxford: Oxford University Press.

Marchesini, Roberto. 2021. *Le radici del desiderio. Alla ricerca delle motivazioni umane.* Bologna: Apeiron.

Maslow, Abraham H. 1954. *Motivation and Personality.* New York: Harper & Row.

Weaver, Jeremiah (ed.). 2016. *Animal Welfare.* New York: Nova Science Publisher.

Chapter 9
Behavioural and Cultural Epigenetics: Social Biologisms Refuted by Developments in Biology

Marco Celentano

Abstract Developments in evolutionary studies, focused on the discovery of epigenetic inheritance and the social-environmental factors influencing it, have led to epochal turning points in the last three decades. Faced with these important changes, we can now consider the genocentric models that guided evolutionary studies from the last decades of the nineteenth century to the early '70s of the twentieth as obsolete. In fact, today we can demonstrate that, in the course of phylogenesis, at least three other kinds of selection, heredity and variation, respectively *epigenetic, behavioural and cultural*, operated alongside the slow processes of genetic variation, producing adaptive or maladaptive effects in a much more rapid way than that implicated by genetic mutations. Thanks to very young research areas such as behavioural and cultural epigenetics, we can now *refute old and new social biologisms*, ancient, ever re-emerging forms of biologistic determinism, *with the tools of biology itself*. We can indeed show that, through the *methylation and demethylation processes*, a lot of socio-environmental factors contribute to inhibit or re-activate gene expression, influencing fundamental aspects of the individual's health, emotional, social and cognitive development with effects that are often transmitted to descendants for several generations. This essay discusses some of the most recent and representative results of these emergent areas of the epigenetic research and eco-evo-devo studies. This analysis reaches the conclusion that it is now anachronistic to hypothesize "human nature" being rigidly codified at the genetic level and substantially unchangeable in its fundamental mental and behavioural propensities.

9.1 Introduction

Within the theoretical framework of the "synthetic theory" developed in the first half of the twentieth century (Huxley 1942), it was excluded that living conditions and social context could, over a few generations, cause hereditary modifications of the phenotype and of its physiological, psychic and behavioural propensities. In fact, the phenotypic conformation was considered as the result of a faithful execution (except for "random transcription errors") of a program entirely written in DNA and scarcely modifiable by the environment. Significant phenotypic changes were

R. Marchesini and M. Celentano, *Critical Ethology and Post-Anthropocentric Ethics*, Numanities - Arts and Humanities in Progress 16,
https://doi.org/10.1007/978-3-030-74203-4_9

considered, therefore, possible only in the long time required by significant genetic or even genomic mutations that were estimated as millions of years (Riedl 1980).

However, different kinds of phenomena verified by the biological and behavioural sciences remained difficult to explain within this theoretical framework: the ability of organisms to adapt to rapid environmental changes, the evident influence of environmental and social stimuli on cognitive and relational development of human beings and other animals, the problem, discussed since the days of Darwin, of explaining the surprising speed with which the evolutionary process took place after the appearance of the metazoans. According to the most accredited theories, in fact, the differentiation of organisms began very slowly but, as of the last phase of the pre-Cambrian, underwent a strong acceleration. In other words, as Steven Jay Gould wrote, for about 2.5 billion years, "the seaweed carpets remained carpets of algae" (Gould 1977: 114; 121), then, within about a billion or at least 600 million years, the evolutionary process led to the appearance of the incredible variety seen in past and present multi-cellular species. In 1867, F. Jenkin had already pointed out that the solution proposed by Darwin to explain this process, the slow accumulation of random favourable hereditary variations produced by natural selection, hardly justify the genesis of a so wide flowering of phyla, species and varieties and the appearance of complex organs such as the eye or the brain in such a short time (Jenkin 1867). All the subsequent reformulations of the Darwinian theory cantered on binomial mutations-selection, "chance and necessity", proposed from then on have been dragging behind this problem.

But as I have briefly reconstructed in the Sect. 2.8 of this volume, the birth of the evo-devo approach and the turning towards an "extended synthesis" of the Darwinism have led, in recent decades, to a substantial overcoming of the theoretical framework of modern synthesis and to a solution of this problem. Developments in these research fields have in fact shown that three other kinds of hereditary transmission have cooperated with the genetic one in determining the effects and times of phylogenesis: epigenetic, behavioural and (at least among "higher animals") cultural inheritance. Three forms of inheritance the effects of which manifest themselves much more rapidly than those implied by genetic mutations.

These changes of perspective have considerable repercussions on at least three issues that have crossed the entire history of post-Darwinian biology:

- the aforementioned debate on the "dilemma" of the speed of evolution;
- the problem of explaining the ability to quickly adapt to environmental changes that all organisms capable of associative learning manifest;
- the *vexata questio* on the degree of influence that, respectively, genes and socio-environmental stimuli have on human experiences, activities and propensities.

As I recalled in the second chapter, the new version of Darwinism emerging from the extended synthesis implies, in fact, that any drastic change in living conditions can trigger "large amounts of [...] hereditary variations" (Jablonka and Lamb 2005: 443), that "some hereditary variations are non-random in origin" and that some acquired information can be inherited. In other words, it implies that evolutionary change can result "from instruction as well as from selection" and that genetic changes can often come as "followers" of evolutionary divergences started at the epigenetic and ethological level (Jablonka 2006; Callebaut et al. 2007; Pigliucci and Müller 2010) and, therefore, that in various cases the "evolution can be very rapid" (Jablonka and Lamb 2005: 442; 443).

In this chapter, I will discuss the results of numerous studies which demonstrate that ontogenetic development of mind and behaviour, in humans and other social animals capable of learning, is decisively influenced, through epigenetic inheritance and cultural tradition, by early and subsequent experiences, social context and living conditions. Mental and behavioural propensities and preferences can therefore develop significant, useful or harmful changes in the rapid times in which social and cultural selection operate. Therefore, to "explain the socio-behavioral status quo in terms of genes" appears, within this new theoretical framework, incorrect and anachronistic (ibid.: 473). In short, according to the emerging perspectives, environmental stimuli, social constraints and cultural heritage channel and largely constrain the gene expression: the exact opposite of what the genetic determinists have always supposed and supported.

9.2 From the "Selfish Gene" to the "Altruistic" One: Moral Innatisms and Evolutionary Psychology

In the fourth chapter of *The Descent of Man* (Darwin 1871), like the third dedicated to a comparison between the human mental faculties and those of other animals, Darwin exposes his conception according to which human morality sinks its remote roots in the social instincts of the higher animals, especially primates. The great naturalist introduces here an approach that could be defined proto-ethological, then developed in the work *The expression of emotions in animals and humans* (Darwin 1872), starting to locate a series of emotional, cognitive, behavioural and expressive patterns that our species shares with others.

In the final pages of the chapter, there is a passage quoted from the letter that Hebert Spencer had written to John Stuart Mill, and that Alexander Bain had published in his *Mental and Moral Science* (Bain 1868): "Our great philosopher, Herbert Spencer, has recently explained his views on the moral sense. He says: «I believe that the experiences of utility organised and consolidated through all past generations of the human race, have been producing corresponding modifications, which, by continued transmission and accumulation, have become in us certain faculties of moral intuition»" (Darwin 1871: 102). Darwin initially seems to align himself with Spencer's moral innatism and intuitionism: "There is not the least inherent improbability, as it seems to me, in virtuous tendencies being more or less strongly inherited" (*Ibidem*). In the following passages, however, he raises perplexities, essentially dictated by the fact that the moral traditions have often preserved and handed down norms which, in his eyes, appear neither useful nor adaptive: "My chief source of doubt with respect to any such inheritance, is that senseless customs, superstitions, and tastes, such as the horror of a Hindoo for unclean food, ought on the same principle to be transmitted" (*Ibidem*: 103).

In other words, it would be necessary to admit, *as Nietzsche was later to do*, that, for millennia, social and cultural selection has in many cases rewarded, preserved and venerated even the "un-adapted", here understood as a melting pot of beliefs and behaviours that damage those who enact them, reducing their vital chances.

The theoretical oscillations between concessions to the moral innatism and critical reservations against it, emerging in this Darwinian text, will cyclically recur in the subsequent biological and philosophical debate, up to our days.

In the last twenty years, in particular, two different models of "moral Darwinism" (Pennock 1995; Wiker 2002; Attanasio 2010; Hodgson 2013; Ruse 2017) have emerged, in which this polarity returns to show itself.

The first model, which I would call "strong", postulates the existence of a series of "innate" instincts, principles and moral judgments, dependent, more or less directly, on our genetic code. This model found some of its best-known supporters in evolutionary psychologists[1] as Jonathan Haidt, Steven Pinker and Marc Hauser, and in the biologist and science popularizer Richard Dawkins.

The second model, which I will define "weak", is instead centered on the hypothesis that "we are not born with any specific moral norms in mind, but with a learning

[1]Evolutionary psychology was born in the early nineties of the twentieth century. Its approach was outlined in the collective volume *The Adapted Mind* (1992), edited by Jerome Barkow, Leda Cosmides and John Tooby. As the title of the work suggests, the first assumption of this branch of psychology is that the human mind and brain are, in their basic architecture and mechanisms, products of the adaptation of the human species to the environment in which it has evolved. The second basic assumption of evolutionary psychology is that the human mind and brain function, according to the model suggested by Fodor, in a modular way (Fodor 1983), that is: through cognitive and physiological circuits which are *mutually independent*. Evolutionary psychology is presented by its supporters as an area of research aimed at finding a mediation between the reductionist and deterministic positions of the genocentric sociobiology and the culturalist approach of the traditional sociology (Adenzato and Meini 2006: XIII). Indeed, it advocates, at least in theory, the overcoming of every innate/learned dichotomy (Cosmides and Toby 2006). However, the positions of its best-known exponents often appear more as a diplomatic and rhetorically "soft" repurposing of sociobiological genocentrism than as a theoretical or experimental overcoming of it. Ultimately, according to their approach, it is possible to explain the origin and motivation of every human psychological attitude, including moral, religious and political ones by discovering the genes they depend on.

These psychologists like to call themselves "Darwinians", but their beliefs diverge profoundly from the Darwin's one. In the concluding chapter of *The Descent of Man*, Darwin stated that "the moral qualities are advanced, either directly or indirectly, much more through the effects of habit, the reasoning powers, instruction, religion, &c.", which means through a social and cultural selection, rather "than through natural selection" (Darwin 1871, II: 404). Therefore, the moral innatism promoted by the main exponents of contemporary evolutionary psychology supports a thesis opposite to that of Darwin. It is no coincidence that it refers to two theoretical models that are *extraneous to Darwinism*: Chomskyan linguistic innatism, which still lacks any adequate genetic and/or neurophysiological evidence from the 1950s to the present day, and the modularistic approach launched by J. Fodor in the 1980s. Moreover, in the new millennium the latter approach underwent extensive criticism and revision by its own author, who has argued that the thesis according to which the human cognition is mostly or totally modular is "devoid of empirical plausibility" and borders on "incoherence" (Fodor 2000).

agenda that tells us which information to imbibe" (de Waal 2006: 166). This leads us to "internalize the moral fabric of our native society" (de Waal 2006: 166; Simon 1990) and rework it according to our experiences. From the 1990s onwards, ethologist and primatologist F. de Waal has offered the most consolidated and documented exposition of it.

We will start our analysis by examining the positions of some supporters of the "strong" model, according to whom moral principles oriented to favour the conservation of "altruistic" behaviour are innate in our species or even fixed on a genetic level. Among these, I will first mention an apparently "unsuspicious" scholar to whom, a few decades ago, no one would have attributed the defence of this model: Richard Dawkins. When he published his bestseller *The Selfish Gene* in 1976, Dawkins seemed to have questionable but *very clear* ideas about our hereditary inclinations: "*If you wish*, as *I do*, to *build* a *society* in which *individuals cooperate generously* and unselfishly towards a *common good*, *you* can *expect little help* from *biological nature*. Let *us try* to *teach generosity* and *altruism*, *because we* are born *selfish*" (Dawkins 1976: 3). These statements did not leave much room for doubt regarding their interpretation: they asserted that our innate tendencies lead us to be *selfish and not generous*. They echoed, renewing its form but not its substance, the position assumed in the previous century by Thomas Huxley in which "altruistic" choices contrast with a natural tendency to protect one's own survival and an unyielding part of our ereditary kit. However, about thirty years later, in the volume *The God Delusion* (Dawkins 2006), he went to reverse this position, though without explicitly admitting it. Indeed, in the sixth chapter of the book, entitled *The Roots of Morality: Why Are We Good?* he wrote: "Where does the Good Samaritan in us come from? Isn't goodness incompatible with the theory of the 'selfish gene'? No. This is a common misunderstanding of the theory - a distressing (and, with hindsight, foreseeable) misunderstanding" (Dawkins 2006: 215). This very self-denial, which Dawkins minimized by presenting it as a simple misunderstanding, had precise historical and cultural causes. In those years, in fact, the hypothesis that some altruistic behaviour has a very strong innate basis in our species had found wide diffusion in a new field of research born in the nineties: evolutionary psychology.

One of the best-known popularizers of this approach had been the cognitive psychologist Steven Pinker. Based on the model of Chomskyan linguistic innatism, and on the neo-intuitionist approach proposed by psychologist Jonathan Haidt, Pinker had launched, in the volume *The Blank Slate: The Modern Denial of Human Nature* (Pinker 2002), the idea of a "Darwinian moral innatism", then re-proposed in the paper *The Moral Instinct* (Pinker 2008).[2] This approach had been then reworked and developed by Marc Hauser, also an evolutionary psychologist, who in the essay *Moral minds. How nature designed our universal sense of right and wrong* (Hauser 2006) had proposed the hypothesis that our "moral sense" is inscribed in the human brain, based on a "universal moral grammar" and determined to a large extent by genetic factors. According to Hauser, it is therefore correct to talk about a very

---

[2]As he himself affirms, Pinker adopts a "weakened", or semi-strong, version of the moral innatism, believing that in our brain there is likely no list of "you must", but only some "if-then" rules.

"moral instinct" which is innate and "immune" to the conditioning of social and cultural factors as authorities and religions (Hauser 2006: xviii).

Hauser's approach, in turn, referred to the Haidt's one which in the article *The emotional dog and its rational tail: a social intuitionist approach to moral judgment* (Haidt 2001) had hypothesized the existence of two separate systems of evaluation that guide the human mind: "moral intuition", phylogenetically older, and the younger "moral judgment", which appeared only after verbal language was affirmed. Indeed, his "intuitionist social model" postulates that it is the phylogenetically older brain functions that are at the basis of the intuitive, quick, automatic and emotional responses we give to questions of a moral nature. According to this approach, therefore, reasoned ethical judgments play, in the mental processing of our ethical choices, only a secondary role, used more to construct a posteriori rational justifications for our evaluations than to identify the conclusions they reach. This way, while not being able to deny that moral judgments are individual re-elaborations of values transmitted by a specific cultural tradition, Haidt tends to minimize the effects of our cultural formation on our emotions, and those of reasoning on the moral judgments, substantially equating both to a posteriori justifications of innate behavioural patterns.[3]

For its part, Hauser, referring to the Chomskyan innatist model, Haidt's intuitionism and John Rawls's neo-transcendentalism, had proposed the hypothesis that there is a "universal moral grammar", based on a "moral instinct", which matures naturally in the brain of each child (Hauser 2006: xvii).

While not going as far as denying the undeniable, that is the influence of cultural differences on the codes of conduct and values internalized by individuals, he postulates the existence of universal "innate principles" of moral judgment only our species is endowed with.

In short, Hauser's theses outline, albeit with some caution, a very strong model of moral innatism, stating that we possess an "innate moral faculty", characteristic and exclusive to our species, that there is a "universal moral grammar", analogous to the universal grammar that Noam Chomsky postulated to explain the learning of verbal language, and finally that these innate devices are impervious to conditioning from social, political and religious authorities.

What kinds of data supports these theses?

Hauser examines the contributions of various disciplinary fields, but the argument he considers decisive does not derive from genetic, neurophysiological, ethological, or historical evidence, but from a study which consisted in administering a "Moral Sense Test" via Internet (see: http://www.moralsensetest.com) to more than 60,000 volunteers from 120 different nations. The test related on some "moral dilemmas" and mental experiments that Hauser himself, along with two of his collaborators, had instructed.

[3]In more recent years Haidt has argued that also political and religious preferences are attributable to genetic factors. See Haidt (2012).

Both Hauser and Dawkins interpret the results obtained from this telematic test—that is, the fact that most of the interviewed people provided similar answers, regardless of their culture of origin, religious affiliation and social status—*as an indication of the universality of basic human moral principles and of their substantial impermeability to social and cultural influences.*

This procedure has raised several criticisms aimed to emphasize an evident inconvenience and disproportion between the poor reliability of the data collection method used by Hauser and collaborators, the kind of data analyzed, and the extreme generality of the conclusions which the author claimed to have been able to reach.

Even a scholar like Philip Lieberman who has made important contributions in both cognitive psychology and studies on human psychic evolution, has attacked the exasperated "adaptationism" that characterizes the approach of these evolutionary psichologists, denouncing the spread of research based on data collected, exclusively or almost, among the WEIRDOs. The acronym indicates the "people of Western cultures, educated, industrialized, rich and democratic" that constitute the majority of "the subjects who played the highly constrained mind-games experiments that form the «empirical» data of *Moral Minds*" (Lieberman 2013: 182) and of its "Toy Experiments".

The result that Hauser and Dawkins interpret as a proof of the existence of an innate universal morality appears, according to Lieberman's pertinent observations, more credible as an indicator of the high degree of *cultural homologation* now achieved in a globalized world, within a circle of people which is westernized enough to be familiar with telematic questionnaires and online games. In any case, it certainly cannot refute the results confirmed by all the countless archaeological and historical studies which had attest to very profound differences developed by past and present human cultures in terms of moral rules and lifestyles in a period that covers (and most likely exceeds) the last ten thousand years.

But, as we have already shown in the previous chapters with specific examples and as this final section will further demonstrate, the developments reached in the past three decades by an emerging discipline such as epigenetics, and the profound renewal it has brought to the field of evolutionary studies finally allow us to dismiss "strong" moral innatism as a dogmatic doctrine now refuted by an increasing amount of experimental results and statistical analyses of historical and empirical data.

9.3 Experience *Creates* Difference: Studies on Monozygotic Twins

Physical and behavioural similarities between monozygotic twins (MZ) have always been among the arguments that the supporters of genetic determinism try to exploit in favor of their own thesis. For example, evolutionary psychologist Steven Pinker, in an article entitled "Why nature & nurture won't go away" (Pinker 2004), tried to prove, drawing on some studies on mono-ovular twins, a thesis characterized by extreme biologistic determinism: "Setting aside cases of extreme neglect or abuse, whatever experiences siblings share by growing up in the same home in a given culture make little or no difference to the kind of people they turn into" (Pinker 2004: 15); "the shared family environment has little to no lasting effect on personality and intelligence" (ibid.).

A thesis that has always been disputed by the psychic and social sciences and that received, the following year (2005), a resounding denial precisely in the research areas to which Pinker most insistently appealed: behavioural genetics and the study on the development of epigenetic characteristics in mono-ovular twins. This refutation was offered by an article published in PNAS, the magazine of the American Academy of Sciences (Fraga et al. 2005). The paper, signed by twenty researchers operating in different countries, was based on a research project coordinated by Manel Esteller and focused on the study of 80 monozygotic male and female twins aged 3–74. It showed that the perfect genetic and epigenetic identity as found in MZ twins immediately after their birth does not prevent the punctual maturing in them, over the course of their lives, of significant epigenetic, attitudinal and behavioural differences. Differences deriving from their lifestyles, from the different environments they have lived in, or different roles they have occupied in a common environment, from biographical circumstances or pathologies which involve the regulation, and therefore the expression, of their genes. The authors of the study found the most significant differences in terms of immune system, organic function, brain micro-conformation and behavioural propensities precisely between monozygotic twins raised in different environments and when different pathologies changed the "epigenetic landscape" of a sibling. "We also established", they wrote, "that these epigenetic markers were more distinct in MZ twins who were older, had different lifestyles and had spent less of their lives together" (ibid.: 10608). Conclusions that, contrary to those taken from Pinker, fully confirmed "the significant role of environmental factors in translating a common genotype into a different phenotype" (ibid.) and the relevance that the life context has for the dynamics of the brain development and personality formation.

For these reasons, according to Esteller and his collaborators, "MZ twins are an excellent example of how to genetically identify individuals can exhibit and therefore provide a unique model for the contribution of epigenetic modifications in the establishment of the phenotype" (ibid.).

Another paper, published in *Science* in 2013, shows, instead, that *even mono-ovular twins grown in the same environment develop, from early childhood, different preferences, attitudes and abilities, and related differences in the respective brain micro-geographies*, if the environment in which they live is sufficiently stimulating (Freund et al. 2013). Particularly interesting is the fact that the study reveals these differences not in human beings, which many scientists consider as the species endowed with the greatest cerebral and behavioural plasticity, but in a model species such as the mouse.

In fact, it offers an experimental account of the development of different behavioural propensities and corresponding differences at the level of neural networks in 40 monozygotic mouse twins raised in a "large enriched environment" which gave them the opportunity to differentiate their activities. Monitored 24 hours a day by a microchip for three months, the twins developed different propensities, attitudes and abilities. The use of imaging diagnosis showed that their brains also underwent different kinds of micro-modifications. The mice had in fact increased the number of neurons in different areas of their hippocampus (an area of the brain that is very important to the learning ability, procedural memory and the encoding of information), and the mice "who explored their habitat more broadly also grew more new neurons" (ibid.: 758) in this area.

Experiments like these show that, to adequately appreciate the incidence of the social and natural environment on the formation of cerebral, psychic and behavioural differences in monozygotic twins, both environmental and biographical differences must be taken into account. They demonstrate that, even in genetically identical individuals, experience creates differences, not only on a cultural but also on a biological level, by modifying their self-regulation systems.

9.4 Experience *Transfers* Difference: Influences of Maternal Lifestyles and Parental Care on Descendants

The developments of behavioural and cultural epigenetics, a research area "which includes both the investigation of the role of behavior in shaping developmental-epigenetic states and the reciprocal role of epigenetic factors and mechanisms in shaping behavior" (Jablonka 2016: 47) already allow, about twenty years after its birth, to corroborate the following statements:

- It is established that a mother's diet, experiences and lifestyle influence embryonic development from its earliest stages and that parental care plays a no less crucial role in the development of the mental and behavioural attitudes of children.
- The effects of these embryonic and childhood experiences persist, in forms that are modulated by subsequent experiences, throughout children's and adults' lives and can be transmitted for some generations to their descendants even in the absence of a reiteration of the external stimuli that caused them.
- These influences, both in human beings and other social mammals, invest all main aspects of development: from the immune system to resistance to stress, from levels of neophobia or exploratory propensity to the development of psychic, cognitive and relational attitudes.
- They induce a series of experimentally verifiable changes in the epigenetic regulation of genetic expression and cerebral microcircuits.

As found in various animal clades, the effects of a mother's diet on her offspring's preferences appear particularly significant in placental mammals, where the zygote lives its prenatal development constantly immersed within the maternal uterus. Bilkó and some collaborators, in a well-known study (Bilkó et al. 1994) fed pregnant, laboratory-grown female rabbits with juniper berries obtaining the development of a clear preference for this food both in offspring bred by their natural mother and in those entrusted to adoptive mothers early. Similar effects on food preferences have produced experiments in which adoptive mothers fed with these berries suckled puppies born of mothers fed with normal laboratory feeds (Jablonka and Lamb 2005). In both cases the preference for that food was then passed on to the next generation.

Does maternal diet also have direct influences on the dietary orientation of human children? Experimental evidence has been accumulating over the past few decades. Often cited are the studies coordinated by the biopsychologist Julie Mennella that demonstrated the transmission of a preference for carrot juice in children of women who had drunk it regularly during pregnancy, or just during its first two months (Mennella et al. 2001). More recently, other studies have clarified that the diet followed by a mother during her pregnancy influences a wide range of organs and systems in offspring and can also affect the biological and health profile of their descendants. Finally, a very recent study deals with a topic to which we will return in the concluding pages of the chapter: the trans-generational effects of undernourishment (Aiken et al. 2016).

But it is now known that the uterine environment not only influences food preferences, but also other temperamental and behavioural traits. For example, in a rodent such as the Mongolian gerbil, the period in the maternal uterus can exert strong inheritable effects on the development of the female offspring. In this animal species, in fact, "a male-biased sex ratio and aggressive female behavior is perpetuated, probably because the mother's phenotype reconstructs a testosterone-rich uterine environment that induces the same hormonal and behavioral state in her daughters" (Jablonka and Lamb 2007: 359)

There are thus many studies on animal models that highlight the direct influence between mother/offspring, or reference adults/offspring and interactions on the latter's psychic and relational development. Lorenz and his collaborators already showed decades ago that it is enough to deprive an anatidae chick of the "greeting" ritual to make it an asocial adult, unable to weave "normal" relationships with conspecifics (Lorenz 1988). More recently, a series of experiments with rats coordinated by Frances Champagne (Weaver et al. 2004; Champagne and Rissman 2011) testified that puppies raised by mothers who donated a low quantity of licking and grooming subsequently showed lower levels of stress resistance, and a higher propensity to neophobia, than those who received greater amounts of care. The females raised this way, when becoming mothers, enacted behaviours similar to those received, obtaining the same effects, which were thus perpetrated from generation to generation (Weaver et al. 2004). Champagne and her collaborators found that these trans-generational effects depended on the fact that the donation and the non-donation of care triggered, respectively, methylation or de-methylation processes of certain portions of DNA.[4]

[4]Methylation is an epigenetic modification of DNA that allows the formation of a bond between a methyl group and a nitrogenous base (one of the bases of which the DNA and RNA nucleotides are composed). In mammals, methylation plays a fundamental role in the development of the zygote, making possible the formation of the chromatin, and therefore of the chromosomes, and then the

They have indeed detected "that increased pup licking and grooming (LG) and arched-back nursing (ABN) by rat mothers altered the offspring epigenome at a glucocorticoid receptor (GR) gene promoter in the hippocampus. Offspring of mothers that showed high levels of LG and ABN were found to have differences in DNA methylation, as compared to offspring of 'low-LG-ABN' mothers" (ibid.: 847). In simpler words: the pups that had received scarce care presented, in the successive phases of their development, a tendency to stress and a neophobia significantly greater than those raised with more care. The females bred by mothers who gave little care also presented "the same epigenome as the mothers and therefore reproduced the same 'unloving' behavior with their children" (Bottaccioli and Bottaccioli 2012: 331).[5] In fact, the methylation process induced by the lack of parental care prevented the development and functioning of an adequate number of glucocorticoid receptors, hormones that allow individuals to adapt to stress, alleviate anxiety and to relax. Individuals who had undergone poor treatment ended up becoming perpetually stressed subjects and females raised in such conditions procreated offspring with similar propensities. Therefore, the greater or lesser donation of parental care was revealed, in this case, as a factor capable of inducing epigenetic modifications that have significant effects from an emotional, psychic and clinical point of view, for both first generation children and their descendants.

Subsequently, to demonstrate even more rigorously that these effects were mainly due to the behaviour of the mother, and not to her genes, Champagne and her colleagues had mothers who gave abundant care bred rat pups born of mothers who gave them few and vice versa. The causal links suggested by the results of the first experiment found full confirmation in the second one: the pups born to "caring" mothers but raised by inattentive mothers developed in their hippocampus low levels of glucocorticoid receptors and behaved more anxiously. Those born to inattentive mothers but bred by mothers who gave wide care showed high levels of glucocorticoid receptors and a more sociable and relaxed behaviour (Iversen 2014).

9.5 Trans-Generational Effects of Traumatic Experiences, Privations and Social Inequalities

Several experiments have proved that also in human beings, "psychic stress has trans-generational effects" (Jablonka 2016: 49). For example: "using blood cells, a study in Gambia found that individuals conceived during the nutritionally-stressful

gene transcription and the differentiation of cells, tissues and organs. The DNA methylation allows the cells to regulate the gene expression by silencing certain genes, i.e. inhibiting their expression. De-methylation is instead a process that has the opposite effect: it fixes, so to speak, the gene in an "ON" position and therefore allows it to perform its functions. Both these processes are strongly influenced by environmental stimuli.

[5]This and all the other quotations from essays contained in this chapter that have not been translated into English are my translations.

rainy season had significantly higher methylation at several important gene loci than individuals conceived during the more plentiful dry season" (ibid.).

More generally, over the last twenty years, the cross results of many studies have shown that for us as for other animals, "stressful or traumatic experiences such as social defeat, a strong or enduring mental shock, physical and emotional abuse, or deprivation of early parental care can have deleterious longterm, trans-generational effects that are mediated by molecular epigenetic mechanisms" (ibid.). In other words, according to the new insights of behavioural epigenetics, "traumatic experiences in our past, or in our recent ancestors' past, leave molecular scars adhering to our DNA. Jews whose great-grandparents were chased from their Russian shtetls; Chinese whose grandparents lived through the ravages of the Cultural Revolution; young immigrants from Africa whose parents survived massacres; adults of every ethnicity who grew up with alcoholic or abusive parents - all carry with them more than just memories" (Hurley 2015).

In 2008, neurobiologist Michael Meaney and molecular biologist Moshe Szyf published, together with other researchers, the results of a study in which they compared the brains of 13 people who had been sexually abused and died by suicide with those of people who died suddenly due to other causes. The former presented an excess of methylation (silencing) of the hippocampal genes, a region of the brain that plays a fundamental role in stress regulation, but also in the fixation of mnemonic data and in spatial orientation (McGowan et al. 2008).

Then, in 2012, Szyf and his collaborators published the results of a study based on the analysis of the methylation rates of most of the genes present in the genome (obtained from blood samples) of 40 people born in England in 1958 (Borghol et al. 2012). These were people who had lived, from childhood, or from a certain phase of their life onwards either in conditions of extreme poverty, or in conditions of great wealth. Overall, Szyf and his colleagues analyzed the methylation status of about 20,000 genes. "Of these, 6,176 genes varied significantly depending on poverty or well-being. The most surprising thing, however, was to notice that methyl changes were more frequently found if the impacting event" that had drastically changed the economic life of those people had occurred "in the early childhood rather than as adults" (Iversen 2014).[6] The authors, presenting the study, wrote: "We aimed to establish whether childhood SEP" (socio-economic position) "was associated with differential methylation of adult DNA" (Borghol et al. 2012: 62). Their conclusions were equally clear: "Disadvantaged socio-economic position (SEP) in childhood is associated with increased adult mortality and morbidity" (ibid.).

In another study, published in the same year, Elena L. Grigorenko, Moshe Szyf and other researchers compared the overall methylation levels of 14 children raised in Russian orphanages with those of 14 Russian children who grew up with their natural parents (Naumova et al. 2012). The experiment confirmed that in the orphans many genes involved in important functions, such as the neuronal communication

[6]This and all the other quotations from essays that have not already been translated into English in this chapter are my translations.

and the brain function development, had a level of methylation, and therefore of inhibition, significantly higher than those found in the control group.

"The study of social and cultural epigenetics is still in its infancy, so we do not have detailed studies of the relations between socio-cultural conditions and epigenetics. It is well established, however, that social inequality (e.g., poverty) in geographically, politically, and economically diverse populations is correlated with an increased risk of cardiovascular diseases, cancer and psychological disorders, and that all these deleterious conditions have epigenetic underpinnings. […] The disposition to develop such deleterious effects can be transmitted to the next generation and contribute to the difficulty of escaping poverty" (Jablonka 2016: 50), thus triggering a "vicious cycle of political-social action" (ibid.). Very clear, in this sense, is the data on the trans-generational effects of social catastrophes such as wars, famines and persecutions. "A 2008 study was first to show in humans that early-life environmental conditions can cause epigenetic changes that persist for life. It investigated long-term effects of the Dutch hongerwinter (hunger winter) of 1944–1945, a seven-month famine imposed by Nazi rationing" (Powledge 2009: 738; 739). The study (Heijmans et al. 2008) showed that individuals conceived during this period, in which the Nazi occupiers reduced daily food rations to less than 700 kilocalories per person, "had, 6 decades later, less DNA methylation of the imprinted *IGF2* gene compared with their unexposed, same-sex siblings" (ibid.). This led, for them, to an increased incidence of diabetes, obesity, schizophrenia and coronary heart disease, significantly higher than that of the siblings of the same sex born in less hard times. A study published by Yehuda and some collaborators (Yehuda et al. 2015) illustrates, instead, the trans-generational effects of different kinds of traumatic events, showing that the experiences of war, segregation, degradation and humiliation influence the disposition of descendants to develop several neurophysiological and psychological pathologies. Indeed, all 22 children of the 32 Holocaust survivors examined were more likely to develop post-traumatic stress disorder (PTSD) than control groups, related to the methylation of the cytosine within the gene encoding the FKBP5 protein, which plays an important role in controlling stress sensitivity.

9.6 Conclusions

Summarizing what has been discussed in this chapter, developments in evolutionary biology and behavioural sciences have now proved that the ability of organisms to adapt to rapid changes in the environment depends, to a significant extent, on epigenetic regulatory mechanisms capable of deactivating or reactivating certain genes, or their influence over certain processes, and that these forms of self-regulation are transmitted to descendants through the epigenetic inheritance. According to these emerging models, biological evolution is based not only on the slow processes of accumulation of genetic mutations and natural selection, but also on these rapid changes in epigenetic regulation, behaviour and survival strategies, which allow living beings to cope with the changes that continually arise in their environmental and social contexts.

This also means that, particularly among animals as the social mammals and birds, the development of individual and collective reactivity, preferences, attitudes and capacities is mainly regulated by inputs coming from the *intraspecific* environment, that is, from the social context in which individuals live. We already knew that this happens through training and learning processes, but we can now state that it is also through epigenetic inheritance. This means that the effects of the experiences of immediate ancestors are transmitted to descendants not only because they take relatives and parents as a model, and/or react to their behaviour, but also because each parent transmits to their children an entire functional and self-regulatory structure that works, above all, at a cellular level and this structure, as we have seen, regulates a lot of aspects related to health, behaviour, cognitive activity and social relationships.

This is all the more true for humans, who live in an environment which, even from a material point of view, is almost entirely anthropic, or built and regulated by humans themselves. This means that the biological history of man is also a *social history*, that its developments, ever since ancient times and even more so today, are channelled towards certain directions by the ways in which human relations are organized within the society. This is true both for aspects relating to emotional and physiological self-regulation, as well as for those relating to psychic, attitudinal and intellectual development.

Furthermore, a wide range of studies attests that every form of traumatic or stressful experience and any kind of social discrimination leave in our minds and bodies molecular and cellular "scars" which cause physical and psychic discomforts and can be transmitted to our descendants. What these studies bring to light is, therefore, a vicious circle, existing for millennia, in which social discrimination produces, through inhibitory, stressful or disabling effects that are transmitted from generation to generation, a continuous strengthening of inequalities. They demonstrate, in other words, that *social discrimination produces biological discrimination* and tends to perpetuate the conditions of their arising and increasing.

But, fortunately, these studies also demonstrate the possibility of the reverse effect. They show in fact that:

- the inclusion of subjects with previous deficits or disorders, caused by epigenetic inheritance or individual experiences, in a social and material environment which is appropriate to them, enriched, stimulating, and non-stressful, can produce considerable compensatory effects and forms of rehabilitation (McGowan et al. 2008; Freund et al. 2013; Ball et al. 2019);
- such improvement effects can concern both the health profile and the management of emotions, the social relationships and the learning abilities (Schneider et al. 2006; McGowan et al. 2008; Freund et al. 2013; Ball et al. 2019);

- in some cases, these positive effects also occur in individuals who have inherited from their parents *genetic* anomalies that hinder learning, emotional management and social relations (Arai et al. 2009; Arai and Feig 2011).[7]

What are the meta-ethical repercussions of these discoveries on theories regarding the innate or learned, genetically or socially transmitted nature of moral imperatives, prohibitions and orientations in human beings?

In short, the thesis of both old and current *genetic determinism* according to which organisms are essentially mere executors of instructions encoded in their genes, and the current human moral and socio-economic systems ultimately the effects of an almost unmodifiable "human nature" produced by natural selection, are now proving, *thanks to the progress and tools of genetics and epigenetics, developmental biology and evolutionary studies themselves, groundless and anachronistic*. What has been discovered in the last decades on the processes of epigenetic inheritance and on the trans-generational effects of personal and historical experiences, allows us to understand that human *social* history has shaped, and is still shaping, not only the psyche and behaviour, but also the biology of human (and not only human) beings, and the self-regulating systems of every individual.

In other words: *biologisms have been refuted by biology itself*.

The data discussed in this essay may be useful to highlight at least three aspects related to this issue:

- the endless disasters produced, since the dawn of "civilization", by the processes of social manipulation of the human propensities and reactivity, implemented by every political and economic regime based on the coercive induction of behavior, the monopoly of information, the gender, class and ethnic discrimination, the thirst for power, conquest and profit;
- the enormous power of mind and behaviour manipulation that the current biological and media technologies and the enormous inequality in the access to the resources that characterize our societies deliver into the hands of restricted economic groups and political institutions. A power of manipulation that is fuelling processes of degradation of the human critical capacity and ethical sensitivity in an increasingly penetrating, incisive and pervasive way;
- the beneficial effects that any attempt to let the forms of human and animal sociality mature from below, creating non-oppressive living environments, capable of offering to everyone a range of life chances, development and expression, could produce if it were socially supported.

Today we can say that the epigenetic landscape that humans inherit from their closest ancestors and develop throughout their lives is, like every other aspect of the human

[7] For example it has been proved that a stimulating environment "can compensate for a learning-deficiency in mutant mice" and "also improves learning in the F1 offspring that inherit the deleterious gene" (Jablonka 2016: 50). Of course, if these forms of compensation are possible for the mice, there is no reason to think that it wouldn't be the same for human beings, all the more so if supported by their social context and by the means it can offer, provided the political wish of the leading class to do so.

world, to a large extent, a *social product*. In our species, moreover, innate behaviour patterns often manifest not as a set of rigid sequences of movements, but as a set of learning programs: "we are born not with any specific social norm, but with a learning agenda that tells us which information to imbibe and how to organize it" (de Waal 1996: 36). That leads us to assume behaviour patterns and reaction norms from the environment in which we live, initially in a completely unconscious way, interiorizing emotional patterns and customs, values and traditions. Epigenetic regulation of gene expression plays a key role in this process because it makes inheritable some effects of experience and propensities. Social learning, and especially early learning experiences, have in turn very strong long-term effects, and may therefore play a role that is even more important than that of epigenetic inheritance in shaping human personality.

Genetic, epigenetic, behavioural and cultural inheritance cooperate, thus, among themselves and with the experience and environmental stimuli to construct the whole neuro-physiological condition, the patterns of reaction, the preferences and propensities which every human being develops over a lifetime.

In short, social life, from the most embryonic of its phases, the one that a fetus spends in the maternal body and in symbiosis with it, moulds individual biological and behavioural expressiveness, inhibiting or enhancing its potential and attitudes and, just like a gardener with their plants, can make them flower or wither.

References

Adenzato, Mauro, and Cristina Meini. 2006. *Psicologia evoluzionistica*. Torino: Boringhieri.

Aiken, Catherine, Jane Tarry-Adkins, and Susan Ozanne. 2016. Trans-Generational Effects of Maternal Diet on Metabolic and Reproductive Ageing. *Mammalian Genome* 27 (7–8): 430–439.

Arai, Junko, and Larry Feig. 2011. Long-Lasting and Trans-Generational Effects of an Environmental Enrichment on Memory Formation. *Brain Research Bulletin* 85 (1–2): 30–35.

Arai, Junko, Shaomin Li, Dean Hartley, and Larry Feig. 2009. Trans-Generational Rescue of a Genetic Defect in Longterm Potentiation and Memory Formation by Juvenile Enrichment. *Journal of Neuroscience* 29 (5): 1496–1502.

Attanasio, Alessandra. 2010. *Darwinismo Morale*. Torino: Utet.

Bain, Alexander. 1868. *Mental and Moral Science*. London: Longmans, Green and Company.

Ball, Natalie, Eduardo Mercado, and Itzel Orduña. 2019. Enriched Environments as a Potential Treatment for Developmental Disorders: A Critical Assessment. *Frontiers in Psychology* 10. https://www.ncbi.nlm.nih.gov/pmc/articles/PMC6414413/.

Barkow, Jerome, Cosmides Leda, and Tooby John. 1992. *The Adapted Mind*. New York: Oxford University Press.

Bilkó, Ágnes, Vilmos Altbäcker, and Robyn Hudson. 1994. Transmission of Food Preference in the Rabbit. *Physiology & Behavior* 56 (5): 907–912.

Borghol, Nada, Matthew Suderman, Wendy McArdle, Ariane Racine, Michael Hallett, Marcus Pembrey, Clyde Hertzman, Chris Power, and Moshe Szyf. 2012. Associations with Early-Life Socio-Economic Position in Adult DNA Methylation. *International Journal of Epidemiology* 41 (1): 62–74.

Bottaccioli, Francesco, and Anna Giulia Bottaccioli. 2012. La rivoluzione epigenetica. *Sistema Salute* 56 (3): 331.

Callebaut, Werner, Müller Gerd, and Stuart Newman. 2007. The Organismic Systems Approach: Evo-Devo and the Streamlining of the Naturalistic Agenda. In *Integrating Evolution and Development: From Theory to Practice*, ed. Roger Amson, and Robert Brandon, 25–92. Cambridge, MA and London: MIT Press.

Champagne, Frances, and Emilie Rissman. 2011. Behavioral Epigenetics: A New Frontier in the Study of Hormones and Behavior. *Hormones and Behavior* 59 (3): 277–278.

Cosmides, Leda and Tooby John. 2006. Origins of domain specificity: The evolution of functional organization. In *Philosophy of Psychology: Contemporary Readings*, ed. Bermúdez, Jose Luis, 539–555. Routledge Contemporary Readings in Philosophy. London: Routledge, Taylor & Francis Group.

Darwin, Charles. 1871 (1981). *The Descenf of Man*. Princenton, NJ: Princenton University Press.

Dawkins, Richard. 1976. *The Selfish Gene*. New York: Oxford University Press.

Dawkins, Richard. 2006. *The God Delusion*. London: Bantam Press.

de Waal, Frans. 1996. *Good Natured: The Origins of Right and Wrong in Humans and Other Animals*. Cambridge, MA: Harvard University Press.

de Waal, Frans. 2006. *Primates and Philosophers: How Morality Evolved*. Princenton, NJ: Princeton University Press.

Fodor, Jerry. 1983. *The Modularity of Mind: An Essay on Faculty Psychology*. Cambridge, MA and London: MIT Press.

Fodor, Jerry. 2000. *The Mind Doesn't Work That Way: The Scope and Limits of Computational Psychology*. Cambridge, MA and London: MIT Press.

Fraga, Mario, Esteban Ballestar, Maria Paz, Santiago Ropero, Fernando Setien, Maria Ballestar, Damia Heine-Suñer, Juan Cigudosa, Miguel Urioste, Javier Benitez, Manuel Boix-Chorne, Abel Sanchez-Aguilera, Charlotte Ling, Emma Carlsson, Pernille Poulsen, Alla Vaag, Zarko Stephan, Tim Spector, Yue-Zhong Wu, Plass C., and Manel Esteller. 2005. Epigenetic Differences Arise During the Lifetime of Monozygotic Twins. *Proceedings of the National Academy of the United States of America* 102 (30): 10604–10609.

Freund, Julia, Andreas Brandmaier, Lars Lewejohann, Imke Kirste, Mareike Kritzler, Antonio Krüger, Norbert Sachser, Ulman Lindenberger, and Gerd Kempermann. 2013. Emergence of Individuality in Genetically Identical Mice. *Science* 340 (I): 756–759.

Gould, Steven Jay. 1977. *Ever Since Darwin*. New York: W. W. Norton.

Haidt, Jonathan. 2001. The Emotional Dog and Its Rational Tail. *Psychological Review* 108 (4): 814–834.

Haidt, Jonathan. 2012. *The Righteous Mind: Why Good People are Divided by Politics and Religion*. New York: Pantheon.

Hauser, Marc. 2006. *Moral Minds: How Nature Designed Our Universal Sense of Right and Wrong*. New York: Harper Collins.

Heijmans, Bastiaan, Elmar Tobi, Aryeh Stein, Hein Putter, Gerard Blauw, Ezra Susser, Eline Slagboom, and L.H. Lumey. 2008. Persistent Epigenetic Differences Associated with Prenatal Exposure to Famine in Humans. *Proceedings of the National Academy of the United States of America* 105 (44): 17046–17049.

Hodgson, Geoffrey. 2013. The Enduring Relevance of Darwin's Theory of Morality. *BioScience* 63 (7): 513–514.

Hurley, Dan. 2015. Grandma's Experiences Leave a Mark on Your Genes. https://www.discovermagazine.com/health/grandmas-experiences-leave-a-mark-on-your-genes.

Huxley, Julian. 1942. *The Modern Synthesis*. London: George Allen and Unwin.

Iversen, Daniel. 2014. L'infanzia dei vostri antenati potrebbe essere scritta nel DNA, ScienzeNaturali.it. http://www.scienze-naturali.it/ricerca-scienza/genetica/linfanzia-dei-vostri-antenati-potrebbe-essere-scritta-nel-dna.

Jablonka, Eva. 2006. Genes as Followers in Evolution—A Post-Synthesis Synthesis. *Biology and Philosophy* 21: 143–154.

Jablonka, Eva. 2016. Cultural Epigenetics. *Sage Journal* 64 (1): 42–60.

Jablonka, Eva, and Marion Lamb. 2005. *Evolution in Four Dimensions*. Cambridge, MA: MIT Press (a revisited edition was published in 2014).
Jablonka, Eva, and Lamb Marion. 2007. Précis of Evolution in Four Dimensions. *Behavioral and Brain Sciences* 30 (4): 353–365.
Jenkin, Fleeming. 1867. (Review of) The Origin of the Species. *North British Review* 92 (46): 277–318.
Lieberman, Philip. 2013. *The Unpredictable Species*. Princenton: Princenton University Press.
Lorenz, Konrad. 1988. *Hier bin ich—wo bist du?* München: Piper.
McGowan, Patrick, Aya Sasaki, Tony Huang, Alexander Unterberger, Matthew Suderman, Carl Ernst, Gustavo Turecki, Michael Meaney, and Moshe Szyf. 2008. Promoter-Wide Hypermethylation of the Ribosomal RNA Gene Promoter in the Suicide Brain. *PLoS ONE* 3 (5): e2085.
Mennella, Julie, Coren Jagnow, and Gary Beauchamp. 2001. Prenatal and Post-Natal Flavor Learning by Human Infants. *Pediatrics* 107: e88.
Naumova, Oksana, Maria Lee, Roman Koposov, Moshe Szyf, Mary Dozier, and Andelena Grigorenko. 2012. Differential Patterns of Whole-Genome DNA Methylation in Institutionalized Children and Children Raised by Their Biological Parents. *Development and Psychopathology* 24 (1): 143–155.
Pennock, Robert. 1995. Moral Darwinism: Ethical Evidence for the Descent of Man. *Biology and Philosophy* 10 (3): 287–307.
Pigliucci, Massimo, and Gerd Müller (eds.). 2010. *Evolution: The Extended Synthesis*. Cambridge, MA and London: MIT Press.
Pinker, Steven. 2002. *The Blank Slate: The Modern Denial of Human Nature*. New York: Viking.
Pinker, Steven. 2004. Why Nature & Nurture Won't Go Away. *Daedalus* 133 (4): 5–17.
Pinker, Steven. 2008. The Moral Instinct. *The New York Times*, January 13.
Powledge, Tabitha. 2009. Epigenetics and Development: Understanding How Epigenetics Works at the Molecular Level Can Be Mind-Boggling. *BioScience* 59 (9): 736–741.
Riedl, Rupert. 1980. *Biologie der Erkenntnis*. Berlin: P. Parey.
Rowe, David. 1994. *The Limits of Family Influence*. New York: Guilford Press.
Ruse, Michael. 2017. *Evolutionary Ethics*. Cambridge: Cambridge University Press.
Schneider, Tomasz, Joanna Turczak, and Ryszard Przewłocki. 2006. Environmental Enrichment Reverses Behavioral Alterations in Rats Prenatally Exposed to Valproic Acid: Issues for a Therapeutic Approach in Autism. *Neuropsychopharmacology* 31: 36–46.
Simon Herbert. 1990. A Mechanism for Social Selection and Successful Altruism. *Science* 250: 1665–1668.
Turkheimer, Eric. 2000. Three Laws of Behavior Genetics and What They Mean. *Current Directors in Psychological Science* 5 (2000): 160–164.
Weaver, Ian, Nadia Cervoni, Frances Champagne, Ana D'Alessio, Shakti Sharma, Jonathan Seckl, Sergiy Dymov, Moshe Szyf, and Michael Meaney. 2004. Epigenetic Programming by Maternal Behavior. *Nature Neuroscience* 7 (8): 847–854.
Wiker, Benjamin. 2002. *Moral Darwinism*. Westmont: Inter-Varsity Press.
Yehuda, Rachel, Nikolaos Daskalakis, Linda Bierer, Heather Bader, Torsten Klengel, Florian Holsboer, and Elisabeth Binder. 2015. Holocaust Exposure Induced Intergenerational Effects on *FKBP5* Methylation. *Biological Psychiatry* 80 (5): 372–380.

The manufacturer's authorised representative in the EU is Springer Nature Customer Service Centre GmbH, Europaplatz 3, 69115 Heidelberg, Germany. If you have any concerns regarding our products, please contact ProductSafety@springernature.com

Printed and bound by CPI Group (UK) Ltd, Croydon, CR0 4YY
28/09/2026
02224086-0006